EARLY WRITINGS IN THE PHILOSOPHY OF LOGIC AND MATHEMATICS

EDMUND HUSSERL

COLLECTED WORKS

EDITOR: RUDOLF BERNET

VOLUME V

TRANSLATIONS
PREPARED UNDER THE AUSPICES OF
THE HUSSERL-ARCHIVES (LEUVEN)

EDMUND HUSSERL

EARLY WRITINGS IN THE PHILOSOPHY OF LOGIC AND MATHEMATICS

TRANSLATED BY

DALLAS WILLARD

KLUWER ACADEMIC PUBLISHERS

DORDRECHT / BOSTON / LONDON

Library of Congress Cataloging-in-Publication Data

Husserl, Edmund, 1859-1938.
Early writings in the philosophy of logic and mathematics / edited by Dallas Willard.
p. cm. -- (Edmund Husserl collected works ; v. 5)
Includes index.
ISBN 0-7923-2262-2 (HB : alk. paper)
1. Logic. 2. Mathematics--Philosophy. I. Willard, Dallas, 1935- . II. Title. III. Series: Husserl, Edmund, 1859-1938. Works. English. 1980 ; v. 5.
B3279.H93E37 1993
160--dc20 93-894

ISBN 0-7923-2262-2

Published by Kluwer Academic Publishers,
P.O. Box 17, 3300 AA Dordrecht, The Netherlands.

Kluwer Academic Publisher incorporates
the publishing programmes of
D. Reidel, Martinus Nijhoff, Dr W. Junk and MTP Press.

Sold and distributed in the U.S.A. and Canada
by Kluwer Academic Publishers Group,
101 Philip Drive, Norwell, MA 02061, U.S.A.

In all other countries, sold and distributed
by Kluwer Academic Publishers,
P.O. Box 322, 3300 AH Dordrecht, The Netherlands.

Printed on acid free paper

Printed in The Netherlands

CONTENTS

EARLY WRITINGS IN THE PHILOSOPHY OF LOGIC AND MATHEMATICS

TRANSLATOR'S INTRODUCTION

The primary intent of this volume is to give the English reader access to all the philosophical texts published by Husserl between the appearance of his first book, *Philosophie der Arithmetik*, and that of his second book, *Logische Untersuchungen* – roughly, from 1890 through 1901. Along with these texts we have included a number of unpublished manuscripts from the same period and dealing with the same or closely related topics. A few of the texts here translated (the review of Palágyi, the five "report" articles of 1903–1904, the "notes" in Lalande's *Vocabulaire*, and the brief discussion article on Marty of 1910) obviously fall outside this time period, so far as their publication dates are concerned; but in content they seem clearly confined to it. The final piece translated, a set of personal notes that date from 1906 through 1908, provides insight into how Husserl experienced his early labors and their results, and into how he saw their relation to work before him: a phenomenological critique of *reason* in all of its forms.

Thus the texts here translated – which obviously are to be read in conjunction with his first two books – cover the progression of Husserl's *Problematik* from the relatively narrow one of clarifying the epistemic structure of general arithmetic, to the all-encompassing one of establishing in principle, through phenomenological research, the line between legitimate and illegitimate claims to know or to be rational, regardless of the domain concerned.

As to the publishing history of these texts, numbers 5–16 were first published by Husserl himself, in the places indicated at the opening of each translation, while the remaining pieces, with one exception, were first published in the *Husserliana* volumes. Numbers 1, 2 and

4 are from Vol. XII. Number 3 is from Vol. XXI (pp. 244–251), and number 23 is from volume XXIV (pp. 442–449), though it, being the exception noted, first appeared in print in *Philosophy and Phenomenological Research*, Vol. XVI, #3 (March 1956), pp. 293–302. The remaining pieces appeared for the first time in *Husserliana XXII*, which is here translated in its entirety – except for the "Editor's Introduction" and the critical notes on the texts. My English translation of numbers 5–11 and number 13, in a less accurate and polished form, have appeared previously, in *The Personalist*, Volumes LIII, & LVIII–LX, and reappear here by permission. Number 11 has also been translated by Richard Hudson and Peter McCormick, in *Husserl: Shorter Works*, (Notre Dame: University of Notre Dame Press, 1981), pp. 126–142.

From a literary point of view, these pieces may be divided into:

I. Complete treatises or essays (4, 6, 7, 11, 17, 18)
II. Reviews (5, 12, 13, 14[a–e], 16, 19, 20)
III. Responses to published criticisms (8, 9, 10, 13)
IV. Fragments and research notes (1, 2, 21[I, II–VIII, X–XII])
V. Letters (3, 21[IX])
VI. Personal notes (21[II], 22)
VII. Encyclopedia articles (15).

However, among the reviews, 5, 13, 19 and 20 are in effect essays, while among those listed above as essays, 17 and 18 are drawn together by Bernhard Rang, the editor of *Husserliana XXII*, from large-scale fragments and highly developed research notes.

In thematic terms, the following main groupings emerge from the texts:

I. The nature of formalisms (algorithmic techniques) and of the formal disciplines (1–10, 13, 21[I–III, XI–XII])
II. The structures and types of mental acts: especially of intuitions, representations and judgments (11–12, 14, 17–20, 21[IV–VIII, X])
III. Intentional objects and "intentional inexistence" (18, 19, 21[IX])
IV. Ontological issues and truth (13, 14, 15, 18, 19, 21[X], and numerous passages from 14[a–e]).

We now turn to the major conceptual developments presented through the texts here translated.

*

Texts 1 and 2 express the assumptions about the nature of general arithmetic (in the form of the calculus) that guided Husserl into his researches for Part II of *Philosophie der Arithmetik*. Part I of that work had undertaken 'analyses' of the concepts *unity*, *plurality* and *number*, by tracing them to their origin in abstraction from adequate intuitions of 'totalities' (*Inbegriffe*). Those analyses are the investigations referred to in the opening sentence of #1 below as "preliminary to general arithmetic." They concern the intrinsic nature of number, not the laws of interrelationships between any numbers merely as such – though in fact Chapter V of Part I begins to move in the latter direction, with its study of the abstraction of the concepts *more* and *less* as derived from the intuition of relationships between the smaller numbers.

At the outset of Part II of *Philosophie der Arithmetik*, Husserl states that "the next philosophical task is that of making psychologically and logically intelligible the origination of a calculational technique based upon those [previously examined] concepts and of investigating its relationship to arithmetical science." (*Husserliana XII*, p. 181) Selections 1 and 2 of this present volume make it clear that up to 1890 Husserl regarded general arithmetic as a science whose domain is the positive whole numbers. It is "the general theory of operations appertaining to the theory of cardinal numbers." By "operations" he has in mind the "combinations of cardinals, or of conceptual and sensible symbolizations standing in their place." {p. 1}[1] He takes this to be in agreement with the majority of investigators, who, in his view, hold that "in the understanding of the domain of cardinal numbers we possess the key for the understanding of the science of arithmetic in general." {p. 2} The task of general arithmetic therefore is that "of defining every conceivable elementary form of numerical construction, and of establishing the laws of their reciprocal connections and transformations." {p. 5}

Now the establishment of the laws of interconnections between "every conceivable elementary form of numerical construction," obviously dealing with an infinitely complex field of relations, essentially requires "a *mechanical* procedure of calculation for equiva-

[1] Page numbers within the "{ }" brackets refer to pages of the English translations to follow.

lence transformations between complicated forms of numerical constructs." {p. 5} And the most important *logical* resources – that is, as Husserl understands "logical" here, the most important *tools* for the realization of *knowledge* (*Erkenntnis*) – in general arithmetic are "the modes of designation peculiar to it." These modes of designation, the well-known symbols characteristic of arithmetic, make *visibly* unmistakable the primary distinctions and relations upon which the calculations depend: known and unknown numbers, the various types of operations, equality and inequality, scope indicators – all with that univocity, brevity and easy distinguishability requisite to a calculational *mechanism* and which today we naturally associate with a 'formalized language'. {p. 6}

Accordingly, the first half of #2 below is devoted to sketching out a *classification* of the types of number designations that enter into the formulae of general arithmetic, including names and variables for individual numbers known and unknown, symbols for the elementary operations such as addition, etc., up through equations and even systems of equations which serve to specify numbers. The systematic interrelations of these types of designations are briefly discussed. But Husserl's real concern is "how to justify logically" both the system of symbolizations used in general arithmetic and the "order of elaboration of knowledge which we find realized in the science as it actually exists." {p. 9} That is, how to clarify the way the use of the symbolism is productive of actual knowing, in the strongest 'scientific' sense. It is not too much to say that *the* problem which chiefly motivates Husserl at this point of his career derives from the fact that, while general arithmetic is an apriori science of purely conceptual truths about numbers in general – but not "analytical" truths, as that term is now used – it does *not* proceed by attending to conceptual connections between numbers described (designated) in the various ways possible. And it does not proceed by *inferences* based upon insight into conceptual connections. As a matter of both fact and essence, that is not its *modus operandi*. Thus, a chasm is opened up between what the science of arithmetic is *about*, its "subject matter," and what is attended to in its primary procedures, those of "formal" transformation or derivation.

The arithmetician begins with the basic operations, such as addition, etc., which are *general* ways of treating numbers. He is interested only in what is true for any numbers, merely as such, subjected to the operations in question. Thus, "Arithmetic is ... an apriori science .... It begins immediately with .... apodictically certain and immediately Evident generalities" which it obtains "by merely bringing certain *basic concepts* clearly before the mind; and those generalities then furnish all further theorems of the science by way of *mediate Evidence* and certainty." {p. 9} Like any science, especially an apriori science, arithmetic is delimited by a basic concept and embraces the objects of that concept as its domain. But it does not deal with these objects *in concreto.* Rather they are dealt with "in a totally inauthentic mode of representation," in terms of determinations that belong to them merely as objects of the conceptual domain. "Out of such conceptual determinations there then arise all of the laws for those objects" dealt with in the science. {p. 10} But these laws and the conceptual determinations underlying them are not what one "pays attention to" in most of the arithmetician's work. When we consider the essential traits of the conscious acts of the working arithmetician we find them to be mainly directed simply upon the sensible features and order of the symbolism in use, and upon the rules for the manipulation of that symbolism. Arithmetic follows, not "the method of concepts," but "the method of sensible signs." "The method of sensible symbols is therefore *the* logical method of arithmetic." (*Husserliana XII*, p. 257) But how is this possible? What is it about calculation with sensible symbols that allows it to produce knowledge of something other than itself – indeed, knowledge (*Erkenntnis*) of the very highest quality?

It will greatly assist us in understanding the nature of Husserl's project with regard to general arithmetic if we carefully distinguish it from the better known project of David Hilbert and others. Husserl's concern is *not* with proving that arithmetic is *complete* in the sense of having a consistent axiomatic base from which every other well-formed sentence in the language or its negation can be logically derived. As is now well-known, the impossibility of completeness for arithmetic has been established. But even if it were demonstrated that general arithmetic is complete, that still would not provide a solution for Husserl's problem. Indeed, that would only raise it again, since such a demonstration would most certainly use a formalized language. The logical question of completeness is not wholly irrelevant to Husserl's concern, but neither is it identical with it.

What Husserl wants to know is *how* – not *whether*, for on that point he has no doubt at all, but *how* – a procedure with a formalized language or algorithm, in which consciousness of the relevant logical relations between concepts and propositions (or of their corresponding objects) simply plays no role, can supply knowledge of a domain left entirely out of consideration in the execution of the procedure. Specifically, how can a process in which one does not think of numbers at all provide us with knowledge of, precisely, numbers and the laws of their interconnections – knowledge, moreover, of the highest epistemic quality, which, because of mental limitations, could *never* be achieved by actual inference or construction focussing upon the relevant concepts and logical relations.

As an attempt to respond to this question, Part II of the *Philosophie der Arithmetik* turns out to be a rather strange piece of writing. It can, I think, be only described as fumbling and inconclusive, though filled with brilliant observations on matters of detail. This is caused by the fact that at its outset its author still hoped to be able to explain how general arithmetic achieves its epistemic results by recourse to inauthentic representations *of numbers*. In the footnote at the opening of Chapter XI, Husserl is lavish in his thanks to Brentano, to whom he professed "to owe the deeper understanding of the eminent significance of inauthentic representing for the whole of our psychical life." (*Husserliana XII*, p. 193n) Of course the whole volume was, with Brentano's permission, dedicated to "My Teacher, Franz Brentano, With Heartfelt Gratitude." By the end of Part II, however, we have come to a very lame conclusion. Namely, that our unavoidable limitation to the use of symbolic number formations in the overwhelming majority of cases where we seek numerical properties, relationships and laws – and in all but a few cases where we seek merely to *refer to* numbers – is what forces us, for the purposes of knowledge, to develop the rule-governed symbol system that makes up general arithmetic. (*op. cit.*, p. 283) This epistemic necessity, this requirement "for the purposes of knowledge," is described by him as the *logical* origin of general arithmetic. The title assigned to this Part II is accordingly: "The Symbolic Number Concepts and the Logical Sources of Number-Arithmetic."

In the "Author's Report" or "*Selbstanzeige*," which he published later in 1891, Husserl indicates that Part II "already clears up the most elemental symbolic methods of number arithmetic, which are based upon the rigorous parallelism between the number concepts and numerals, and between the rules for compounding concepts to form judgments and the rules for compounding symbols to produce formulae." (*op. cit.*, p. 287) What is *not* cleared up is the role of *calculation*, or formal transformation and derivation. Does it also work by some sort of parallelism, or what? Husserl does not know at this point, and closes the "*Selbstanzeige*" by saying that "The higher symbolic methods, of a totally different type, which constitute the essence of the general arithmetic of number, are reserved to the second volume [of the *Philosophie der Arithmetik*], where that arithmetic will present itself as one member of a whole class of arithmetics, united through sharing identically the same algorithm."

In candid statements from an 1891 letter to Carl Stumpf we find a much clearer explanation of his change in view. This important letter was written at the point where the galley proofs for Part I of *Philosophie der Arithmetik* had been completed. He hoped to complete the pages for Part II in the next nine weeks. But he confessed to his mentor and friend that he "possesses as of yet no coherent outline of the part dealing with *arithmetica universalis*," i.e., of Part II of the book, and can only hope that it will come to him 'in a rush', such as he had experienced with certain other writings. Then comes a startling admission that the guiding conviction of his *Habilitationsschrift* (of 1887) has proven false. The concept of cardinal number cannot serve as the foundation of general arithmetic. "By no clever devices, by no 'inauthentic representing', can one derive negative, rational, irrational, and the various sorts of complex numbers from the concept of the cardinal number." {p. 13} "General arithmetic" finds *application* to cardinals, ordinals, and other domains, but "no common concept underlies these various applications of arithmetic from which that science can be derived." {p. 13} The question then is: What constitutes its content? What sorts of conceptual objects are its propositions about?

Some of those propositions obviously can be interpreted in terms of cardinal numbers and their relationships, but many cannot – e.g., those involving such symbols as "$\sqrt{-1}$" – even though proofs involving them are perfectly correct and reliable. Husserl had originally treated signs only in relation to designated concepts. Then he noticed that one can have two formally identical proofs, one of which seems to deal with "impossible" concepts and the other with "possible" ones. This led him to see that the symbolic proof *process* actually used in arithmetic is indifferent to whatever concepts might be correlated to the signs involved. From this he concluded: "The *arithmetica universalis* is no science, but rather is a segment of formal logic. Formal logic itself I would define as a symbolic technique (etc., etc.), and designate it as a special – and one of the most important – chapters of logic as the technology of knowledge." {p. 17, cf. p. 50} He recognizes that this result calls for "important reforms in logic," and admits that he "knows of no logic that would even do justice to the very possibility of a genuine calculational technique" (*ibid.*) We have noted above that this lack remained *after* he finishes Part II of *Philosophie der Arithmetik*, and that he promises a remedy for it in a projected (but never completed) second volume.

The complaint about the inadequacy of current logics in relation to symbolic calculation is repeated at the end of selection 4 in the present volume, "On the Logic of Signs (Semiotic)." {p. 50} In fact, this fascinating piece is apparently Husserl's first systematic effort toward a 'logic', in his special sense, for symbolic calculation. Here we shall devote extensive comments to it, for it provides a framework for many of the conceptual developments found in the other papers in this volume. It is difficult to date with precision, but its concerns and much of its very wording seems to me to place it together with the Stumpf letter, Part II and the "Foreword" to *Philosophie der Arithmetik*, and the long critical review of Schröder. (selection 5) This locates it in the period from middle or late 1890 through the spring of 1891 – obviously a time of incredible labors, leading to fundamental reorientations toward the nature of the deductive sciences, as noted in the 1900 "Foreword" to the *Logical Investigations*, but already in the 1891 "Foreword" to *Philosophie der Arithmetik*. (*Husserliana XII*, p. 7) The impact of works by Schröder upon him during this period was very great – as he himself indicates (though not mentioning Schröder by name) in that 1900 "Foreword," and as I have elsewhere explained.[2]

[2] Dallas Willard, *Logic and the Objectivity of Knowledge: A Study in Husserl's Early Philosophy*, Athens, OH.: Ohio University Press, 1984, p. 120.

* *

"On the Logic of Signs" begins {p. 20} by stating the issue which, we have indicated, baffled Husserl at the time: How can we deal cognitively with a domain of objects (concepts) that is not really before us, and how can the most secure of all sciences be based precisely on work with such absent 'concepts'? The obvious answer is, by means of *signs* or *symbols*. Everyone knows that arithmetic does whatever it does by operating on signs. To learn arithmetic, especially in its middle to upper reaches, is to learn how to manipulate a sign system. With reference to cognition more generally, signs enable us to carry on intellectual engagement with subject matters not standing before us as the work proceeds, and perhaps not even thought of. But *how* do they do that?

In working toward an answer Husserl starts from the fundamental fact that some objects which come to consciousness do so in the course of bringing something else to mind, while others do not. The latter 'just stand there' without carrying the mind beyond themselves. Thus, "Contents [read: *objects*] can be given to us in a two-fold manner. First, in the *authentic* manner; namely, as that which they are. Second, in the *inauthentic* or *symbolic* manner; namely through the mediation of *signs* – signs which are themselves authentically represented." {p. 20} The focal phenomena of this entire investigation, and one seldom out of Husserl's mental horizon hereafter, is this characteristic 'deflection' of intentionality to 'an other'. This "relationship [of 'deflection'] between the sign and the designated must be specifically noticed" if the concept of sign is to be possible and if we are to be able purposively to invent and use signs. {p. 21} And, on his view, that relationship *is* directly given to us in select cases.[3]

The basic nature of the sign 'relation' empowers the sign for its special role in consciousness and cognition. The referent is somehow 'there' with the sign. "If something is not directly given, but is given only as mediated by signs, then the complex of these signs, or the sign composed of them, *stands in place of it*" {p. 30}, so far as our mental engagement with it is concerned. But a sign can deputize in this fashion in *knowing*, as distinct from mere consciousness, only if it is rigorously univocal and sufficient by itself to pick out its object.

[3] {p. 330} and see the VIth "Logical Investigation," subsection 6.

If it meets these two conditions then it is, by Husserl's definition, *an inauthentic representation.* {p. 30} Thus, every inauthentic representation is a sign. But not every sign qualifies as an inauthentic representation, because some do not meet the two-fold condition required to function in *knowledge* in his strongly normative sense.

In what follows the "signs" discussed are "inauthentic representations" in the sense just explained. On the way to answering his question as to *how* signs are capable of mediating knowledge of their referents, Husserl provides a brief exploration of certain distinctions between signs. At least nine distinctions are discussed, not all of equal importance for his question. First, he distinguishes *extrinsic* or external signs from *conceptual* signs. The former is a sign "which has nothing at all to do with the specific concept [character] of the designated." Proper names are cases in point. A "conceptual sign," by contrast, is a characteristic of the thing which it is used to designate, e.g., being red. Univocal and multivocal, as well as simple and composite signs are distinguished in the ways one would expect. {p. 22} The distinction between *direct* and *indirect* signs is also obvious. But it requires comment because of the vital role it is to play later on in Husserl's doctrine of the degree of fulfilment which a representation may have. This, in turn, is crucial to his account of *Erkenntnis* and its 'objectivity'.

The indirect sign," he says, "is a composite sign in which the partial signs are not at the same 'level' as each other, but are rather superimposed upon each other and refer to each other." {p. 23} That is, the indirect sign of an object deflects consciousness to it only by first deflecting consciousness to a further *sign*, which then (again, directly or indirectly) deflects consciousness on toward the object. The number of intermediary signs that can be interspersed between a sign and its ultimate object is left open at this point. Proper names are taken as examples of direct signs, while "any general name is indirect," because it designates its ultimate objects, in its extension, by first deflecting the mind to properties or "signitive marks" belonging to them. {p. 23} A very great deal of John Stuart Mill's theory of "names," with which Husserl was quite familiar, survives in certain parts of these classifications.[4] He then proceeds to classify signs {p. 24} into the identical and the non-identical, the equivalent and

[4] See Mill's *A System of Logic*, Book I, Chapters 2 & 3, London: Longman, 1970, or any of the many other editions of this work.

the non-equivalent, those for representation contents and those for psychical acts (especially *judgments*), and finally the natural and the conventional (*kunstliche*). On the whole, these distinctions obviously are not organized around any common principle, but all can be used in describing the sign systems of mathematics.

Now the signs which make up the algorithms of *arithmetica universalis* fall on both sides of some of these distinctions, but not of all. Specifically, they are extrinsic (and sense perceptible), *not* conceptual, they are conventional and *not* natural, and in the most important cases they are indirect and *not* direct: "All higher level mathematical signs are indirect, are multi-tiered signs of signs of signs .... We will see how *indirect, purely external,* altogether *univocal* signs .... play a role of the highest importance in arithmetic." {pp. 23–24} They are also conventional, not natural, even though the system as a whole has developed blindly and not purposively. That is, their power to 'deflect' to a further sign or an object is based on a purposive use of them to designate what they designate – and, indeed, for the very purpose of enabling us to represent and judge about numbers and number relations. Once established, however, these conventional signs work just like natural ones. "The same natural laws underlie conventional signs and natural ones." {p. 24}

It is important to understand that Husserl neither gives nor seeks any ultimate "explanation" of what makes 'signing' possible. Rather, he wants to clarify what it is, and this allows for observations of it arising in various ways, e.g., with as well as without *purposeful* adoption of the sign as designation of its particular object. "There would simply be no higher mental life – much less, then, science," without the purposeful employment of external signs to aid us in thinking and knowing. {p. 29}

With this classification of signs before him, Husserl proceeds to give an account of how they accomplish what they do for knowledge. It comes in two parts: a "psychological" and a "logical." The former opens with a description of the psychical activities which make up normal thinking. This description is given in much the same form in many of his writings – e.g., in selection #11 {p. 167} – and serves as one of the enduring points of reference for his work. He finds that the usual rapid flow of thought contains 'representations' of an extremely 'inauthentic' type. "Words or letters, accompanied by indistinct and unclear phantasms ... [bearing] isolated and fragmentary qualities, the rudimentary beginnings of higher psychical activities ... all without a fixed content or continuity ... – *these* are our thoughts." {p. 31; cf. pp. 167–168} Yet we normally proceed with assurance and to a correct outcome. "The signs and fragments of experiences stand in place of the genuine concepts, but *that* they do so is unnoticed by us."

Given this description of actual thinking, the question of how what is thus really present 'in' Experience (*Erlebnis*) does what it does for consciousness and knowledge beyond 'Experience' reasserts itself. And Husserl now replies that the ultimate subject matter is here 'available' to us – no doubt in some highly attenuated sense – in virtue of a property, or at least a disposition, that actually belongs to the sign as an authentic representation. There is, thus, an actual or actualizable connection here between sign and ultimate subject matter. {p. 32}

Thought bearing upon the concept of *the sphere* is offered as an illustration. With the word "sphere" a ball is imaged, "in which the shape alone is specifically attended to." This accompanying image, with a property crudely approximating to the intended concept, thereby symbolizes it. And it may disappear again leaving only the word, "but its appearance nonetheless suffices in order to secure us in a confident grasp of the subject involved." {p. 32} In general, there adheres to any representation we may have "a more or less extensive complex of memories: of words, of sentences, of images with absolute or relative properties usually noted, which are intimately connected to the [inauthentic, symbolic] representation by means of association, and from amongst which, depending upon the direction of interest, now these, now those, are reproduced." {p. 32} And we are so assured of the possibility of "unfolding before ourselves the full content of the verbal signification" that we take ourselves to be directly in touch with the relevant subject matter.

This story is somewhat complicated by important cases of symbolic thought where we cannot, and know we cannot, "fully unfold the verbal signification." {p. 33} Here the inauthentic representation is not a mere matter of abbreviation or convenience, but of necessity. The corresponding authentic representation, bringing the object 'itself' before us, is not a possibility in these cases, either at the time or permanently. A past event, for example, is forever lost to authentic presence. But also God, Africa, 'external' things, and other minds fall here, as do the non-existent objects of contradictory concepts. However, Husserl holds that nothing new in principle is required to understand how symbols relevant to such cases work. We must, indeed, learn the use of symbols – learn *what* 'symbolizing' or 'signifying' is – from the type of cases earlier described. {p. 36} But the 'tendency toward' – the 'reaching', the *intentio* – is obviously present even where it absolutely cannot be carried out to completion in the ultimate subject matter. The signs still have meaning, still are 'about' some specific subject matter.

Husserl is concerned to stress the fact that the role the sign or inauthentic representation plays is not one that depends upon our awareness that it is so doing. This phenomenological point is fundamental to later critiques of supposedly "scientific" accounts of various matters: e.g., Formalist accounts of logic, or physicalistic or naturalistic accounts of the life-world. The sign-relationship and the symbolic nature of our consciousness is, mercifully, lost to our awareness on most occasions, though we always *could* reflectively bring it before us. But this convenient forgetfulness can lead to dangerous denials. Our practice of forming judgments is not usually guided by awareness of the inauthentic character of our representations, but proceeds blindly as components of our experience naturally associate with one another. The train of associations, the natural disposition of signs to call forth certain others, "reproduces now this, now that group out of the associative complex belonging to the concept" at issue. {p. 36} Both our judgments and our inferences proceed *as if* they were following the "authentic and true concept of the relevant facts," even though they really are moved only by the associative tendencies of the fragmentary contents described above. Thus we have symbolic or inauthentic judgments and inferences as well as representations.

But this sketch of psychological laws that govern the movements of thought with symbols does not provide us with all we need to understand arithmetic. When judgment and inference enter the scene, they bring with them the questions of *truth* and *justification*. No mere explanation of what brings judgment and inference about can deal with the question of *right*, as Husserl very well knows. And when we ask the question of what right we have to proceed in thought as we naturally do, the first answer must be, according to him, "None whatsoever!" Normally "we proceed *without* any justification; we are guided, not by a motive to knowledge, but rather by a psychological mechanism." {p. 37}

Nevertheless, the question of truth does persist. Of course a logically unjustified procedure may well lead to true results. Indeed, they most often do – "A very interesting fact metaphysically," Husserl observes. {p. 37} But rather than take up the question of why that should be so, he attempts to find in the "inner structure" of the strangely reliable natural processes of judgment and inference "the grounds why they are suited to produce truth (though not knowledge)." He holds it must be possible to describe "a parallel logical process" – that is, a process purposively designed to secure truth. And by finding this "logical process" *in* the natural, unreflective process, we shall see that the latter goes *as if* it had been rationally devised to conform to the logical process. We see "why the non-logical process must nonetheless have the effect of a logical one, and that is the logical illumination of which we spoke above." {p. 38} Thus, in the cases where the processes are indeed so formed as to guarantee truth, we see that they *must* yield truth, and on the basis of this insight our result becomes, not just a judgment that happens to be true, if indeed it is true, but knowledge, assured truth, truth with insight into why it must be so.

What do we find 'in' the psychological process that could make it *logical* in the relevant sense? Husserl distinguished two cases. In the first our judgments are authentic, but we are basically judging about surrogates and not what the surrogates are 'standing in' for. The justification of the procedure stems from the fact that the surrogates include in themselves "precisely those constituents and properties of the genuine concepts upon which the judging interest is directed." {p. 38} The validity of the judgment is guaranteed for the authentic concepts, the true subject matter, inasmuch as they "possess, among others, precisely the properties of the surrogates that were especially attended to and judged of."

Hence the *logical* schema involved is: "A judgment applies to an object X provided only that X has the property A; G possesses the property A; thus the judgment also holds true of G." X is our surrogate, bearing the property A, and G is our intended subject matter. Not that we actually think in such forms. As already noted, we judge through the blind, associative powers of surrogates, but do so *as if* we were dealing with authentic concepts. And these logical forms present in the judging, though not consciously guiding it, enable us to see, upon reflection, why such a procedure, in the appropriate cases, leads to no error. We *see* that the appropriately structured non-logical process *must* yield the same result as the logical one – except that, in the latter, insight into the justification is given, which therefore constitutes the result as *knowledge*. {p. 39}

But in a second group of cases inauthentic judging and inferring "is based upon mere external symbols." Here it is the order of the signs themselves that associatively guides us in forming inauthentic judgments from the structure of other judgments. "Sensible signs (e.g., names) of the representations are combined with sensible signs of acknowledgement and rejection; sentences originate; systematic sequences of sentences symbolize deductions; and the judgment process consists, then, in allowing an external progression along the chain of signs to surrogate for genuine inference." {p. 39} Now the question is, once again, "What is the basis for that status as truths which is enjoyed by the results of these natural mechanisms?" In answer, Husserl points out that the association of ideas that produces the terminating sentence in these cases is keyed to *form*: "the external manner of linkage and arrangement among the premisses." Innumerable possible inferences have the *same* form, and many forms have a systematic type which "stamps itself upon the memory" in such a way that when in speaking and judging we "move a few stages into the familiar pattern, reproduction leaps ahead to offer us the form of the conclusion sentence." {p. 41}

This, then, is the psychological mechanism for the second group of cases, and in it we upon reflection find, once again, the parallel *logical* structure that forces the mechanical procedure to produce a correct result. The functioning of the psychological mechanism requires that the linguistic designators be univocal, at least in the context, and that the premisses involved determine the conclusion sentence in a purely formal and univocal fashion. {p. 42} Otherwise the appropriate associations could not arise. But this means that the same structure *could* have been adopted for purposes of achieving knowledge, i.e., if we were *planning* a process where the conclusion would have to be guaranteed truth. That such "formal" inferring is a tool of the highest importance for scientific progress is, Husserl says, something that will be demonstrated "by our own theory of arithmetic." {p. 43}

Now Husserl clearly recognizes that these very general reflections on how the structure in the psychological mechanisms governing sign use can be "logically illumined," yielding devices or guidelines that can be purposively utilized to attain assured truth, do *not* "provide a *logical* illumination for those algorithmic modes of procedure which have on such a wide scale become the vehicles of progress in the exact sciences." {p. 44} He has dealt only with the "natural judgment" as it employs and rests upon symbols. Natural thought processes often do in fact lead to error, and they therefore require careful regulation. But the same is true of the elaborate technical symbol systems that are essential to theoretical science. The algorithm of general arithmetic did not emerge as the result of a logically controlled purpose. "The system of arithmetic as a whole, with its wonderous structure, is not the product of intentional foresight, but rather is a natural development." {p. 46} Consequently, while it "works," and we quite well *know* that it does, we do not know *why* it works, why its results *must* be true – when, indeed, they are – and hence arithmetic could not present itself to Husserl and others of his day as a body of *knowledge*, as a system of logical *insights*.

"It is only," he holds, "in cases where the procedure itself is a logical one, and where we have logical insight into the fact that, such as it is and because it is so, it must lead to truth, that its result becomes, not a mere *de facto* truth, but rather a knowledge of the truth." Pre-logical operations with signs may attain truth without being based on logical understanding, but "*logical sign operations* ... are followed on the basis of knowledge [of why they work], and therefore do not merely provide us with truth, but rather with assured truth." {p. 47} "A utilization of symbols for scientific purposes, and with scientific success, is still not therefore a *logical* utilization." {p. 48} And this is precisely what we find in mathematics. Logical reflections on symbolic techniques can, therefore, prevent highly troublesome errors. But it also can invent novel methods and even establish rules for devising new methods. {p. 51}

At this point in his career, however, Husserl simply is in no position to realize his goal of a "logic of signs" – a logical illumination of how, in general, inauthentic representations can be used to provide "knowledge" in his strong sense of the term. The remaining papers in this present volume are largely to be taken as research reports dealing with various issues that must be dealt with *before* he can do that. We note that at this juncture he seems to have no idea of the manifold structural parallelisms that are indispensable to his later thought: those between the categorial structures of being in general, the logical structures of concepts, propositions and theories, and the sense perceptible structures of algorithmic symbol systems.[5] Similarly, he has no grasp of what he is later to call "the most fundamental of epistemological distinctions, the distinction between the real and the ideal." (*op. cit.*, subsection 51) Consequently, he does not even manage to articulate what later becomes *the* question for him: that concerning how the logically illumined structures discussed above are supposed *to fit together with* the psychological process of representing, judging and inferring, or concerning "the relationship ... between the subjectivity of knowing and the objectivity of the content known." (*LI.* p. 42) Thus he really cannot proceed to take on logical clarification of *arithmetica universalis* at this point. His initial "solution" raises too many problems that must first be resolved.

Succinctly, he must elaborate a general ontology within which the act/object nexus can be made sense of: one with a clear view of whole and part, property and instance, quality and relation, necessity and contingency. Only this will allow him to give the mind and language a 'substance' and character of its own – which it certainly must have – *without* at the same time creating a Cartesian or Lockean "inside" from which there is then no escaping. At this point he really does not know what to do with the categorial (the properly ontological) – or even with the mental, though he has picked up a few ideas about it, and some vocabulary, from Brentano. This is why in "The Logic of Signs" the representation is treated strictly as an *object*, never as an act. Many years of struggle with the nature of the "representation" (*Vorstellung*) lie just ahead of him, at some points in which he will state that it is better to avoid the word altogether {p. 146} and will experiment with new terminology – especially "*Repräsentation*" – in hopes of dispensing with it. (selection 11 below) But the exposition just examined is in terms of the qualities, powers and dispositions of the sign itself. The sign or "inauthentic representation" is treated as if it simply did things *to* or *for* consciousness. This is in some sense correct, of course, but no place is made for the grasping, interpreting, apperceiving of the 'sign' by or in the mental act, which becomes so

[5] Subsections 48 & 62 of the "Prolegomena to Pure Logic," which is volume I of *Logical Investigations*, translated by J. N. Findlay, New York: Humanities Press, 1970. Hereafter this edition is referred to as "*LI.*"

important later. The sense of our doing something to or with an object (the sign) "itself present" to confer upon it its signitive character steps onto the stage by 1894, as we shall see, and moves toward full flower in later works, as the 'stuff' of the mental act is increasingly filled out in Husserl's view. Nonetheless, to have circumscribed the basic sign phenomena, and to have in effect indicated that it has virtually an unrestricted range, so far as the objects which can manifest it are concerned, is a great step toward Husserl's ultimate goals, with results which remain prominent in his later writings.[6] The next major steps are the 1894 paper, "Psychological Studies in the Elements of Logic," and then *Logical Investigations* itself. But a side trip into the world of the Formalist logicians was mandatory.

* * *

In the context of Husserl's early work (see *LI* p. 53) there were really only two accounts of logic that had any wide acceptance. These are the psychologistic and the formalistic (or algorithmic), the latter of which grew into what we today call "symbolic" logic. The former, often associated with John Stuart Mill, but carried to astonishing lengths by lesser known writers such as G. Heymans, F. A. Lange (see "Prolegomena" chapter IV), and Th. Elsenhans {p. 248}, was simply *immolated* in the fiery criticisms of Gottlob Frege and Husserl – by Husserl in his "Prolegomena to Pure Logic" – though it is now rising from the ashes, once again, in certain quarters of contemporary Anglo-American philosophy. Only in a very few places does Psychologism come up for discussion in the papers of this volume, though some incisive comments on it are contained in #13 and in #14[c] {pp. 249–259}. Therefore we will not deal with it at any length here, but refer the reader to the "Prolegomena" itself, and to Chapter IV, Section 3 of my book earlier referred to, *Logic and the Objectivity of Knowledge.* Section 2 of this same chapter explains Husserl's critique of the algorithmic 'logic', which was usually taken to be a logic of "extensions" or classes, but we must

[6] See especially *Logical Investigations*, I (subsections 9–14, 20 & 23), III (subsection 15), V (Appendix to subsections 11 & 20), and elsewhere. The striking similarity between the views of C. S. Peirce and Husserl on the basic sign phenomenon should not go unnoted.

also say something about it here, since nearly all of Husserl's writings in this connection are contained in the present volume, specifically, selections 5 through 10 and 21[I–III]. We shall address the question of how, on Husserl's view, the formalist logicians such as Boole and Schröder fail, no less than the psychologistic logicians – though for different reasons – to provide a *logic* for the formal disciplines and for the sciences where they play an essential role.

We should note that in certain circles of early and middle Twentieth Century philosophy, Husserl was regarded as having decisively refuted the extensional and algorithmic interpretations of logic, as well as the psychologistic ones.[7] But of course "symbolic logic" won out culturally, if not philosophically, and his painstaking critiques remained buried in these early papers in obscure journals. To be sure, *he* did not forget them. Although he seems to have thought that the "formalists" had their hearts, but not their heads, in the right place, he continued to regard the results of his early critiques of them as so definitive that "formalistic" issues are hardly mentioned in the *Logical Investigations* or elsewhere subsequently. In 1903 he speaks off-handedly of "the follies of extensional logic laid bare [by him] over twelve years ago" {p. 199}, and in his last writings he characterizes the philosophical use of "symbolic logic" practised toward the mid-century as "the idolization of a logic that does not understand itself."[8] What, then, were Husserl's criticisms of the algorithmic 'logic' of Schröder and others in the late 1800's? First we must say that they had nothing to do with the formalization of deductive techniques as such. Husserl regarded formalization as an absolutely essential move in the progress of logic as a discipline. Toward the end of the "Prolegomena" (in subsection 71) he has strong praise for the new mathematical treatments of logical theories. Indeed, he attacks those who "remain stuck in the common prejudice that the essence of mathematics lies in number and quantity," and asserts that it is the

---

[7] See, for example, W. M. Urban, *Language and Reality*, London: George Allen & Unwin LTD, 1939, pp. 320–321 and elsewhere. Bernard Bosanquet, H. H. Joachim and H. W. B. Joseph were similarly appreciative. See the latter's "A Defense of Freethinking in Logistics," *Mind* XLI–XLII, 1932–1933, for a line of criticism remarkably similar to Husserl's in approach and outcome.

[8] P. 189 (cf. p. 141) of E. Husserl, *The Crisis of European Sciences and Transcendental Phenomenology*, translated by David Carr, Evanston: Northwestern University Press, 1970. See also my "Husserl's Critique of 'Extensional' Logic: 'A Logic That does Not Understand Itself'," *Idealistic Studies* IX, #2, May 1979, pp. 142–164.

philosopher who is out of bounds in attacking the mathematical form of treatment in logic, which in fact is "the only scientific one." (*LI.* p. 244) And this was not an attitude new to him in 1901, but is clearly present in the papers here to follow. {See for example pp. 93n and 127; cf. 244}

Given all of this, however, the leading proponents of algorithmic logic, of the 'algebra' of logic, show an astonishing capacity to misunderstand exactly what it is that they have and have not accomplished. Selections 5–10 and 21[I & III] below are entirely given over to demonstrating this point, and Ernst Schröder, as the most distinguished representative of the formalist tendency, is the one upon whom Husserl centers his often burning critique.

Schröder presents his "algebra" as formulating the rules or methods by which the human mind arrives at knowledge: *exactly* what Husserl is interested in! At the very opening of his book he describes the subject matter of logic (and of his treatment there published) as "thinking, insofar as it has knowing as its goal."[9] But when we look at this "algebra" of *deductive* inference, which is all that Schröder offers us, we do *not* find a logic in the sense that Schröder himself claims. {p. 57} We do not find an account of why valid thought processes in general are valid, but a calculative mechanism which, for the most part, dispenses with deductive *inference* altogether, and puts 'in its place' the manipulation of symbols in terms of formalized rules. Establishment of the very equivalence of the symbolic process with the deductive one which it 'replaces' is, correctly, no part of the algebra offered; but certainly it must be a part of what any genuine *logic* of signs would have to provide. Calculation and deduction are not only distinguishable, they are separable as well, and beg for a *true* "logic" that would put them into intelligible connection. {pp. 55–57} At most, what we get from Schröder is *one* symbolic technique which some deductive *thinking* might be translated into, thereby increasing our power to identify possible valid transitions, while insuring that validity would not be lost through the symbolic transformations permitted or their conceptual correlates.

But, as noted, the account of how the algebra itself serves the purpose of *knowledge* is no part of the algebra, and is not given along with it by the "Formalists." Moreover, deductive thought can occur

---

[9] *Vorlesungen über die Algebra der Logik*, Vol. I, Leipzig: B. G. Teubner, 1890, p. 1.

in its very best of forms without being routed through the symbolism. Calculation is not deduction; but, if it were, the logical calculus would still not be a *theory* of deduction – a point which Schröder and his associates completely miss, according to Husserl. The 'laws' of this calculus "are not rules to which everyone does and must conform, so far as one infers correctly .... Let the practice of deduction draw ever so much use from such methods ..., the logical *theory* of deduction ... is not even touched upon." {p. 57}

The still further move which takes the algebra to be a calculus of *classes*, only deepens the self-misunderstanding that Husserl finds among the "Formalist" logicians. If the calculus were a calculus of classes, it would still be only a calculus, and the above objections would still apply. But in fact the character of the algorithm *as* a calculus has no essential connection with classes or sets. Selection 6 below was developed by Husserl precisely to show that you can have a satisfactory calculus, that operates entirely as a calculus – one, moreover, involving precisely the Schröderian formula – where the interpretation given is in terms of "conceptual contents" (or "intensions," as we might say today). {p. 95 & 123} A strictly formal language as such is not determined by its interpretation, and is as friendly to an intensionalistic interpretation as to an extensionalistic one.

Moreover, the calculus *restricted* to the class interpretation ceases to be a *logical* calculus, for then it is incapable of expressing the various types of judgments and their interconnections which actually show up in natural epistemic progressions – most of which have absolutely nothing to do with classes, and have no need of being replaced by statements about classes in order to serve as vehicles of knowledge in the strongest of senses. On Husserl's view, the "natural" form of the universal judgment is: "Provided something is an *S*, it is a *P*." {p. 102} This form is the one present in the spontaneous movements of thought, and it refers only to conceptual objects: that is, objects *as* falling under certain concepts. Even statements about classes or sets can only refer to and judge about *their* objects *as* falling under the concept *set*. His own "calculus of contents (intensions)" is developed from this one judgment-form alone {p. 109}, and it can, he thinks, stand as a *truly* logical calculus precisely because all judgments are about conceptual objects, whereas not all are about classes.

Of course his *calculus* is, no more than Schröder's, a *logic*, a theory *of* epistemic progression – whether or not involving algorithms. Nor is *calculation* in even his "direct and authentic logical calculus" identical with the deduction that occurs in thinking. He constantly emphasizes the distinction between inventing techniques (algorithmic or not) for use in deduction and developing a *logic* of deduction, which explains, among other things, why such techniques work. "Nothing has so much contributed to the widespread disdain of the calculus as the vague confusion of these two objectives, so essentially different, on the part of those who have invented and advocated the calculus." {p. 101} In addition he notes the further confusion of the calculus with a *language* {pp. 68–69}, a criticism that later resurfaces in Wittgenstein. Now Husserl takes the distinctions which he finds here to be so obvious that the treating of an algorithm – and especially a "calculus of extensions" – as a logic, etc., must, in his opinion, involve some form of self-deception. {p. 54} No doubt this way of talking, which re-emerges at various points throughout his career, did not endear him to the formalist logicians. He also comments on "a peculiar fact" which shows up in both "the logical and the mathematical domains: One can be an outstanding technician in logic, while being a very mediocre philosopher of logic; and, again, one can be an outstanding mathematician, while being a very mediocre philosopher of mathematics. (Boole provides an outstanding example of both.) It almost seems as if the mental activities required in the one area and in the other were heterogeneous in the extreme, since it is so extraordinarily seldom that they are found united in one person." {pp. 57–58}

Perhaps he is right. In any case it is still true today that you will be hard pressed to get one who has just completed an algorithmic process – a formal derivation, a semantic tableau, etc. – to ask, much less answer, the question: How does *this* show that *a general form of inference*, or even a particular argument, of the appropriate corresponding type, is valid? They will usually operate from the assumption that the algorithmic process is all there is to know, or perhaps that the symbolic display just is the same thing as the general form or the argument. The role of logical symbolisms (today, including their computerized embodiments) in the production of knowledge, and the epistemology of logic itself as a body of knowledge, is in no better condition today than when Husserl concerned himself with it. Perhaps worse.

* * * *

But if help for Husserl's concerns cannot come from the "Formalists," where are we to turn? The answer which Husserl adopts – with what degree of self-awareness it is hard to say – is that we must turn back to the painstaking analysis of the very 'substance' of the mind and its mental acts and their components. We must examine the mind itself if we are to understand how it works – through algorithms or otherwise – to achieve knowledge. This might seem obvious, but it might also be thought that it had *already been done.* Did not Descartes, Locke, Kant, Mill and others do just this? The turn to 'language' and the resultant 'Analytic' movement in philosophy that occurred, mainly in the English-speaking countries, during Husserl's lifetime assumed that they had done it, and no doubt that *was their intent.* But Husserl was convinced that they had failed to carry through to success with an idea that was basically right. Nonetheless, to his way of thinking it remains the only way forward, and so he sets out firmly on the pathway toward what later came to be his Phenomenology – "Phenomenology as descriptive theory of the essence of pure Experience."[10] The first genuinely coherent steps in this pathway are represented by the 1894 article "Psychological Studies in the Elements of Logic" (selection 11); and the first compendium of results from following it is his VIth "Logical Investigation," titled "Elements of a Phenomenological Illumination of Knowledge," which culminates the period of his work that produced the papers in this volume and stands as his first full answer to the question that opened "On the Logic of Signs." {p. 20}.

Husserl's choice is, I think, clearly manifested in the title of the 1894 article. This crucial paper, the centerpiece of the present volume, is to be studied along with selections 17 and 21(IV–VIII). If one has understood Husserl's main problem after the collapse of his initial research program, its title and the direction of its inquiry will make perfect sense. Here we have two "*Psychologische Studien zur elementaren Logik*" – two studies of mental states and process, with the goal of attaining an elementary or basic logic, keeping in mind the sense of "logic" established in previous discussions. I have tried

[10] *Ideas I,* subsection 75; cf. p. 1 of *The Idea of Phenomenology*, translated by W. P. Alston and George Nakhnikian, The Hague: Martinus Nijhoff, 1973: "... phenomenology as the general doctrine of essences, within which the science of the essence of cognition finds it place."

to make the point of the enterprise stand out by translating the title: "Psychological Studies in the Elements of Logic."

By now we know that when Husserl speaks of "logic" he is referring to a theory of why the processes that bring us to knowledge *must* result in knowledge, in assured possession of the truth. The "pure" logic which emerges a few years later is only one essential part of a logic in this more inclusive sense. Such a theory is to be sought in "psychological" studies precisely because the transition from prior mental states *to* a condition of knowing is – and who but a few over-zealous 'sociologists of knowledge' could deny it – a transition that occurs through the mental processes of individual human beings. Something in certain states is to *guarantee*, to *necessitate*, something about certain others. In particular, certain operations with inauthentic representations or "signs" are to make knowledge of their 'absent' subject matter possible – which means to guarantee that, given the appropriate conditions, things actually are as they are represented, and that at least in some cases they could be *found* (authentically represented) to be so. {cf. pp. 166 & 168–169}

The first part of this treatise therefore deals with necessary connections between types or elements of mental states. It is titled, "Concerning the Distinction between Abstract and Concrete," but the issue back of this distinction really is that of *necessary connection*, for the 'abstract' turns out to be, precisely, the necessarily dependent. Thus, he opens with the observation that the totality of consciousness is always a unity in which each element exists in combination with every other element, and proceeds to note that the *manner* of unification varies significantly from case to case. Especially, some elements are *independent* or separable from others, while in other cases this is not so, the elements in question being *dependent* or inseparable. {p. 139} He mounts a general, seemingly disinterested, inquiry into the relation of dependence (inseparability) between mental elements and states, no doubt in the hope of eventually discerning the necessary connection between the operations of consciousness with symbols and the knowledge to which certain ones among them necessarily lead or can lead. Here we see his characteristic method – which shows up in his earliest publication (1887), and later becomes such a major part of "phenomenological" work – as he works by varying (largely in imagination, in "thought experiments" as it were) certain components of Experience wholes to see what the effects are on the other components and on the whole itself, and especially to see when changes in one component bring about changes elsewhere "as an Evident necessity." {p. 140}

Dependent 'contents' exhibit, in such cases, conjunctive variations with a *necessity* that is open to *insight.* Such variations are based upon what have been traditionally called "synthetic apriori connections." Husserl illustrates this type of dependency by reference to tone sensations, lately studied to such exhausting lengths by his associate Stumpf. "The intensity of a tone is not something indifferent to the quality; not something which is, so to speak, external to it ....The two contents interpenetrate. They exist within each other, not outside of each other." {p. 141}

If we were to check these descriptions from our own experience, now, we would have to keep in mind that he is here talking about what we should call tone *sensations*, and not tones themselves – though the findings are later extended to "objects in general," whether specifically mental or not, and whether real or Ideal. {p. 179n} They are, in short, later extended to all "Ideal law" connections, whatever the domain, and have no restriction to the 'mental' as such.

The "definition of concepts" now follows: A *dependent* content is one where we have Evidence that modification of at least one among the contents given with, but not included in it (not "analytic" to it), *must* yield a change or suppression of that content itself. "Any content of which this is not true is *independent.* In this latter case, the thought of the content itself remaining intact while all simultaneous contents are suppressed contains no absurdity." {p. 142} Husserl adds a helpful part/whole formulation for *dependent* contents: Of them he says that "we have ... the Evidence that they, being such as they are, are conceivable only as parts of more inclusive wholes." He also concedes that independence, while absolute in some cases, also admits of degrees for various types of contents. {pp. 143 & 144} Of special importance, in conjunction with his doctrine of "indirect" signs, discussed above, is his statement about *immediate* and *mediate* parts. Our inauthentic representations, it will be recalled, contain as parts the signs (or other inauthentic representations) which they depend upon in order to be representations of their ultimate subject matter. Now in 1894 this point is subsumed under his general conception of dependent parts. Thus, while no application is explicitly here made by Husserl to his problem of inauthentic or symbolic thought, a number of incidental points are established that are of great significance for the solution of that problem, as well as for the upcoming second of the "psychological studies." These include the discovery that the relation of concrete to abstract can itself be given in intuition, which means that we can (in select cases) *see* – or 'intuit' – necessary connections present to us. That is what the talk of "Evidence" and "Insight" is about. Also, being *abstract* is found *not* to consist in being the object of a special mode of consciousness, and hence not to require being the object of representation or concept as opposed to intuition. {p. 147} The abstract and the concrete can, in general, be equally intuited or represented or neither. Their status is ontological, not epistemological. Bearing these points in mind we turn to the second study.

Now its purpose is to achieve clear concepts of intuition and Representation, and of some of their contrasts and interrelations. He notes the coincidence in common usage between "intuition" and "perception," but only to draw on psychological findings that question it. "We think we *see* the equality of the side areas of a perceived cube, although the areas *really* seen are unequal. The perceptual judgment does not raise the factual content [the 'sense data'] of the perceptual representation to the level of objective consciousness. Rather, it finds in that content – more precisely, in certain of its Moments – its occasioning cause." {p. 149} (Later on he will speak of the Moments *founding* the judgment.) Most perceptual representations are not "authentic," but are inauthentic representations involving "deputization by parts, images, signs, and the like." They are *not* intuitions. "By such means the represented is in truth not *set here before us* at all. Rather, an intuition is a 'setting before' (*Vorstellung*) in a more literal sense, where the object is actually put before us in such a manner that it is itself the substrate of the psychical activity." {p. 150} For an act to be an intuition its content must "become immanent." {p. 150}

Certainly Husserl has far to go before clarity is achieved with such language, but he does here mean this – whatever else – by "immanent": that there is no element in the object of the act which fails to be really present to it exactly as it is intended. Intuition is "the simple state of being-turned-to a Present (*Präsent*) content ...." {p. 327} The "intention" of the act finds complete "satisfaction" in the content upon which it is directed. This is possible, but also may be lacking, in a wide range of acts, including act of conceptualization and phantasy, as well as of perception. {pp. 150–151} To repeat, 'abstract' contents are, in general, as susceptible to intuition as are concrete ones. {p. 152} We also find that representations are sometimes formed by substituting genuinely intuited contents for *supposedly* intuited ones – such as the ideal forms of geometry, which no one can literally see – and by utilizing signs as representations of what they designate. {pp. 153–154}

After his survey of cases, Husserl proceeds to specify the concepts *intuition* and *Representation*. He notes that there are two groups of "psychical Experiences" (*psychische Erlebnisse*) that are called "representations" (*Vorstellungen*). Members of one group "have the peculiar character of not including their objects in themselves as immanent contents (and thus, as present *within* consciousness). Rather, in a certain manner which still must be more precisely characterized, they *merely intend* their objects." "Merely intended" applies to a content (object) that is "not given *in* consciousness, but ... [is] aimed at, minded, referred to 'with understanding', by means of some contents which are given in consciousness." {p. 154} Note that "merely intentional" does *not* mean "non-existent." Obviously the *sign* structure discussed above re-emerges here. {p. 20} The contents 'within' consciousness are, he now says, "Representants" (*Repräsentanten*) of the objects "merely intended." They serve in this capacity without any conceptualization of whatever relationships actually obtain between the *Vorstellung* and the object intended. *Vorstellungen* in this first sense Husserl now baptizes "*Repräsentationen*," which we are translating as "Representation" with capital "R." Other "psychical Experiences" also called "representations" do not merely intend their objects. "Rather, they *really include* those objects *within themselves* as their immanent contents. *Vorstellungen* in this sense we call '*intuitions*'."

Contrary to what might be regarded as the most obvious of readings, the talk of being "really included" or "present" within consciousness or the representation should *not* be taken to refer to a part/whole or even predicate/subject relation. Rather, as we have noted with regard to "immanent," it refers primarily to a way of being given to consciousness. In the language of some years earlier, already discussed, it refers to objects "being given in the authentic manner; namely, as that which they are." {p. 20} The precise relationship between the authentic givenness of an object (content) and some type of ontological containment of it in the mind or act of consciousness is not yet clear for Husserl, and I suspect that as of 1894 he still partly thinks of ontological containment as what guarantees or makes possible the way of being-given that he here calls "intuition." By 1908, on the other hand, he has a firm hold on the ambiguity of transcendence (and immanence) as it applies to objects of mental acts. The immanent may be what is literally a constitutent, really (*reell*) contained in, the mind or act.[11] But it may also be what has "self-givenness in the absolute sense." (*op. cit.*, p. 28) It is, further, clear to him in 1908, and perhaps much earlier, that immanence in the latter or epistemic sense does not presuppose immanence in the former or ontological sense, especially as it concerns Ideal objects. But the 1894 paper seems to me to allow the two immanences (and the corresponding transcendences) to run obscurely together. However, the epistemic sense is nothing but the "authentic" manner of being given which, we have seen, emerges some years earlier in Husserl's writings.

In sum, a "representation" where all intended aspects of the object are 'present' just as intended is an *intuition*, while one where at least some aspects intended are *not* so present is a *Repräsentation*. But a number of further clarifications are now added. To be an intuition an act does not require development from prior Representations of the same object. "The child has intuition of the very first contents to which it attends." {p. 155} Nevertheless, the passing of Representations over into the corresponding intuition is among the most important of epistemic phenomena, and it plays a crucial role in Husserl's ultimate understanding of "the logic of signs" and of the power of inauthentic thought. "If a Representation goes over into its correlative phenomenon, e.g., into an intuition immediately intended by it, then the immediate psychical Experience of the fact that the intuited *is also* the intended shall be designated as consciousness of the *fulfilled intention*." {p. 156} That the concepts of *truth* and *being* are essentially connected with the possibility of appropriately fulfilled intentions, is a point of huge significance that is not clearly worked out in the paper, and possibly not until the *Logical Investigations* or even later.

Husserl further finds that intuition is not to be restricted to the con-

[11] *The Idea of Phenomenology*, pp. 27–28.

tent immanent *at* an instant in time. An object such as a melody is heard in one act elongated *through* time. It is fully given in the elongated act as a whole, even though all of the tones are never present at one instant. A similar, but not identical, point is to be made about physical *things*. "The limitation of the concept of intuition to momentary acts would too much impair its common and useful employment." {p. 158; cf. 317-318} The materials here translated under the title, "Intuition and *Repräsentation*, Intention and Fulfilment" {pp. 313ff.}, contain, relevant to these issues, some of the finest descriptive work in all of Husserl's writings. It is a pity that for one hundred years they have not been available as a standard for what phenomenological work should be.

Now a number of profoundly important descriptive insights are developed in response to the question of whether or not the same *kind* of 'representing' act is involved in the intuition and in the Representation. "Is it the intent of our definition that in both cases essentially the same psychic function, called 'representing', is carried out," with the distinction between Representation and intuition to be located elsewhere in the total configuration of the respective acts? {p. 159} The answer will be a definitive "No!" On the way toward this answer, however, we deal with the subordinate question of whether there "corresponds to each content [object] of consciousness an act which intuits it." {p. 159} The answer here also is negative, and intuition is marked off from two other ways in which objects can be *directly* present to consciousness. First, "background" objects, present "to howsoever determinate a degree," are not properly said to be intuited. "Only what is separately and specifically noticed can be designated as intuited," and "background" objectivities do not meet this requirement. But even this special sort of noticing is not enough. "Often several partial contents are isolated off to themselves, and stand out from a common background, and yet we ... 'look at' only one." {p. 160} In attending to a melody, for example, certain background noises are also very determinately heard and even recognized. In this case, "We have heard the noise, but not intuited it." {p. 160} The "declarations of inner experience" mark intuition as "a peculiar type of engagement with ... a content [object] that is separately and specifically noticed." Thus we have three distinct modes of presence singled out, corresponding to objects in the "background," to objects "separately and specifically noticed," and to objects attended to with an additional descriptively characteristic emphasis. The latter, which – in a fully developed consciousness at least – always presupposes the other two, is what Husserl here calls "intuition." These distinctions undergird many of his later investigations, including his critical discussions contained in the present volume. (See for example the discussion of Cornelius. {pp. 186–188})

But neither intuition nor the other forms of presence noted constitute that "psychic engagement with a content" found in the Representation. It is obvious, according to Husserl, that not every psychical act is to be designated an intuition of its object. E.g., the aesthetic disapprobation of a painting is not an intuition, although it is based upon an intuition. {p. 161} Similarly with the Representation. Consider the cases where "the same content [object] is at one time the basis of a pure intuition, but at another time the basis of a Representation ..., [or] where certain figures or arabesques first effect us purely aesthetically, and then there suddenly flashes the understanding that they could be symbols or word signs." {p. 161; cf. 331ff.} We should note that in this quotation we come upon another one of those descriptive reference points which serves to anchor vast stretches of Husserl's work. He urges us to "fix upon" the dawning understanding *as* it "suddenly flashes," before this "new state of affairs comes to explicit logical consciousness through thoughts and words – which often can happen only after some span of time." He also describes related cases of the occurrence of signs or words with and without significational consciousness, as well as the somewhat different "distinction between simply taking a *concretum* X just as it itself is, and then grasping it as Representant for any arbitrary X." {p. 161} In the range of cases, of variable types, thus marked out, "what we have is obviously no mere distinction of content." More content is not what constitutes the 'shifts' noticeable here. "If anywhere at all, it is here that the witness of inner experience is clear, and yields the conviction that there exists in the two cases in question a different mode of receiving the content into consciousness, a different manner of psychical engagement with or in the content." {p. 163} Consequently, words heard are not, when functioning meaningfully, to be described as intuited. Sign consciousness does not consist in intuitions of the symbols plus something more. "The immanent content of a Representation is, as such, not intuited." {p. 164; cf. 349, 390n & 413} Many types of the higher order and more complicated acts require an elaborate interweaving of intuitions and Representations. But the two types of acts remain radically distinct in the way they bear upon contents or objects. With all this established, Husserl concludes his 1894 paper with some observations on work still to be done. He acknowledges that "the solution to the problem of *what kinds* of conscious states they [intuition and Representation] are is not yet attained." {p. 165} Whether they are simple or composite, and how they relate to other known phenomena, remains in the dark. He emphasizes the fundamental significance of research into these two functions "for the psychology of knowledge and logic," and especially as concerns the Representation, since all logical activities are based, precisely, upon Representatives. He expresses astonishment that there can be such a thing as the Representation: "a psychical act [that] can reach out beyond its own immanent content to another content which is not truly encountered at all ..., [where], while we are engaged with the Representing contents, we take ourselves to be employed about the Represented objects themselves." {p. 167} Here, he thinks, "lie great, unsolved puzzles ..., the most obscure parts of the theory of knowledge ...," having to do with "the possibility of knowledge in general." {p. 167} Especially scientific knowledge, in the understanding of which both the older and the newer logic "totally leave us in the lurch. Logic, the 'theory of science', must concede, if it will be honest, that all science is a mystery to it." {p. 168} The by now familiar complaint still stands.

* * * * *

However, the significance of the "Psychological Studies" for the development of Husserl's views is so great that a few additional comments about its historical position in his work may be useful. He himself states, in his "Personal Notes" of 1906, that this text "is a first sketch of the *Logical Investigations*, especially of the IIIrd and the Vth." {p. 491} How are we to understand this claim?

It will be recalled that the IIIrd Investigation is titled, "On the Theory of Wholes and Parts." It clearly corresponds to the first "Psychological Study," ostensibly devoted to the distinction between abstract (non-independent) and concrete (independent) contents. {p. 139} Now in the Introduction to the IIIrd Investigation Husserl remarks that this distinction between *contents* "first showed up in the field of descriptive psychology of sense-data" (*LI*. p. 435), which of course is exactly what we have in the first Psychological Study. But in the IIIrd Investigation Husserl finds that contrast between non-independent and independent which can be *viewed* in sense contents to be but "a special case of a universal distinction" which "plays an extremely important role in the field of *objects as such*," and which receives its systematic treatment "within the *pure* [*a priori*] *theory of objects as such*." (*LI*. p. 435) This pure theory is, of course, to be expounded in the IIIrd Investigation, though the realization that the first "Study" was in fact dealing with a distinction applicable to objects of every type goes back at least to 1897. {p. 179n}

When in the IIIrd Investigation we come to Husserl's results regarding the distinction in question, we note a radically new element: that of the *species* of the contents (objects) concerned, and of the Ideal laws ("laws of essence") governing those contents *through* their species. Thus: "Non-independent objects are objects belonging to such pure Species as are governed by a law of essence to the effect that they only exist (if at all) as parts of the more inclusive wholes of a certain appropriate Species." (*LI*. p. 447; cf. p. 460) Husserl claims that this is "what we mean by the terser expression that they are parts which can only exist as parts, that cannot be thought of as existing by themselves." The color of this paper, for example, "is by its pure Species predestined to being as a part." Independent objects, by contrast, fall under no such species law, and "may, but need not, enter into more comprehensive wholes."

So it is a new understanding or appreciation of "the most fundamental of epistemological distinctions, the distinction between the real and the Ideal" (*LI*. p. 193), that allowed Husserl to move from the stage of the "Studies" to that of the "Investigations." "One must," he remarks in 1900, "clearly grasp what the Ideal is, both intrinsically and in its relation to the real, how this Ideal stands to the real, how it can be immanent in it and so come to knowledge." (*loc. cit.*) Once this is done, the descriptive distinctions which emerge in his 1894 explorations take on their proper and profound significance for the interpretation of representation and knowledge. Beings or objects of all types – sense-perceptible or not, individual or universal, mental or physical – exhibit the same structures of dependence and independence devolving upon Species laws, and thus they all reside within frameworks of "synthetic a priori" or even categorial ('formal') interrelationships that open up to "insight" with "Evidence" when properly approached. "Ideas [essences] themselves can, in a corresponding, if somewhat modified sense, be spoken of as 'independent' and 'non-independent'." (*LI*. p. 448) Essence laws will prove to be the key to answering Husserl's fundamental question about the knowledge function of 'signs' or 'inauthentic' representations.

The crucial relationship of "foundation," which plays such an indispensable role in all of his further work, is provided its systematic exposition in terms of dependence and independence in Chapter Two of the IIIrd Investigation, and the relation of whole and part itself is then explained in terms of foundation. (*LI*. pp. 474ff.) In these expositions also, everything turns upon synthetic unity under Ideal law. Husserl deftly avoids what in Anglo-American philosophy came to be known as "Bradley's Regress," by basing unity, not on further *parts* of some type – which would always require yet further parts to unite *them* in turn into the whole – but upon the *already* unified species under which the parts of a whole fall.

The conceptualization of the whole/part structure under Ideal law is applied to the analysis of linguistic meanings in the IVth Investigation, which sketches out a *pure theory of semantic forms*. (*LI* pp. 493f.) It is then further applied to the analysis of the conscious act as such, with special emphasis upon acts of representation and their fulfilment in intuition, in the Vth and VIth Investigations respectively. So the second Psychological Study, "Concerning Intuitions and Representations," is in fact expanded to encompass both the Vth and the VIth Investigations. This was because Husserl came to see, first, that the 'act', the basic unit of intentionality, has its essential integrity independently of fulfilment of the intentionality resident in it, and second, that fulfilment in intuition – being a matter of two or more specifically qualified acts in a whole arising through a peculiar type of "synthesis of identity" – required an eidetic analysis separate from, though dependent upon, that of the mental 'act' itself.

The Vth Investigation deals with laws of essence governing the make-up of the mental act as such. Its title reads, "On Intentional Experiences and Their 'Contents'." (*LI*. p. 531) Not every element of consciousness, nor every element in an act, is of or about an object. Those elements in conscious life which are of or about objects are called by Husserl "intentional Experiences" or "acts." The act is a certain type of whole, grounded internally – and to some degree externally – by the ideal laws which govern the species that define it and its parts. In the Vth Investigation Husserl painstakingly analyzes these laws and parts, and devotes exhaustive studies to such special issues concerning them as the sense in which every act of consciousness is 'founded' on a representation, the relationship between judgments and representations, the contrast between intentional and immanent 'contents' of acts, and what is and is not to be made of the talk of 'intentional' or 'mental' inexistence of objects. The crucial papers in this volume on Twardowski and on Intentional Objects, as well as the draft of a letter to Marty {pp. 473ff.}, lay out the reasoning back of his position on this latter topic more fully, I think, than do the "Logical Investigations" themselves.

These inherently interesting and often misunderstood matters are not just discussed for their own sakes, however. They *must* be correctly interpreted in order finally to clear the way to an answer for Husserl's basic question of how what is present in or intuitively before consciousness allows us to have knowledge of what is neither present in nor standing directly before consciousness. And this answer is given – in a fully developed, though later to be supplemented form – in the VIth Investigation.

Briefly, his answer is that what is present in or intuitively before consciousness in inauthentic or symbolic representing, judging or inferring, is laden with *dispositions tending toward other experiences – indeed, usually many layers of other experiences – through which the same object may be brought before the mind in increasing fulness of confirmed detail*, up to the point where it is given as fully as possible, in a manner essentially prescribed by the type of object it is. We do not, of course, bring the validity of an argument or argument form, the truth of a theory, or the loyalty of a friend, before us for understanding, confirmation or disconfirmation, in the way we would the species of a tree or the existence of a rodent in the hedgerow. But every representation or merest thought or mention of an object specifies an 'approach' to that object through further relevant acts, culminating, ideally, in the fullest possible presence of "the thing itself," or allowing us to determine that there is no such object, or that it is quite otherwise than it seemed. The fulfilment of a representation is thus a synthesizing totality (whole) of acts, organized under Ideal law, in which the object more or less absent at the outset becomes progressively more present. To *know* the possibility of fulfilment, with respect to symbolic representations and judgments – as with our algorithmic formulae – is, indeed, to know that the absent subject matter is as it is represented, however "inauthentically." For there would be no such possibility if it were otherwise. The elaborate discussions of fulfilment, 'disappointment', 'levels' of knowledge, intuition, Evidence and truth, in the IVth Investigation, are Husserl's attempt to provide the outlines of how this dispositional solution to his basic problem about knowledge on the basis of non-intuitional representations, judgments and inferences works.

In subsection 10 of this Investigation he sums up his point about knowledge with inauthentic representations, though he is there specifically dealing with our consciousness of sense-perceptible objects: "All perceiving and imagining is ... a web of partial intentions, fused together in the unity of a single total intention. The correlate of this last intention is *the thing*, while the correlate of its partial intentions are *the thing's parts and aspects*. Only in this way can we understand how consciousness reaches out beyond what it actually experiences. It can so to say mean beyond itself, and its meaning can be fulfilled." (*LI*. p. 701; cf. {p. 167}, quoted above) We re-emphasize that the process of "reaching beyond" and achieving the object itself in intuition will vary, depending on the type of object in question. But the basic, Ideal law structure of parts and wholes and founding relations between acts progressively closer to intuition, and finally – in some cases – reaching the objects "themselves," is the key to understanding the possibility of knowledge in the absence of the object known.

The papers collected in this volume do not, individually or collectively, offer a solution to the problems confronting Husserl's attempts in the 1890's to understand cognition, scientific or otherwise. However, we do find here and there in them major parts of the solution that came to be systematically elaborated in the *Logical Investigations*. Three of these parts deserve special mention here because of the light they cast upon his growing insights into intentionality. These concern: 'defect' as the root of 'reach' (*intentio*), the so-called intentional or 'inexistent' objects, and the nature of truth.

The Intentional 'Reach': We have noted that the 'reach' of Representations (or of acts of consciousness generally) beyond what is 'in' them was a matter of astonishment to Husserl. He repeatedly returns to this phenomenon in his works, and these papers contain some fundamental advances in his effort to understand it. Of course he always works from the idea that only what *is* in the conscious act can enable us to 'go beyond' it to what is not. (His version of "internalism," as we now call it.) The basic sign structure is present here. But what is the factor that makes it work? Selections 17 and 21[IV–VI] are especially relevant to this question, and the main point which emerges from them is the role of a certain *lack*, or *deficiency*, or *restriction* which always manifests itself *upon* the content that serves as sign or inauthentic representation.

Intention is described by Husserl as "a tensed interest that is incited by a given content but is not directed upon it .... It has an Ideal direction upon a content not given, tinged in conformity with that content's [object's] special character; and it does not have the character of the satisfied interest – which, rather, is the corresponding positive feeling." {p. 453} He devotes careful explorations to how perceptual contents, e.g., the beginning of a melody {p. 314}, as well as images, words (perceived or imaged), verbal significations, concepts and conceptual contents (universals) *all* carry with them, for the ordinary mind, a penumbra of associations which establish determinate "lacks" that constitute the intentional bearings of the act involving them beyond what is actually given in it. {p. 336} "If a familiar melody is broken off in the middle, or is incorrectly played further, if an unusual letter strays into a printed word, if an easily observable detail in the familiar arrangement of objects in our room is changed, and so forth, then we *directly* have the feeling of a certain lack. We feel ourselves to be restricted in a certain way – not restricted in general, but rather restricted by what is intuited. The feeling seems to emanate from it." {p. 337}

It is the *determinacy* of the lack that brings forth, or is disposed to bring forth, quite specific further objectivities. The relevant 'lacked' imparts its own special coloration or "tinge" to the lack and to the symbol bearing it. As in the case of our earlier discussion of tones, the segments of experience which are interwoven through the penumbras of associations "do not simply exist along side" of one another. {p. 314} Because of this "character of the unfinished, the deficient, we 'feel ourselves drawn onward', the movement begs continuation." {p. 314} A "quasi-quality" or "Gestalt" emerges, with dispositional links to all sorts of further objects, which in turn lead on to others. The comments in this passage apply directly to the hearing of a melody, but they are paradigmatic of all forms of 'the reach beyond' the given. On the other hand, any perfect fulfilment eliminates the respective intentionality *qua* 'reach beyond'. As he later says, "The intuition fulfils the intention which terminates in it as not itself again being an intention which has need of further fulfilment, but as offering us the *last* fulfilment of our intention." (*LI.* pp. 762–763)

Intentional Objects: Of course the object which *would* offer the ultimate fulfilment of the reach may not exist in a particular case. Or is that indeed possible? Would that amount to saying that there was a reach which was a reach for nothing at all? And how then could it even be *a reach*? This presents us with "the paradox of so-called objectless representations," which Husserl submits to a brilliant analysis in selection 18, given the title "Intentional Objects." He determines that the apparent contrast between intentional and real *objects* of conscious acts is not a classification of objects at all, but of representations. The object, whether real or not, is never resident in a representation of it. {p. 353} And "a merely intentional object is a non-existent one, a true object an existent one." {p. 356} "There are no 'different modes' of existence and validity." {p. 366; cf. 368–369} To say that an object is real or intentional is to say something about how the corresponding representation does or does not fit into the various contexts of judgments possible for it. In particular, it is to say something about the truth or falsity of the corresponding existential judgments. {pp. 369, 375} "It is the signification alone that is the inner and essential determination of the representation, whereas the *objectual relationship points to certain frameworks of truths or judgments* which the signification fits into." {p. 376}

Truth: The tie of mind to being is, then, for Husserl – contrary to many who read him through Meinongian lenses – not representation or intentionality alone, but truth. However, "the discussion of the relationship between representation and object inherently leads to discussion of the concept of truth." {p. 381} Truth, and the Experience of truth in fulfilment, is treated in numerous passages in this volume, but above all in the second part of the selection on "Intentional Objects." {pp. 378ff.} Husserl accepts a version of the "correspondence" theory of truth, but rejects all 'similarity' accounts of what the correspondence amounts to. To understand what truth is, and what kind of correlations it involves, we must "look at cases where we can intuitively encounter the agreement between representation and object: hence, where the 'object itself' is given, and simultaneously the representation which represents it – both given in *one* act in which the agreement is a lived experience." These are the cases of "Evidence, in which we grasp the truth itself, and thus do not merely represent or hold-to-be-true, but rather live through and behold (*erleben und erschauen*) truth." {p. 384}[12]

In such cases "We have insight into the law that the signification *in specie* and the fulfilling sense *in specie* 'agree'." {p. 384} Thus, the prior correlation here in the *Experience* of truth is a matter of an *identity* between the senses of a "mere thought" and of a perception of what the thought is about. Reflectively "living through" or Experiencing *this* identity, however, makes it possible to compare the objectivity itself to our consciousness of it, and it is here that truth (or falsity) is found. Truth is a relationship which shows up between consciousness and its objectivity when the objectivity is found to be as it is thought to be. It is "an adequation with the 'thing itself'." (*LI.* p. 763) But while this is where we *find* it and come to know what it is, truth is in essence a relation between the Ideal constituents of acts of consciousness – propositions or "judgments in the logical sense" – and what those propositions are inherently about, regardless of whether or not they are ever actually thought by, or occur to, anyone. Indeed, it is a simple fact that not all truths which refer to objects "come to knowledge and are thus contents of our actual judgments." {p. 378} Although the full-blown doctrine of the Ideal content of the judgment is not yet at his disposal, he sees that truth and falsity are, as he later says, "essentially properties of the contents of such [actual] judgments" (*LI.* p. 165), and he does not allow truth and falsity to depend for their existence or nature upon our acts of consciousness.

On the other hand, knowledge of truth is what, for the most part, mediates our knowledge of existence. Most of our judgments of existence are made and sustained without confirmation. But: "If an intuition of the object subsequently occurs, then there results an identification of what was 'represented' and affirmed *as* existing with the object intuited. In such a way we obtain intuitive knowledge of existence, we grasp the correspondence between 'representation' and 'object'. And conversely: because we have frequently lived through such correspondence, we are capable of linking the merely

[12] Cf. *The Idea of Phenomenology*, pp. 29–30.

'represented' up with the representation ... or objective assertion of such correspondence ...." {p. 333} Not intuition, but the rigorous correlation of being with truth is, accordingly, the foundation of most of our knowledge of what is and how it is.

* * * * * * *

Having now emphasized major conceptual points from the selections here translated, I leave the reader to ponder the riches of Husserl's descriptions, analyses and arguments. I have followed certain conventions of usage and spelling in order to keep some of his distinctions before us which might otherwise shift or be lost as one reads:

"Authentic" is almost always used to translate the term "*eigentlich*," which is associated by Husserl with the special type of situation where the object of an act is "itself present," or at least is not represented merely symbolically ("*uneigentlich*").

"Evidence" and "Evident" with capital "E" translates the word "*Evidenz*," and its variants, the meaning of which – closely related to that of "*eigentlich*" – can only be established from these and other texts

"*Erlebnis*" is in most cases translated by "lived experience," while "experience" translates "*Erfahrung*," except in a few cases where the context clearly indicates "*Erlebnis*."

"Idea" with capital "I" always translates "*Idee*," never "*Vorstellung*." "Ideal" is usually capitalized to emphasize the objective status of what it qualifies, namely, the "*Ideen*" and their laws.

"Illumination" and its variants, with capital "I," translates "*Aufklärung*" and its variants, in opposition to "*Erklärung*" ("explanation"). In a few passages where usage demands it, "Elucidation" (with capital "E") is used instead of "Illumination."

"*Kunstlehre*" is often not translated, though it is sometimes associated in the text with the English term "technology." It is, I think, not translatable except by lengthy paraphrase. It refers to the theory of a technique – which, etymologically considered, is a techno-logy, but is not what "technology" currently conveys.

"*Kunstliche*" is translated as "conventional" *or* as "artificial," depending on context, but sometimes the nuances are unclear.

"Moment" with capital "M" translates "*Moment*," which Husserl generally uses for a non-independent individual, in opposition to a "fragment" ("*Stücke*").

"Reel" is used to translate "*reelle*," marking individual *psychical* existents, as opposed to "real" ("*reale*") individual existents, subject to both space and time.

"Reflexion" with an "x" indicates reflexivity, self-reference. "Reflection" is reserved for a process of consideration, meditation, thoughtfulness.

"Representation" and its variants, *with capital "R,"* never translate "*Vorstellung*," etc. It translates the special term "*Repräsentation*" and its variants, which were used by Husserl to mark a very important distinction in the analysis of the mental act.

"Representation" and its variants, *without* capital "R," always translate forms of "*Vorstellung*."

"Sense" always translates "*Sinn*," except where its other meanings will be very clear from the context.

"Signification" always translates "*Bedeutung*."

"State of Affairs" always translates "*Sachverhalt*," except for a few cases where it translates "*Sachlage*."

I periodically re-introduce the German term in brackets to aid the reader's memory of my usage.

Numbers standing alone in "[ ]" symbols in the right-hand margin always refer to the pages of the '*Husserliana*' volumes used in translating.

I am especially indebted to Norbert Purzkar for working through the text and translation of selections 1–16 and 22, and to Kerstin Plack and Kathleen Chapman for the same on the remaining selections. Arthur Szylewicz labored over the translations of 17–20, 21[I–XII] and 22. I have not always followed the advice of these fine scholars, and they are not responsible for errors remaining.

Rudolf Bernet and Reto Parpan gave invaluable assistance on several matters, as did Philip Buckley and Reggie Lilly. Mrs. Ingrid Lombaerts kindly invested a huge amount of time and talent in the typescripts for this volume, and saved me from innumerable errors. My heartfelt thanks to her. And finally, without the excellent editorial and interpretive work of Bernhard Rang (on *Husserliana XXII*) and Lothar Eley (on *Husserliana XII*) I could had done nothing.

I have compared the French translations by Jacques English where these were available.

The Faculty Research and Innovation Fund of the University of Southern California has provided financial assistance.

My colleagues and students at the School of Philosophy in The University of Southern California, and especially the Directors, John Hospers, Martin Lean and John Dreher, have been very generous and supportive through the years it has taken me to complete these translations. I am grateful to all of them, and to David Malone and Marshall Cohen, Deans of the Division of Humanities at USC.

It has cost my immediate family, Jane, John and Becky, a very great deal for me to do this work. In addition, Becky assisted with "data entry," and John saved the entire project from loss by his computer expertise. Jane always makes it possible and delightful for me to keep going.

There is no way to acknowledge adequately the greatness of W. H. Hay, as a human being and as a teacher. This is a frustration which I share with many of his students. So I will only say, "Thank you Bill," for – along with many other wise kindnesses – suggesting long ago that I should read Husserl's *Logical Investigations*, and for spending many hours discussing them with John Atwell and me.

A final, no doubt presumptuous, word. If you have already read a great deal of Husserl, and regard yourself as familiar with his views, try to "bracket" what you believe you know about him and his views and just follow the line of these early texts, doing the phenomenological work on the cases discussed, and relevantly similar cases, as you go along. A famous student of Husserl long ago advised readers of his works to read them backward, starting with the last writings and proceeding toward the earlier ones. This frankly seems to me to be contrary to the very laws which govern the understanding of an author's contributions; and I am convinced, for whatever that is worth, that it is impossible thoroughly to understand Husserl's later works without first coming to terms with the painstaking expositions of detail that are contained in the early writings but are either abridged to the vanishing point or just omitted in the later works. Failure to proceed in this way seems to me partly to explain why Husserl has so largely ceased to be a living force in philosophical research today and lives on only in a supressed form in the works of people such as Heidegger or Derrida or Ricoeur, who reject his basic aspirations and arguments for a realistic philosophy of knowledge. In any case, the "official" way has been tried, and the result stands before the philosophical public. It might at least be interesting to see what the effects of reading Husserl in the way I have suggested would be.

THE CONCEPT OF GENERAL ARITHMETIC[1]

After completion of the investigations preliminary to general arithmetic,[2] we now turn to the logical analysis of the domain itself. We will do well at this point to call special attention to the presupposition from which we began, as something self-evident, and which dominated the entire process through which our investigations developed. It was simply this: that arithmetic has to do with cardinal numbers (*Anzahlen*) – "numbers" in the most underivative and narrowest sense of the word – and with their reciprocal interrelationships. Accordingly, we must take *general* arithmetic to be the general theory of operations appertaining to the theory of cardinal numbers. Wherever we have spoken of "operations" we were dealing with combinations of cardinals, or of conceptual and sensible symbolizations standing in their place. With respect to the starting point and the germinal core of our developments toward the construction of a general arithmetic, we are in agreement with mathematicians that are among the most important and progressive ones of our times: above all with *Weierstrass*, but not less with *Dedekind*, Georg *Cantor*, and many others. Indeed, we even feel ourselves to be in a certain harmony with investigators who, like *Kronecker*, hold the cardinal concept to be a mere derivative from the concept of the ordinal number. In fact it must be realized that the progression of logical developments from

[1] 1890. First published in *Husserliana XII*, The Hague: Nijhoff, 1970, pp. 375–379.

[2] The reference here is to the 'psychological' investigations of Part I of *Philosophie der Arithmetik (Husserliana XII*, pp. 5–178). This and the following piece should be understood as fragments from Husserl's efforts to solve the problems of Part II of that book. (D.W.)

a starting point in the ordinals (assuming its soundness) to general arithmetic would have to run completely parallel and be analogous to the one followed by us up to now, and that all of the problems which will yet arise, and all of the difficulties which have to be cleared up, remain *mutatis mutandis* the same for that way of thinking as for ours. And thus it is in general with any manner in which general [375
arithmetic is presented (however the presentation may be conceptually founded), provided only that it considers the domain of "positive whole numbers" to be the first and most underivative domain, the sole foundation of all remaining domains of numbers. *Which* concept it thereby assigns to this expression, "positive whole number," makes no essential difference for the formal progression of further developments, with all of its enigmatic obstructions. We can here dispense with giving the grounds of this circumstance, which is noteworthy in itself, and with giving evidence for our claim; especially since we would otherwise have to presuppose as already cleared up questions whose illumination is, precisely, the main goal of our further inquiries. Moreover, this chapter will no doubt present an abundance of confirmatory examples. It is enough to indicate that we find ourselves in full harmony with a number of the leading mathematicians – and in wide-ranging agreement with yet others – in our view held to this point concerning the foundations upon which arithmetic rests. And precisely the same holds true if we glance over into the sphere of professional philosophers. Indeed, here we can even claim that of those who have devoted any special inquiries at all to the questions of the philosophy of mathematics, probably most of them have thought that in the understanding of the domain of cardinal numbers we possess the key for the understanding of the science of arithmetic in general. Among many others I mention *Mill, Bain, Sigwart, Wundt,* and *Dühring*. Certainly we must not be silent concerning the fact that yet others, mathematicians as well as philosophers, are of a different view. The one group of them supposes, for example, that it is not the concept of the cardinal, but rather that of continuous magnitude (whether simply that, or of "directed" or vectoral magnitude) which underlies general arithmetic; whereas others, once more, denying *any* general foundation, see in this discipline a true formalism, devised for the purpose of solving problems from the domains of magnitudes of the most diverse kinds. Undiverted by such disagreement, we will strive to adhere consistently to the path we have in fact adopted. The

dialectical conquest of the grave difficulties which many times seem to bar that path will of itself lead us to present and criticize those other theories; and we take seriously the question of the viability of our own provisional theories and of the extent of the domain of application which they govern. Moreover, it goes without saying that it cannot [376] be our intention to provide an exhaustive logico-critical presentation of the general arithmetical domain. Our developments will be more suggestive than detailed, in the course of which we want to investigate specifically a few main points that, because of their obscurity or difficulty, have up to now resisted logical understanding. In older times the mysteries of the calculus were very much complained of. A quite comprehensive literature, to whose development the most outstanding of philosophers and mathematicians have contributed, testifies to a serious effort to suffuse those mysteries with the light of knowledge. This fringe area between philosophy and mathematics has been called "the metaphysics of the calculus" (probably following the example of *D'Alembert*). Not as if it had anything to do with first philosophy, with metaphysics in the primary and literal sense. But, of course, wherever a range of curious and maddening puzzles mocks at human knowledge, there the name "metaphysics" must not be lacking. In our time a complete transformation of official scientific opinion concerning this class of problems has come about. They are considered as definitively disposed of, as products of an outdated and limited mode of conceptualization, which must clear away and disappear of themselves once the point of view is transformed and elevated. One's own mode of conceptualization is regarded as so clear and irreproachable that it seems justifiable to look down upon older ones as either aberrant whims or childlike stupidities. What magisterial reprimands had even the great *Cauchy* constantly to endure, because he (in repeated and hopeless attacks upon the difficulties) could not gain the upper hand. Indeed, there has been no lack of isolated voices that have warned against precipitancy, and have found the courage to characterize the presumed clarity of the moderns as a huge deception. But they have not prevailed. And this is probably because, more fortunate in critique than in construction, they did not master the issues; but especially because, condemning where they did not understand, they improperly restricted the range of the calculus – which was at least secured by an immense number of fortunate applications. Disregarding these exceptions, it almost appears as if the present age – which, after all, does so much pride itself on its logical insight into the essence of mathematics, and which [377] so haughtily scorns the unclarity of its ancestors – has lost all sensitivity to fundamental difficulties. And this holds as much for the mathematicians as for the logicians involved here. I can boldly state: In the logical respect, the modern treatments of general arithmetic, in spite of their explicitly logical aims, represent a decisive step backward in relation to those which derive from the more remote periods of mathematics. In those times there was at least manifested a vivid sense of serious difficulties which had to be overcome, although there was no genuine capability of making them clearly precise, much less of completely resolving them. But our time is so full of its illusory clarity that it has ceased to sense the difficulties, much less then to aim in addition at their solution. There is a remarkable incongruity, moreover, between the rather high regard in which the modern epoch holds itself in this domain of fundamental questions, on the one hand, and that division of opinions already touched upon, on the other – which certainly is not always felt and understood in its true magnitude and significance, but nevertheless presents itself to critical inquiry as being very considerable. The result of our deliberations, which are to be undertaken with utmost carefulness, will be that our time, if any, has cause to be modest, and that its favorite theories must be rejected *in toto*, partly in consequence of logical errors of the most serious type, and partly because those theories are inadequately limited in scope. It shall wholly retain to itself only one glory: to have brought the technical development of the calculus to a scope and a perfection such as earlier times had never seen and had been surmised only by their most brilliant representatives. To be sure, in precisely this perfection of modern arithmetic there also lies, as will be shown, one of the chief sources of error. It is all too easy for the practicing mathematician to succumb to the temptation of confusing the methods of development for the extension of arithmetical technique in the most consistent way possible with the methods of a logical development of arithmetical science itself. The arithmetic of today is an unrivalled technique, with respect to the extension and the practical reliability of its rules. But it is no science, provided that [378] by "science" we understand, precisely, a system of knowledge.

Let us now pass over to our subject matter. According to our explanation of the concept of general arithmetic, this discipline has

the task of defining every conceivable elementary form of numerical construction, and of establishing the laws of their reciprocal connections and transformations. It aims to provide a system of rules that makes possible a mechanical procedure of calculation for equivalence transformations between complicated forms of numerical constructs. Its most important logical resource are modes of designation peculiar to it. In earlier times general arithmetic was described as a "calculus of characters"; and, in the use of letters (instead of numerals) expressed by that phrase, its distinction over against common or numeral arithmetic was likewise sought. That was an error. Apriori there certainly is nothing which stands in the way of studying general number relationships also by use of numerals. Of course numerals do evoke within us the representations of determinate numbers of our number system. But why should we not be able to abstract from this, and regard the numeral sign *hic et nunc* present as a mere instance standing for a wholly arbitrary number? The indicated sums, products and other types of combinations of any two numbers thought of through numeral expressions would then be able to stand as symbolic place-holders for any such combinations in general, and could furnish the basis for considerations of general arithmetic. However, there still lies in numeral expressions a great disadvantage from the standpoint of general arithmetic. All too easily one gives in to the temptation of illegitimately generalizing properties of number combinations that have validity only for the instance accidentally selected. In the mode of treatment proper to general arithmetic, none of the operations indicated can be actually carried out; for the form of combination, which is there the only object of interest, and concerning which theorems are supposed to be discovered, would immediately vanish. The number which is the finished result of the operation shows no trace of its origination. But the powerful habit of calculating with numbers has the effect that in the course of the transformations one now and then involuntarily evaluates a result. Painstaking attention no doubt could prevent this, but it is a defect that such attention is required at all. If the operat- [379]
ing of general arithmetic is itself to become a process of mechanical calculation, then it must not be intermingled with devices that are intrusive and troublesome. An especially great drawback to numeral signs resides, finally, in their failure to make apparent in the expression the distinction between the numbers which are given and those

that are sought. One therefore selects other signs which make it easy to escape all of these inconveniences: the characters of the various alphabets a, b, c, ... α, β, γ, in which the later characters, x, y, z, ... ξ, η, ζ, etc., are reserved for the designation of unknown numbers. That these signs also satisfy to the fullest the other requirements which must be imposed upon an appropriate symbolism for a calculational mechanism – such as univocity, brevity, easy distinguishability – is clear.

In addition to these general signs for numbers general arithmetic further requires several types of other signs. There are the operation signs for the indication of operational combinations. They combine simple into complex signs. Further, there are the relational symbols "=" and "$\gtrless$" for symbolizing the existence of an equality or inequality between two simple or composite expressions. This way of talking about the 'equality' of two expressions that are differently formed is, of course, wholly improper. One should prefer to speak of an equivalence, since it is after all only the numbers symbolized by means of the expressions that are equal.

Finally one needs, in addition, the parentheses, in order to make stand out clearly, in complicated combinations of operations, the relationship and sequential order of the operations involved.

ARITHMETIC AS AN APRIORI SCIENCE[1]

If we take up a textbook in arithmetic that is composed on scientific principles, we find, disregarding applications of numbers, the following procedure adopted: First the "basic operations" are dealt with, such as "the four species," after which there follows (as a rule) the elementary operations of the "third level": potentialization and its two inversions. These operations are defined and their laws established, which follows an order grounded in the nature of the facts. In the domain of each operation one first considers two-term combinations, and then those of several terms, which latter can be regarded either as complexes or as two-member combinations containing in their members some further combinations. Finally there are taken into consideration, combinatorially, the various possibilities for mixing the different operations. After this there follow investigations of a wholly different sort: One treats numbers defined by means of "definitional equations" – first, solutions of the simplest equations, then of those that are ever more complicated or difficult. The unlimited multiplicity of numerical determinations which fall within the domain generated in this way (and one is easily convinced that it fails by far to encompass all conceivable formations) appear, accordingly, as grouped in the following manner:

1) Certain types of number formations – and always formations of one number out of two numbers – are singled out as the elementary ones. They are called basic operations or basic combinations.

[1] From *Husserliana XII*, pp. 380–384. No date given in *Husserliana*. The date 1891 is given in the French translation by Jacques English: E. Husserl, *Articles sur la logique (1890–1913)*, Paris, 1975.

2) The innumerable formations which are left over from 1) presuppose the basic combinations in a twofold manner:

A) In the one manner we have mere compositions from the elementary formations, originating in that within an elementary combination [381] of two indeterminate numbers the one, or even both, is thought of as replaced by an always elementary combination. With the terms of this latter, then, we can do the same thing, and so on. One obtains number formations of an arbitrary and ever-increasing degree of complication, which presuppose exclusively the concepts of the basic combinations. If one calls the number formations that are realized through a single combination *simple*, and thus calls a combination of combinations (or, more precisely, of results of combinations) *composite*, then the formations considered under this heading are to be characterized as composites of basic combinations.

In following out the sequence of combinations we have the representation of the gradual generation of the intended number out of its components. With respect to this we can also speak here of constructive number formations. If we consider further that the basic combinations with simple and given components are thought of as irreducible elementary formations in general arithmetic, by means of which the *results* of combinations are regarded as given and engendered, then the extension of the concept of constructive number formation to these elementary formations immediately imposes itself upon us. It is they that produce the respective number, insofar as there can be talk of such a thing at all in general arithmetic. Of a "production" in the authentic sense we here must, of course, have no thought, where there are joined numbers that are indeed assumed to be known, but are indeterminate. For general arithmetic, the number that is the sum of a and b, for example, is regarded as produced and given once we write down a + b.

B) Number determinations of the second group are of an entirely different kind. If a, b, c, . . . are any given numbers, then any arbitrary combination of these numbers – whether it is simple or composite – constructively defines for me a new number P. If, now, in one or several formations of this kind we think of a single one of the components (e.g., a) as unknown, and of the number P, however, as given – or, otherwise expressed, if we think of the numbers previously sought as components, but of some one among the earlier components as the number sought – then we obtain (disregarding the few exceptions where basic combinations result) totally new types of number formation, which we can designate as *inversions* of the constructive [382] types. These are the well-known definitions of numbers by means of equations or systems of equations.

Now the issue is how to justify logically this selection and classification of number determinations, as well as this order in the elaboration of knowledge which we find realized in the science as it factually exists.

The arithmetician begins with an explication of a series of basic operations, and an inquiry into the laws governing them. If his procedure is not arbitrary, then the first question must concern the concept of basic operations.

In order to respond to this question some preliminary reflections are necessary: Arithmetic is, in the view of most thinkers, an apriori science. This entails that it does not begin with singular facts, in order subsequently to rise to probable generalizations through induction; but rather it begins immediately with assured – and indeed, with apodictically certain and immediately Evident – generalities. It obtains these by merely bringing certain "basic concepts" clearly before the mind; and those generalities then furnish all further theorems of the science by way of *mediate Evidence* and certainty. What, then, if this is correct, are the immediately Evident certainties underlying arithmetic, and what are the concepts from which they issue?

Arithmetic is supposed to establish the laws for the domain of numbers. Each science, and in particular each apriori science, thus has its domain. It is delimited by a general concept and embraces the objects of that concept. Such concepts are, besides the concept of *number*, also those of *set*, *magnitude*, *spatial extension*. There are apriori theories of sets, magnitudes and space. But the objects with which these sciences, as apriori, have to do are never (either in the research or in the formulation of the theorems) the objects of the relevant concepts as given *in concreto*. Rather they are objects of those concepts in a totally inauthentic mode of representation, which either confers upon them no sort of distinguishing determinations at all, or at most only the ones which issue purely from the nature of the respective domain. To be sure we often enough hear, in these sciences, of "determinate" objects (determinate sets, magnitudes, etc.). But then what is intended are either objects which are thought of in the inauthentic mode of representation, as *somehow or other* [383] determinate, but are not at all genuinely set before us with any degree of determinateness; or else objects are meant which, *via* certain relationships and combinations, are thought of as generated from objects that are represented as just described – and consequently as receiving a conceptual determination in relationship to them. Out of such conceptual determinations there then arise all of the laws for those objects. It lies in the "nature" of the respective domain that its objects, as such, found certain relationships; that, within certain general forms and by means of objects given (or, rather, by means of objects *as* given, and otherwise thought of indeterminately), new objects can be defined; and that, from the conceptual forms of these new constructs, the existence of this or that determinate relationship between relevant objects can be grasped by direct or indirect Evidence. Each such relationship provides us with a law.

And thus we find it actually to be in arithmetic. It lies in the nature of the number domain that for any two arbitrary numbers a and b there obtains, in exclusive disjunction, one of the relationships $a \gtreqless b$. Likewise there lies in the nature of the number domain certain forms for constructing numbers by means of numbers. We find, for example, that two arbitrary disjunctive numbers can be combined in thought into *one* number that includes in itself the units from the two numbers and those units alone. Also, that for two arbitrary numbers there is a number which is their sum, is something we know even with apodictic Evidence. It is clear that for the formation of the concept of sum we have no need to consider concrete numbers. With the concept of number we have the concept of a collective combination of units of whatever sort. We need only define these units in such a way that the units colligated are identical with those of the arbitrarily given number (i.e., in indeterminate representation), and we then have the concept of *sum*. In a similar manner there emerges within the domain of pure numbers (i.e., numbers that are thought purely *as* numbers, each thought just so and not otherwise than as is signified by the expression "a certain number") yet many other ways of constructing numbers from numbers – either directly or as mediated through formations already defined, and possibly with the aid of the relational concepts $\gtreqless$. Of all of these constructions it is to be said that they are grounded in the "nature of the domain of number." But by this nothing further is meant than what has just been explained: that for their conceptualization we require only the pure [384] numbers, provided that all of the determining concepts which we come to use are grounded in conceptual representations of numbers as such.

Finally, one also is easily convinced that all arithmetical theorems have the form of relations of equality or inequality, setting the various number determinations in relationship to one another or to simple numbers.

We now want to take a closer look at the manner in which apriori science emerges in the domain of numbers.

LETTER FROM EDMUND HUSSERL TO CARL STUMPF[1]

My Dear Professor:

Forgive me for answering your cordial new-year's letter only today. After losing two precious weeks at the turn of the year through a burdensome illness, zeal for my work allowed no breathing space until the beginning of this week. I was by this date to have finished a section of my book *Philosophy of Arithmetic*, and that required every minute. Therefore only now do I convey my heartfelt thanks for your good wishes.

We were very happy to hear that you and your dear ones continue to be well, and that, except for Rudi, you have been spared by the influenza. We too are among those rare exceptions. Hopefully, diptheria and influenza have not too greatly weakened poor Rudi. You did well (as I heard from Mrs. Cantor) to hold him out of school for the winter. I also was delighted to learn of the completion of your second volume of the *Tone Psychology*. I suppose you could have meant no other page-proofs in your last card. *Vivat sequens*! I hope my turn also will come in the not too distant future. About 200 pages of proofs[2] are now finished, and I still have 150–200 pages to complete in the next 9 weeks. If heaven does not now deny me the strength, and if the evil demon of nervousness also is subdued from

---

[1] From the Winter of 1890 or 1891. I think 1891 is the more probable date because of the indicated stage of proof-reading and composition (surely for the *Philosophy of Arithmetic*). First published in *Husserliana XXI*, pp. 244–251. See p. 459 on the date of the letter. Dr. Ullrich Melle has given invaluable advice on the translation. [D.W.]

[2] For *Philosophy of Arithmetic*,(Halle-Saale: C. E. M. Pfeffer, 1891); 2nd edition in *Husserliana XII*, (The Hague: Nijhoff, 1970). [D.W.]

here on, I hope then to bring it all to a happy conclusion. I confess that for this I possess as of yet no coherent outline of the part dealing with *arithmetica universalis*. However, the guiding ideas have been ripened and secured through years of painstaking deliberations. I unfortunately do not have the gift of first coming to clarity in the process of writing and rewriting. So long as I am unclear, the pen does not budge. But once I have come to a clear understanding, everything moves along rapidly. My *Habilitationsschrift* was likewise put into a finished text within a few weeks, without previous outline.

So I also hope now – if I retain the happy disposition which has turned me into a new man this winter – to be able to lay out in one thrust, in the course of the next 9 weeks, the theory developed over [245] a period of years. The results which I have obtained are striking enough. The opinion by which I was still guided in the elaboration of my *Habilitationsschrift*, to the effect that the concept of cardinal number forms the foundation of general arithmetic, soon proved to be false. (The analysis of the ordinal number already made this clear to me.) By no clever devices, by no "inauthentic representing,"[3] can one derive negative, rational, irrational, and the various sorts of complex numbers from the concept of the cardinal number. The same is true of the ordinal concepts, of the concepts of magnitude, and so on. And these concepts themselves are not logical particularizations of the cardinal concept. The fact is that "general arithmetic" (including analysis, theory of functions, etc.) finds application to the cardinals (in "number theory"), as well as to the ordinals, to continuous quantities, and to *n*-dimensional manifolds (time, space, color, force, continua, etc.)

Since, now, no common concept underlies these various applications of arithmetic, from which that science can be derived, what constitutes its content? What sorts of conceptual objects are its propositions about? An odd question! In recent decades the number has been repeatedly defined as a sign. Does arithmetic perhaps deal with mere signs? Is it thus to be a mere game with symbols? Why not, says *Helmholtz* – if only the 'game' has the immense advantage of allowing for all of these "applications"? But who will explain to

[3] On the immense significance of this comment for Husserl's development, consider the footnote on the first page of Chapter XI of the *Philosophy of Arithmetic*. It is unlikely that this footnote would have been published in 1891 if this letter were from the winter of 1890. [D.W.]

us how, in general, a mere game with signs can admit of applications. *Helmholtz* has not remotely done so. He falls into the same errors as his predecessors: Cardinal (or else ordinal) numbers are regarded as the foundation of arithmetic, but – as mere signs – are themselves explained by means of the series of number signs. Since, now, a moment's thought makes it clear that such fundamental concepts are not to be explained away in terms of names, how can he accept such a theory of arithmetic?

Arithmeticians who – now with hesitation, and now decisively – explain numbers as signs, allow themselves to be guided merely by the study of algebraic formalisms. These mathematicians (*Grassmann* above all) have brought to Evidence the possibility of deriving [246] the whole algorithm of arithmetic and analysis by means of mere sign definitions ($1 + 1 = 2$, $2 + 1 = 3$, etc.; $a \cdot a = a^2$, $(\sqrt{a})^2 = a$, etc., all understood in the sense of mere equivalences of signs upon paper). This occasioned their identification of number and sign. (*Helmholtz* stated this with full certainty on p. 20 of his "Zählen und Messen."[4]) A deficient understanding of the basic concepts of the abstract domains to which arithmetic applies then led to the further identification of cardinals, ordinals, and magnitudes with mere symbolizations. Of the fundamental difficulties contained in their theories, these investigators were not even aware. I will briefly discuss them.

There is nothing to be wondered at in the fact that a system of signs and operations with signs can replace a system of concepts and operations with judgments, where the two systems run rigorously parallel. Upon this fact syllogistic and the logical calculus rest, as does the much more subtle system of ordinary arithmetic. To calculate is not to think (infer), but rather is to derive sign from sign systematically, in conformity with set rules. The sign which is obtained at the end is interpreted, and thus the desired thought results. So we have a method of deriving a judgment from given judgments, not by means of actual inferring, but rather through a rule-governed procedure in which, from the arithmetical symbols for the data, the symbol for the result is obtained in a mechanical fashion. (Thus, calculation must not be confused with speaking in a mechanical fashion, which so commonly occurs in place of thinking. This – with few exceptions – is mere association of ideas. The former, by

[4] In: *Philosophische Aufsätze. Eduard Zeller zu seinem 50. jährigen Dr. Jubiläum gewidmet*, Leipzig, 1887, pp. 17–52. [D.W.]

contrast, is a tool for the deduction of *new* conclusions.) In every case, one might then suppose, to the conjunction of signs there must at least potentially correspond a conjunction of thoughts, just as it actually occurs in those examples cited. Senseless signs and relationships among signs – of what use should they be to us? That a class of general representations, and the related judgments, permits certain systematic methods of deduction which can be comprehended in uniform rules – this alone appears to ground the possibility of devising calculational mechanisms by means of a parallel symbolization. One follows the signs and their rules, and thereby spares and replaces thinking and the rules of thought. But it seems absurd that signs which designate nothing, signs without any interpretable sense, should accomplish something.

But how do things stand in this respect with the system of signs of arithmetic? If it deals with discrete magnitudes, then "fractions," [247]
"irrational numbers," imaginary numbers and, in the case of the cardinal numbers for example, the negative numbers also lose all sense. In relation to continuous quantities (e.g., weight), fractions and irrational numbers are meaningful signs; but the imaginary and negative numbers are not. And so on. Nevertheless, in the "theory of number" (dealing exclusively with whole numbers) the integral and differential calculus, and possibly its irrationals, are applied – with success! Thus everywhere. Each theorem obtained is an actual identity, if one calculates it out. How is this possible?

A closely related question formed the starting point for my own investigations. Since I originally considered signs only in relation to designated concepts, in the case of numbers such as $\sqrt{2}$, $\sqrt{-1}$ and the like, for example, I had to take them as representatives of "impossible" concepts. Thus I first sought to get clear on how operations of thought with contradictory concepts could lead to correct theorems. I found no other possibility than that of the indirect or apagogical procedure of proof. But that is not under discussion here. Finally I noticed that, through the calculation itself and its rules (as defined for those fictive numbers), the impossible falls away, and a genuine equation remains. Indeed, the procedure of calculation runs its course another time with the same signs, but now referred to valid concepts, and again the result is correct. Thus it is not a matter of the "possibility" or "impossibility" of concepts. Even if I mistakenly imagine that the contradictory exists – even if I hold the most absurd theories concerning the content of the corresponding concepts of number, as many great mathematicians have done – the calculation remains correct, if it follows the rules. So it must be an accomplishment of the signs and their rules. Then I studied "formal" arithmetic, as well as ordinary calculation, and in this way came to a result that harmonizes with Helmholtz's formally, but is conceptually based upon a totally different foundation. Perhaps I would have come to the same result more quickly, had I not been so totally frightened away from the 'sign theory' by Helmholtz's attempt, with its outright incoherencies.

The question which I stated above is answered thus: The sign system of *arithmetica universalis* divides into a certain sequence of levels, comparable to that of a system of concentric circles. The lowest level (the innermost circle) is occupied by the signs $1, 2 = 1 + 1$, $3 = 2 + 1$, etc.; the next by fractional signs; and so on. The signs [248] of the lowest level, and they only, are independent. Those of the higher levels are formally dependent upon those of the lower levels, and ultimately upon the lowest. To each domain there belong rules of calculation ("formal laws"). Those of the higher domains are dependent upon those of the lower, and include them formally. The rules of calculation are, then, so formed that each "equation" (in whatever way it may be set up, i.e., by means of whatever domain levels) is satisfied as an identity with reference to *the* signs and *the* domain of rules which it actually involves. Thus, for example, if an equation between whole numbers (or their signs) is proven with the aid of all the parts of the *arithmetica universalis*, it nonetheless incorporates into itself nothing of this route of proof: it is an identity with reference to the signs which it contains. It is an identity in the sense of the signs and sign rules of the lowest level.

Thus, if we are dealing with the domain of the cardinal *concepts*, and if I can prove that they and the elemental relations of the cardinal numbers admit of an adequate symbolization by means of the signs of that lowest range (that of the totality of whole numbers), then each correct sign equivalence ("equation") between the signs of this level represents a correct arithmetical proposition. But from this law it follows: For the discovery of the laws of number, in the genuine sense of the word, I can use the whole algorithm of the *arithmetica universalis*. (Notwithstanding that all of the remaining signs of the *arithmetica universalis* are 'senseless', admitting of no interpretation.) And so for all levels.

After all of this I can state: The *arithmetica universalis* is no science, but rather is a segment of formal logic. Formal logic itself I would define as a symbolic technique (etc., etc.), and designate it as a special – and one of the most important – chapters of logic as the technology of knowledge. In general, these investigations appear to push us toward important reforms in logic. I know of no logic that would even do justice to the very possibility of a genuine calculational technique.

These results are of the greatest importance, of course, for the logical understanding of all mathematical sciences, insofar as the *arithmetica universalis* plays a role in them. But even pure positional geometry [*Lagengeometrie*] appears to me in a new light; for, even though it does not use the algorithm of arithmetic, it too has found it necessary to invent special algorithms in various of its problem areas. Using intuition and actual construction one simply never gets very far; and hence there immediately develops from elementary iterative constructions, by means of an appropriate symbolization, [249]
a calculational mechanism with which one then can formulate and solve certain classes of problems. To be sure, the algorithms here are of the earlier type – of the same type as ordinary arithmetic. Each step of the calculation can, if one wants, be immediately transformed into intuition or into a symbolic representation utilizing it. In the common treatment provided by modern geometry, now the old algorithm, now new ones, are used, as happens to be convenient.

Little has been done for the logic of geometry. It is incredible that even the logical justification of the old method of Cartesian coordinates (and that justification is by no means something obvious) has still not been demonstrated. Here too the arithmetical investigations were of great use to me. In fact, for each case of applied arithmetic, the results of those investigations compel us to demonstrate, through analysis of the relevant concepts and their relations, that the area of application can be symbolized by signs of the *arithmetica universalis* – or by signs of certain domain levels in it.

Thus, if one has discovered the symbolizations for a domain of magnitudes, then the theory of that domain is merely a matter of calculation and of interpretation of the results attained. If the available symbolizations are not sufficient, then one can attempt to devise new ones that will accomplish what is desired. I was in this position with regard to continuous and discrete manifolds of more than two dimensions. Upon discovery of the symbolization, Euclidean geometry of the *n*th order also was theoretically complete. The theorem of the sum of the angles and the Pythagorean theorem were obtained. Thus does the method of signs solve the basic questions in the philosophical theory of geometry.

But enough of this! All the more so as this latter theory belongs to the *second* volume of my work. And here I am already writing on my fourth sheet of paper! Excuse my imposing for so long on your patience. The words flood from my pen with such rapidity and ease, and I all too gladly seize the opportunity to express myself. If all is not said in so precise and well-ordered a manner as it could or should be (looking back over it, I am not very well satisfied), then please forgive me. The ideas here sketched out in haste will soon present themselves in a more carefully developed form and nicely printed; but I am already horrified to think of the penetrating and critical eye which you will cast upon these pages.

I had it in mind to write to you of yet so many things that would please you or be of interest, and now I must be far too brief. Above all: The Hayms are at present wonderously friendly toward us. They [250]
have invited us to join them at supper with an intimate little group, and also have spent an evening with us. In general, they are very kind – even affectionate – to us.

In my course on the philosophy of mathematics I have eight students, seven of whom are quite prompt and listen eagerly.[5] I was unwilling to lecture in ethics for two students only, and hence canceled out. You will scold me for this, but in the interest of my health and work it was a very wise choice.

1. I have announced logic lectures for the summer (afternoons, 3 to 4 hours). Our class schedule reads:
2. The History of Philosophy, Entire: 5 hours by Uphues.
3. The History of Modern Philosophy: 4 hours by Erdmann.
4. Psychology: 4 hours by Vaihinger, *and* 5 to 6 hours by Erdmann.
5. Logic: by Uphues and myself.
6. Ethics: by Haym.
7. On Materialism and Spiritualism: 1 hour, Erdmann.
8. On the Body and the Soul: Vaihinger.

[5] Following their wish, I lecture mainly on questions about the logic of space, and provide a detailed criticism of the Riemann/Helmholtz theories. [marginal note]

9. Pedagogy: by Uphues.

Now I would call that a rich menu! Accordingly, I of course could lecture on nothing other than logic, which suits me very well for the present. Probably I will yet announce lectures on the philosophy of mathematics, for 1 or 2 hours.

Erdmann is preceded by a very fine reputation. He is said to be a very charming and open person, and likewise for his wife. The *Pischels* and the *Meyers* (the latter of whom are quite popular) rave on and on about them.

Our friendship with the *Arnims* is, if possible, more cordial than before. One or two times a week we spend the evening together, usually reading. Lately we have been reading Mörike's *Mozart's Trip to Prague*. If you are not familiar with this charming novelette, you *must* read it.

I had a letter from *Brentano* at the turn of the year. I seemed to hear no joyous, elevated mood speaking from it. He complains that there is so little critical interest in his book on ethics.[6] I greatly lament the fact that I am at this point unable to undertake publication of a review in some journal! I did let him know that I was thinking of dedicating my book to him – which, to my joy, he now has accepted without any objection. [251]

With that I finally come to a close. With many cordial greetings from our house to yours, I remain

Yours very truly,

E. Husserl

PS. While I was lifeless and dull, I had you quite within my reach. Now that I have once again joined the human race – mentally rested and refreshed – and could so genuinely enjoy the blessing of your presence, you are far away. I now feel this very deeply.

---

[6] Brentano, *Vom Ursprung sittlicher Erkenntnis*, Leipzig, 1889. [Added in *Husserliana* edition, p. 150.]

ON THE LOGIC OF SIGNS (SEMIOTIC)[1]

How is it that one can speak of 'concepts' which one, nevertheless, does not authentically (*eigentlich*) possess, and how is it not absurd that the most certain of all the sciences, arithmetic, is to be based upon such concepts? We respond by means of a general analysis belonging within the domain of logic.

Concepts, or contents in general, can be given to us in a two-fold manner:

First, in the *authentic* manner; namely, as that which they are.

Second, in the *inauthentic* or *symbolic* manner; namely, through the mediation of *signs* – signs which are themselves authentically represented. Thus, for example, any intuitive representation in sensation or phantasy is an authentic representation, provided it does not serve us as a sign for something else. But if it does so serve, then in relation to this latter it is a symbolic representation.

The word "sign" in our definition is to be taken in the widest conceivable sense. Thus we do not restrict ourselves to external, sensuous signs, which we affix to things for the purpose of distinguishing and recognizing them more easily. Of this kind are proper names, such as "Peter" or "John," and likewise the names of *abstracta*. But general names also are signs. Each general name is a sign for a general representation, and this, in turn, is a sign for any object that falls under the corresponding abstract concept. Thus, given this mediation through the concept, each general name is a sign for any of the objects which it comprises in virtue of its "connotation." Furthermore (as

[1] First published in *Husserliana XII*, pp. 340–373, dated 1890. The content suggests to me very late 1890 or, better, 1891, after the discoveries announced in the letter to Stumpf. [D.W.]

is already clear from the last example) we take any conceptual mark (*Merkmal*) – so far as it serves, precisely, as a distinguishing mark – to be a sign. Any characteristic whatsoever, whether it be an absolute or [341] relative one, can upon occasion serve as the signitive mark for objects that possess it. This obviously is the source of the equivocation in the term "mark": In its original signification it meant the same as "sign." But then it was constricted to cover characteristics *as* signs, in order finally – in an extended sense – to signify the same thing as characteristic in general. However, a characteristic does not in all circumstances serve us as a mark, even though any characteristic *can* serve that purpose upon occasion. At one time we are interested, for example, in the characteristics of aluminium as such, so far as they enrich our knowledge of this metal. But at other times just those very same properties, confirmed as belonging to an as of yet unknown body, can serve as the signitive mark to determine that that body is, precisely, aluminium.

Thus, as the sign for something (for a content in general) any and everything can serve which marks it out – which is suited to distinguish it from other things, and by which we then are in a position to re-identify it. But for us this re-identification comes into consideration not merely as a psychological mechanism that functions without our needing to be aware of it – one which, upon the occasion of the one representation, calls the other back into consciousness without our having, even in some general sense, to be conscious of the fact that it is precisely the former which has recalled the latter and mediated its re-identification. In order for the concept of *sign* to be possible, and in order for us to be able purposively to invent and use signs, the relationship between the sign and what is designated must be specifically noticed. And we have on innumerable occasions actually confirmed by experiment that external and sensuous (as well as conceptual) marks are suited to direct our thoughts toward the contents which possess them. The concept of the sign is, precisely, a relational concept. It refers us to something designated.

Signs can be categorized in various ways. We distinguish:

1) *Extrinsic* signs from *conceptual* signs; or signs in the narrower sense from marks. An extrinsic sign is one which has nothing at all to do with the specific concept of the designated, with its content. The name of a person, for example, stands in this relationship to the person himself. It designates, but does not characterize. [342]

A conceptual sign is any inner or external property (*Merkmal*), in the common sense of the term, when that property serves as a sign. Both sorts of properties are dependent upon the specific concept of the designated. The former are determinations which are included as partial contents *within* the representation of the designated content. The latter are relative determinations, which characterize the content as one term of certain relations resting upon it. It is, moreover, noteworthy that absolute properties also, when closely considered, represent relative determinations. When one describes an object unknown to us by designating it, among others, as one that is red, then it is not red as such, but rather being-red (i.e., the relationship between thing and color, well known to us in its peculiar character) which serves to characterize the object. But the external properties are also relative in this manner, without prejudice to their *specific* relational content. With these properties, then, we have a multiple relation: the relation of the designated thing to other things and, moreover, the relation of that thing to the relation itself, through which the relative attribute gains its status *as* an attribute. If, for example, the first relation is a relation of similarity between *A* and *B*, then the latter relation is that which makes possible the attribute "similarity to *B*." We can verbally express the sign-representations for absolute and for relative properties or marks as follows: In the first case, "Something which is red in color," and in the other case, "Something which stands in the relation of similarity to B."

2) *Univocal* and *Multivocal* signs, where again we must distinguish between signs that are univocal or multivocal accidentally, and those that are such in their very nature and definition. Any proper name, for example, is univocal by definition, but it contains an accidental multivocality when several persons find themselves with the same name. On the other hand, any general name is accidentally univocal when *de facto* only one object exists that falls under the concept connoted by it, whereas in its nature and definition it is multivocal.

By reciprocal combination and limitation of multivocal signs, composite univocal signs can be formed. This is a procedure which constantly aids speech in relation to general names, resulting in a huge savings of signs as well as in a significant gain in the knowl- [343] edge disseminated by the reformulated signs.

In conjunction with this we refer to the division of signs into:

3) *Simple* and *composite*. This, however, is not to be confused with

another, and an especially important, division that intersects with it: that into 4) *direct* and *indirect* signs. Sign and fact can, namely, be either linked directly or linked indirectly, under mediation of other signs. The indirect sign is a composite sign in which the partial signs are not at the same level as each other, but are rather superimposed upon each other and refer to each other. Z is a sign of object G in virtue of the fact that Z is a sign for Z_0 which is a sign for G; or in virtue of the fact that Z is a sign for Z_1, this a sign for Z_2, this again perhaps a sign of a Z_3, and so on, until finally the sign Z_n directly designates G. Any proper name is a direct sign, any general name is indirect. In fact, the general name designates the object under the mediation of certain conceptual marks. The adjective "red" directly designates being-red (including the *abstractum* red as a metaphysical part), which, precisely, can then serve as the signitive mark for the object itself – although other conditions must be added in order to make the designation univocal. All multivocal signs which connote a determinate range of multivocality are indirect; for such a connotation can only come about through a general mark or property, which thus mediates between the sign and the designated.

In the case of indirect signs it is necessary to distinguish: that which the sign *signifies* (*bedeutet*) and that which it *denotes* (*bezeichnet*). With direct signs the two coincide. The signification of a proper name, for example, consists just in the fact that it names precisely this determinate object. With indirect signs, on the other hand, there are intermediaries between sign and fact; and the sign designates the fact precisely through these intermediaries, which therefore constitute the signification. The indirect sign Z has the signification that it directly denotes Z_1, this directly denotes Z_2, etc., and finally Z_n directly denotes G. Accordingly, the signification of the general name, for example, consists in this: that it denotes some object on the basis [344]
of and by means of certain conceptual properties which the object possesses.

All higher level mathematical signs are indirect, are multi-tiered signs of signs of signs. One easily sees that this division also intersects with the first two mentioned. It is especially worthy of mention that the mediating partial-signs can be univocal as well as multivocal, external as well as conceptual (and in some cases a mixture of the two). We will see how indirect, *purely external*, altogether univocal signs (besides the indirect, mixed signs) play a role of the highest importance in arithmetic.

5) *Identical* and non-identical, *equivalent* and non-equivalent signs. Two signs are identical if they denote the same object, or objects of one and the same extension in the same manner. The one is the mere repetition of the other, e.g., "Horse" and "Horse," "five" and "five." Identical multivocal signs do not in general denote identically the same object, but always objects of identically the same extension, as determined by the signification of the sign.

Two signs are *equivalent* provided that they denote the same object, or else objects of one and the same extension, in different manners, whether by different external or by different conceptual means: e.g., a pair of synonymous names such as *King* and *rex*; Wilhelm II = present German emperor; $2 + 3 = 5 = 7 - 2 = +\sqrt{25}$.

Definitions in the sense of a genuinely formal logic present us with special cases of sign equivalences. A definition is a sentence which expresses the signification of one external sign by an equivalent sign of that type. A direct external sign has no signification expressible in signs, and hence it cannot be defined. Thus we have, for example, proper names, names for *abstracta*, the sign "1" and the like.

One can, finally, divide signs into those for *representation-contents*, and those for psychical acts and, above all, *judgments*. Most words in our language are independent or dependent signs for contents. Judgments appear linguistically in the form of sentences. [345] Mathematical judgments appear in the symbolic form of equations, inequations, congruences, and the like.

Signs for representational contents divide into those for *absolute* contents and those for *relations*, including true relations (*Beziehungen*) as well as combinations (*Verbindungen*). The latter customarily find linguistic expression in syncategorematic words, such as "and" and "but." In arithmetic one distinguishes signs for numbers from the signs "=," "><," and from the operation signs "+," ".", etc.

Toward a Division of Signs into Natural and Conventional

The same natural laws underlie conventional signs and natural ones. In conventional signs there emerges as a novel aspect the influence of the will, guided by the drive toward knowledge, and also the capacity of governing the course of judgment formation by the will into conformity with, precisely, those epistemic interests.

The invention of conventional signs actually is already found at the lowest level of human development. The psychological preconditions required by it – the capacity for understanding the function of signs, and the power of will over the underlying psychical dynamics – are so very simple and so widely realized that we should not be amazed when even animals make themselves understood through signs, to a certain extent. A sensible expression, for example, which for one individual presents itself as a natural sign, can simultaneously become mediator of 'understanding' for another individual. The knowledge of this result may, then, give occasion for using the natural sign purposefully as a means of understanding. Thus, through a frequent and mutual usage there develop signs with a fixed, conventional signification. Similarly, also, with the origination of conventional surrogates: e.g., with the first and simplest of which we know, the number signs. In most languages the word "five" means the same as "a hand."

A further distinction between signs is that into *formal* and *material* signs. It is of fundamental significance for logic. It is characteristic of the state of formal logic that ideas about the distinction of form from matter remain unclarified up to now. Two wholly heterogeneous distinctions [346] have always been run together here: the distinction between the judged content and the act of judgment, on the one hand, and that between the terms of the relation and the relation, on the other. The form of the judgment is confounded with the form of the relation. The old explication of the judgment as a relation or combination of representations rests entirely upon this confusion. By far the most of our judgments bear upon relations; and, because of this, judging is identified with relating. However, this view did not proceed with the necessary consistency, and the constituent parts of the relation were assigned now to the form, now to the content. In the judgment, "God is just," "God" and "just" were assigned to the matter; in the judgment, "All men are mortal," the "all" (as with quantity signs in general) was assigned to the form, on the view that the quantity goes with the mode of the judgment. *Brentano*'s epoch-making investigations have put an end to these erroneous theories. According to his results, any judgment is an affirmation or a denial of a representation content. If, now, one is guided by the principle of the distinction between form and matter which has been dominant up to now –

according to which all that goes into the mode of the judging, and thus into the judgment act, is assigned to the form – then the judged content would have to be regarded as the matter of the judgment, and the acknowledgement or rejection as the form. However, for formal logic that other principle of distinction, according to which the formal is grounded in the mode of the relation, is certainly of greater significance. In fact, those uniformities of assertion structure which give rise to well-distinguished classes of assertions are almost always correlated with forms of relations, and only the fabric of inferences associated with those forms permits a formal mode of treatment in the broader sense – i.e., an algorithmic.

We want now to discuss in more detail the distinction which we have in view: In any composite thought we distinguish matter from form. The matter is represented by names, while the form is represented by syncategorematic expressions, whether simple or composite. The names serve (and this is their peculiar function) for the designation of the absolute contents, the relational *terms*. By contrast, the syncategorematic expressions have the function of expressing the relation between the absolute elements of the thought (*Gedanken*). [347]
We take the word "relation," here as elsewhere in this work, in a very broad sense. We count among relations both relations in the narrower signification, which belong within the primary contents, and also relations which are mediated through psychic acts. We are interested, in the latter respect, only in the judgment and the relating acts. To grasp a relation from the "point of view" of the one and of the other term is a specific psychical activity which belongs within the genus of representing. If this activity is combined with an acknowledgement or rejection, then we obtain the most important class of judgments, in which the matter is articulated and ordered. If, for example, we represent a similarity of *A* and *B*, then the relating activity yields the representation of *A* as having the relative attribute, "similar to *B*"; but the acknowledgement yields the judgment, "*A* is similar to *B*." A relation of magnitude between *A* and *B* provides the relational representation of *A*'s being greater than *B*, from which the judgment, "*A* is greater than *B*," arises; and so on. The relational terms *A* and *B* belong to the matter, and the supplementary expressions, "being greater than . . . ," "is greater than . . . ," and so on, belong to the form. To the form belongs, further, the distinction of position, which marks out the subject term and the predicate term of the relation – I

say of the *relation*, and not of the "judgment." Through the relating activity the relational terms lose their equivalence: the one becomes the main term, the subject, to which there is assigned, as an attribute, the standing-in-relation-to-the-other-term. The predicative term is a constituent of the grammatical predicate. If I say, "Gold is yellow," then gold is the subject term, and the abstraction (the "metaphysical part") "yellow" is the predicate term; but the grammatical predicate is the adjective yellow, i.e., "having yellow" or "being yellow." The acknowledgement bears upon the attribute *as* attribute of the gold. It presumes that articulation of the judged content which is introduced by the relating activity. The distinction between subject and predicate thus belongs wholly to the judged content, and not to the mode of judgment. However, corresponding to our principle, we do not count the whole predicate as matter, but rather the predicate term alone.

Obviously the distinction between matter and form is a relative one. *Any* represented content can serve us, upon occasion, as the term of a relation. Thus a represented relation, a relational judgment, [348]
a chain of inferences, and so on, also can belong to the matter. In any deduction the individual judgments are constituents of the matter. For a deduction is a composite judgment. But in such cases the standpoint from which things are considered will always make clear what is matter and what is form. It is, then, always a matter of *which* relation is assumed.

From the standpoint of the isolated judgment, for example, everything in the assertion which expresses the judged relation belongs to the form, while all that is taken as a relational term belongs to the content.[2] If one of the terms is composite, then, with reference to this composition, the combining element belongs to the matter, and to the form belongs the mode of combination. In the case of inference the premisses and the conclusion are the matter, and their arrangement – in so far as it is of significance for the bearings of the propositions upon one another – is the form. Only at a second level does the form of the separate propositions, and at a third level the form of *their* matters, enter into the form of the inference: insofar as the process and content of the deductive activity is conditioned in part upon them.

[2] We will then attribute the form of the matter *indirectly* to the form of the judgment, if it enters into the basis for the judged relation, e.g., as in: If A is b, then C is d.

Given the manner in which we have defined matter and form, we must say that an existential proposition, "*A* exists" (in which *A* is a content that is simple, or non-articulated, or one involving no attributes) has neither matter nor form. In order to take all propositions into account, we could perhaps state our definition as follows: To the matter belong the contents or substrata of our logical activities, while the activities themselves belong to the form. Logical activities are the judgment and the representing activities sub-servient to it: above all, relating, conjoining, distinguishing, and so on. In the proposition, "*A* exists," "*A*" would then be the expression of the matter, and "is" that of the form. In the expression "similarity of *A* and *B*," "*A*" and "*B*" would belong to the matter, while "similarity" and the "and" and the "of" would belong to the form; for the latter indicate a relating and articulating activity which is presupposed in the judgment, "There is a similarity between *A* and *B*."

That *our* distinction of matter and form really is of value to a formal logic is best proven by the sciences in which a truly fruitful and comprehensive range of derivations by means of formal mechanisms [349] has occurred. These are the sciences dealing with numbers, magnitudes and extensions. Therein we find, in every case, that the symbolism does not distinguish judgment act and judged content, but rather we have signs for terms of relations and signs for relations. And of the latter there are two kinds: the one expresses the existence or non-existence of a relation, thus implying a judgment; while the other does not do this, but rather merely indicates the formation of a composite, relational representation. Thus, for example, the arithmetical symbols $=$, $\gtrless$, $\neq$, $\equiv$, etc., and the geometrical signs $\sim$, $\cong$, etc., are of the first sort; for "$=$" signifies "is equal," "$>$" signifies "is greater than," and so forth. The arithmetical operation symbols $+$, $\bullet$, etc., are of the second sort. Special signs for acknowledgement and rejection are not required by the formal mode of procedure.

Further Division of Signs into Natural and Artificial

It is not our task here to portray in detail what immense significance the inauthentic representations, as well as symbols in general, possess for our entire psychical life. They begin to take hold on the earliest levels of psychic development, and accompany that development –

ever expanding, and fulfilling ever more inclusive and more complicated functions – up to the highest levels of development. Indeed, we may claim still more: They do not merely *accompany* psychic development, but rather they essentially *condition* it, making it possible to begin with. Without the possibility of external, enduring marks of reference (*Merkzeichen*) as supports for our memory, without the possibility of symbolic representations serving in place of authentic representations that are more abstract or too difficult to keep distinct and to operate with (or, indeed, serving in place of representations that *as* authentic are altogether denied to us), there would simply be no higher mental life – much less, then, science. Symbols are the great natural instrument by which the limits of our psychical life, originally so narrow, are broken through, and by which the essential imperfections of our intellect are, at least to a certain degree, rendered harmless. Through characteristic detours, sparing of higher thought, they enable the human mind to accomplish things which directly, in the workings of authentic knowing, it could never bring about. Symbols [350] serve the economy of mental achievement as tools and machines do the economy of mechanical achievement. With free hand the most excellent draftsman does not draw a circle so well as the school child does with a compass. The most inexperienced and feeble man accomplishes incomparably more with a machine (provided only that he knows how to handle it) than the most experienced and strongest man without it. And it is not otherwise in the mental domain. Take the symbolic tools away from the greatest genius and he becomes less capable than the most limited of minds. Today a child who has learned to calculate can do more than the greatest mathematicians could do in antiquity. Problems which for them were hardly conceivable, and wholly unsolvable, are today solved by a beginner, without special effort and without special merit. And as tools of labor present a series of levels, ever increasing in complication up to the most wonderful of machines – mirroring the progress of mankind in mechanical productivity – so it is also with symbols in relationship to mental productivity. Upon the conscious application of symbols, the human intellect raised itself to a new and truly human level. And the progress of intellectual development runs parallel with progress in symbolic technique. The magnificent development of the natural sciences, and that of the technology based upon it, constitute above all else the pride and glory of recent centuries. But it certainly seems that no less a claim to fame belongs to that remarkable symbol system – which has not yet received its clarification – to which the sciences and technology owe most, and without which theory as well as practice would be helpless: the system of general arithmetic, the most wonderous mental machine that ever arose.

Among signs, the "*inauthentic*" *representations* play an especially important role. According to our definition, any content which is not given to us as what it is, but rather is given only indirectly by means of some sort of sign, is one that is inauthentically represented. It is, accordingly, clear that the concepts *sign* and *inauthentic representation* do not by any chance coincide. Each inauthentic representation is indeed a sign, but it is not conversely true that each sign is an inauthentic representation. If something is not directly given, but is given only as mediated by signs, then the complex of these signs, [351] or the sign composed of them, *stands in place* (*vertritt*) of it. But not every sign has this deputizing function, nor, indeed, is every sign qualified for it. For the thing is only given indirectly, through the sign, when the sign is univocal, and when it is by itself alone sufficient to mark *(kennzeichnen)* the thing out, whether externally or conceptually. Only then can the symbol serve as deputy of the thing.

Moreover, one must not here force upon the concepts of *univocal sign* and of *deputization* a rigorous, logical sense. For the psychological possibility of deputization, all that is required is univocity of the sign, in the psychological sense. In and for itself, and from the logical point of view, the sign may be multivocal; but under the *hic et nunc* actual circumstances, with the present controlling direction of interests, it is univocal, and therefore is suited for deputization. But where we wish to use inauthentic representations (possibly with a distinct consciousness of their function) to attain *knowledge*, there we have of necessity to free ourselves from all contingent and variable "circumstances" and, accordingly, assign to the symbols a well-defined logical signification that gives them a rigorous univocity. In order to allow the distinction in question, between inauthentic representations and signs, to stand forth more fully, we will formulate the definition of the former concept as follows: Any sign (simple or composite, external or conceptual, and so on) which functions as *deputy* for what is designated is an inauthentic representation.

This deputization can, now, be either a temporary one, or a (more or less) enduring one. The inauthentic representations can, that is:

1) Serve as mere intermediary for the production of the authentic representations corresponding to them. In this manner there function, for example, conventional markings, mnemonically designed sequences of words, mechanically learned verse, and the like.

2) But the inauthentic representations can also completely take the place of the authentic ones as surrogate representation. Here again two cases are to be distinguished:

A) The inauthentic representations serve merely as convenient [352] replacements for the authentic ones in carrying out some higher level psychical activity. As "higher" psychical activities in this respect we count: representing in phantasy as opposed to sensation; representing of more abstract as opposed to concrete contents; representing of relational as opposed to absolute contents; representing in acts of higher order as opposed to that in acts of lower order (and, accordingly, also the representing of a multiplicity as opposed to that of a single content); and the representing of psychical acts as opposed to the representing of primary contents (and, accordingly, also the representing of relations *within* primary contents). Wherever it is possible, then, the contents – but above all, primary contents and primary relations – requiring a lower psychical activity function as surrogates for those which are (briefly put) higher. In the course of any rapid flow of thought, inauthentic representations of the type here considered prevail to an extraordinarily high degree. Words or letters, accompanied by indistinct and unclear phantasms, and, in and with these latter, isolated and fragmentary qualities (*Merkmale*), the rudimentary beginnings of higher psychical activities, and so on – all without a fixed content or continuity, now contracting into the mere word images, and now approximating (in this or that respect) to the clarity of a genuine representation: – when closely examined, *these* are our thoughts. And so perfectly and assuredly do they surrogate for the concepts really intended, that in the majority of cases we do not notice the distinction between the two at all, in spite of the huge gulf separating them. The signs and fragments of experiences stand in place of the genuine concepts, but *that* they do so is unnoticed by us. Now, how is it possible that such highly inadequate surrogates – which are in part foreign in their intrinsic natures to the true concept of the subjects concerned – can nonetheless stand in place of those subjects, and can provide the foundation for the judgments, acts of will, and so on, which are directed upon them? The answer is: In virtue of the fact that the deputizing signs (changing from moment to moment in relation to the same fact) either include in themselves, as a partial content, precisely the property upon which the momentary interest bears, or at least possess the aptitude to serve as the beginning or connecting point for psychical processes or activities which would lead to this property – or even to the full concept involved – and which we can arouse and produce wherever it may be required. [353] If, for example, we are concerned with the concept of a sphere, then, like a flash, there appears with the word the representation of a ball, in which the shape alone is specifically attended to. This accompanying representation, whose property crudely approximates to the intended concept and thereby symbolizes it, may then disappear once again, leaving only the word remaining. But its appearance nonetheless suffices in order to secure us in a confident grasp of the subject involved. Often there will suffice the word alone, plus a lightening-like judgment of recognition. In the course of the further flow of thought there then stands out from the store of memory this or that Moment, which is precisely the one we need: E.g., the geometric definition – whether as a mere sentence with the recognized sound complex, or in the form of an inadequate "intuitionalization" (perhaps three or four lines of equal length, proceeding out from one single point, a crudely inexact phantasm) – or the mode of production by rotation of a circle, or some theorem or other, and so on. All of this is reproduced with that degree of approximation to the relevant genuine concept for which, precisely, we have need, either immediately and directly, or in successive steps. We are presupposing that the requisite reproduction processes run their course with unconditioned trustworthiness. Just so soon as they do not do so, as soon as memory deserts us, understanding also ceases immediately, and the symbols do not fulfill their purpose, the flow of thought falters, and we then even find that we are lacking the true concepts.

In this manner there thus adheres to each actual representation a more or less extensive complex of memories: of words, of sentences, of images with absolute or relative properties usually noted, which are intimately connected to the representation by means of association, and from amongst which, depending upon the direction of interest, now these, now those, are reproduced. It is not meant that the interest should or could direct itself upon something unconscious (namely, the "unconscious" contents stored away in the 'box' of the memory). The interest goes, of course, only upon content which is really present; but that psychical act constitutes the psychological ground for the reproduction of a content associately connected to the one currently present, which, earlier united with it, rests upon a [354] homogeneous interest.

In the process of the rapid flow of thought, signs (as was already mentioned) act as surrogates without our noticing it. We suppose ourselves to be operating with the genuine concepts. But even when, forced to reflection, we notice the true state of affairs – as when, suddenly becoming uncertain, we reflect on the signification of a word – we are, as a rule, still satisfied with mere surrogates. Some sort of vestigal residue reproduced, and a vivid recognitional judgment associated with it, suffices us. They assure us of the possibility that in any instant we would be capable of unfolding before ourselves the full content of the verbal signification. We feel ourselves to be confidently in touch with the matter at hand, and proceed onward in the expectation that the mechanism of reproduction will function well.

B) The class of surrogate representations that we have dealt with up to now is characterized by the fact that the authentic representations which the surrogates replace are at our disposal at any moment. Where the controlling interest can only be satisfied through those authentic representations themselves, they quickly reappear from memory. It is also clear that the prior existence of these authentic representations forms the presupposition of those inauthentic representations which here serve as substitutes. It is wholly otherwise in these respects with symbolic representations of the second class. These do not subserve a mere convenience of thought. They are not signs or abbreviations for original authentic representations that are easily reproducible in any moment. The symbols here, quite to the contrary, refer to facts whose authentic representation is denied us – whether at the time, or permanently. In many cases authentic representation at least has a *psychological* priority over the symbolic. This is true of many phantasy representations, and with all memory representations. With the former it is still possible that the relevant objects may subsequently return to authentic representation – as when I think of a picture in the next room, and then go in and look at it. By contrast, objects of the latter type remain forever inaccessible to would-be authenticizations. A past event which I recall is something that no power on earth calls back into actuality. It is clear [355] that this distinction holds true in general. On the other hand there is an immense number of cases where the symbolic representation has priority over the authentic one. And also here the cases divide, once again, into two groups – depending upon whether the intended authentic representation can subsequently come to actualization, or whether it remains forever denied to us. Examples which easily present themselves may illustrate this. We read with understanding a geographical report of a journey without having seen at any time landscapes, human beings, animals, etc., of the kinds described – much less, then, the very ones described themselves. But it may actually happen that we one day travel to the same place, and subsequently become acquainted with them. This is easiest for us in the cases where the objects described are of a type otherwise well-known to us. The single properties are given; their grouping and interrelationship is easily accomplished in phantasy, conforming to familiar models; and the image of a "something" which resembles the projected phantasm serves as an adequate substitute for the subject at issue – which then also, in case it does itself show up sometime, can be recognized. If we are sufficiently familiar with representations which arise in this manner, and which are very rich in content, then there immediately ensues a surrogation of them by means of substitute representations that are more convenient, poorer in content, or even wholly external (that is, by means of a symbolization of the second level), answering to our mental drive toward economy, which has been repeatedly mentioned. Let us now consider examples from the other group. Already with concepts such as *Africa*, *the earth*, etc., we have concepts that belong here, although they are closely related to the above examples. Further, we have here the concept of *man* in the sense of physiology and psychology, and – taken in an analogous sense – the concept of an animal and of a plant, the concepts of the sciences, and so on. An extremely large complex of inauthentic representations, ordered together by means of various judgments, and with the possibility of unlimited expansion (but within the framework of characteristic properties): *this* forms the sum of that which the one most familiar with such a concept can make present to himself, or can indirectly designate as belonging to it. Naturally we also do not, in such cases, operate with those complexes themselves, not even to the extent in which they stand at our disposal. Rather, we operate with concise surrogates, and thus with indirect symbols which, through the mediation of especially characteristic properties (those which form the kernel around which all else crystalizes) and external signs, des- [356] ignate and replace the intended concepts. The greatest distance from genuine representations is attained in those symbolic representational constructs such as God, external thing, real space, real time, other minds, and so on – furthermore, in the construction of contradictory concepts such as wooden-iron, round-square, and the like. Whereas with examples of the previous class it is always conceivable that a, so to speak, quantitative extension of intellectual capacities would make a genuine representation (e.g., of Africa) possible, it is clear that, with the examples now cited, no heightening, of whatever kind, of those capacities could lead to the concepts intended. This is clear for some cases because an Evident judgment guarantees that the properties involved are incompatible; and for others because that which is intended, by means of negative determinations belonging within the conceptual content, is explicitly thought of as extra-psychical, and therefore as non-presentable (*unvorstellbar*). Beyond this, some of these cases contain as properties concepts of infinite increase – and, indeed, not merely in the sense of the unlimited, but rather in the sense of an actual infinite (as, e.g., the concept of God, the concept of infinite perfections). These are concepts the authenticization of which would presuppose an actual infinity in psychic capacity that is totally incomprehensible to us.

The psychological understanding of the surrogate representation of the type here considered requires no new principle in comparison with the previous class. Only this one thing deserves mention: that to the symbolic representations which are derived from relevant authentic ones there belongs, in the nature of the case, a psychological priority over those symbolic representations which do not originate in that manner. We must already be quite familiar with that practical equivalence between authentic representations and their symbols which makes possible the use of the latter in place of the former, before there can occur a formation of symbolic representations which are not founded through prior authentic representations. That inauthentic representations of this type must gain more and more in widespread use and importance with the development of language requires no special explanation. The formation of higher-level concepts went hand in hand with the development of language. Any composite of conceptual properties linked by the indeterminate concept of a "something" (or an equivalent substitute for it) could, then, serve [357] as a symbolic representation on the basis of the known relationship between symbol and subject matter.

We now wish to discuss a few points of greater importance, which concern inauthentic representations as a whole. We begin with a distinction which we have already pointed out. One must, namely, not confuse the fact that surrogate representations are being used with being aware of that use. The latter is absent in many, if not most, cases: The fragmentary contents and signs surrogate, but *that* they do so, we do not notice. Even there where the symbol relation belongs within the content of the inauthentic representation, that relation itself usually disappears, in that a surrogation of the second level sets in. One speaks, for example, of Bismarck. That my representation of him is inauthentic, is a fact well known to me; and its symbolic character even belongs within its full content. But in the rapid flow of thinking there surrogates for it, in turn, an abbreviation – perhaps a fragmentary phantasy image – and the *sign*-relationship gets lost. Inauthentic representations are the basis of our common practice of forming judgments. If it thus is certain that we as a rule operate with signs without specific awareness of the fact that we are doing so, then it is clear that for our practice of judgment formation the circumstance that the signs are signs cannot become effective as a motive of *knowledge*, even though the judgments are directed toward authentic concepts, and not toward the symbols. Accordingly it surely is not *logical* motives – i.e., motives toward *knowledge* – that guides judgment activity in practice, but rather blind psychological laws. We thus do not perhaps operate with signs instead of with the facts because we have made an induction, and because abundant experience has taught us that signs and facts stand in such a relationship that a procedure for forming judgments which is based upon signs always proves correct for the correlative facts. No. We proceed without reflection, and thus without induction also. The true situation is far simpler. We follow in judgment the train of the association of ideas, which, depending upon the course of our interests, reproduces now this, now that group out of the associative complex belonging to the concept. And our judgments and inferences, fixing upon these [358] – now richer, now poorer – fragmentary contents (and sometimes, as we shall see, fixing continuously and exclusively upon symbols), proceed as if they nevertheless were based, everywhere and always the same, upon the authentic and true concept of the relevant facts. And, indeed, they so proceed simply because we just do not notice that we are operating with surrogates instead of with the full concept. It is, therefore, the same with our judgments as with our representations: instead of the authentic ones we have symbolic judgments, but we do not notice that they are symbolic.

However, we must not rest satisfied with the illumination of the factual psychological state. With the judgment a perspective comes into consideration which is lacking with the representation: namely, concerning the two-fold question about justification and about truth. As to the subjective aspect of this question, it reads, in our case: With what *right* do we operate in our practice of judging in the manner described, using symbols instead of the true concepts? The answer lies in the above explanations. We proceed *without* any justification; we are guided, not by a motive of knowledge, but rather by a psychological mechanism.

But this does not settle the second or objective aspect of the question, the one about *truth*. Indeed, a logically unjustified procedure can quite well lead in the end to true results. One such case – and in fact quite a strange one – here lies before us. It certainly would be quite well conceivable apriori that a psychological arrangement of our nature should force our practical (extralogical) judgment always or predominantly to error, and only exceptionally to truth. In reality matters stand just the opposite. On the average we get along very well with judgments in terms of surrogates (and the overwhelming majority of all judgments are of this type). This is a very interesting fact metaphysically. Here one could, extending a *Hume*an reflection, go on to say that it conforms to the general wisdom of nature to secure by a mechanical instinct an activity of soul so essential to the survival of the human race. This activity then remains on the average free from error in its effects, holds valid right from the beginnings of life and thought, and is independent of the laborious and fallacious proofs of reason, which first become possible at a more mature period of development. Recently it is perhaps preferred to explain this [359]
teleological trait of our nature by means of *Darwinian* principles. – Yet all of that is irrelevant here, where the metaphysical is not in question. What we do and must strive for is a logical illumination of the state of affairs. "How so?" one will ask. "A logical illumination

of a procedure recognized to be non-logical? Is that not a contradiction?" It will not be difficult to make clear the soundness of our aim. If a characteristic type of judgment process, although not guided by motives of *knowledge*, nevertheless leads to correct results, then we still must seek and find in its inner structure – in case that really is something transparent to us – the grounds why it is suited to produce truth (although not knowledge). In other words, it must be possible to state a parallel logical process which explains the mechanism of this type of judgment process logically, and conceives of it, so to speak, as if it had been rationally devised by means of that logical process. With the aid of the logical process we attain to an understanding of why the non-logical process must nonetheless have the effect of a logical one, and that is the logical illumination of which we spoke above.

We therefore ask: How does it come about that in the common praxis of judging we can dispense with authentic concepts? We intend judgments about them, but there nonetheless underlies our judgments only those surrogates – so fragmentary and so variable in many ways (and, indeed, varying in relation to the same subject matter). How is it possible that our judgments are independent of these surrogates, and hold good of the true concepts upon which alone they are directed?

For the purposes of a response we have to distinguish two main classes of cases: 1) those cases where the process involves, both in the single steps as well as in their linkage, an authentic judging: one which has a symbolic character – though it is not specifically noticed – only in virtue of the fact that the matters judged about consist of surrogates instead of authentic representations; 2) those cases where the judging itself is inauthentic – and, indeed, in such a manner that external signs (possibly sentences and systematic complexes of sentences) enter as surrogates for judgments and inferences.

For the first class the following is the simple solution to our puzzle: [360] It is correct that these judgments involve only such uncertain, unclear, variable surrogates. But these surrogates include, in any given moment, precisely those constituents and properties of the genuine concepts upon which the judging interest is directed. As objects of special attention they become, not unclearly and uncertainly represented, but represented with just that degree of clarity which the judging requires – howsoever obscure the remaining constituents of the surrogate may be. That the surrogates change from moment to moment may be true; but in doing so they follow the changes in the interest of the judgment. If, now, we inquire into the value of these judgments for knowledge, then it is clear that they must be valid for the authentic concepts, inasmuch as those concepts possess, among others, precisely the properties of the surrogates that were especially attended to and judged of – as they would, in general, also have validity for any and every thing to which those properties belong. Thus, from the logical viewpoint the following schema is presupposed: A judgment applies to an object *X* provided only that *X* has the property *A*; *G* possesses the property *A*; thus the judgment also holds true of *G* in just the same respect. The sign "*X*" stands for our symbolic representation – possibly a phantasm in which the property *X* is exclusively attended to and judged of – the remainder being wholly obscure. Precisely the same property is also shared by the intended subject matter (the '*G*'), and therefore the judgment also holds true of it. Not a trace of such logical reflections is to be found in the natural course of our thinking. Our practical activity of judging is, indeed, not a logical activity. We produce judgments on the basis of surrogates, and treat them without further ado – unencumbered by the question of justification – as judgments about authentic concepts. But here we recognize why doing things this way leads to no error. We see that the non-logical process must yield the same result as the logical one, with the one difference (essential for theory, but not for practice) that the latter also guarantees insight into the justification, while the former does not.

Up to this point we have investigated only the basis for the objective truth of isolated judgments of the type in question. But it is clear that for the understanding of inferences relevant here nothing essentially new is to be added. If isolated judgments about surrogates are equivalent to those about the related authentic concepts, then a process of inference consisting of judgments of the one kind will also [361]
be equivalent to one consisting of judgments of the other kind.

Let us now turn to the second group of cases, where our judging is based upon mere external symbols, while we nonetheless advance unhindered in the process of judgment. This is possible by virtue of the fact that the judging here is itself not, as previously, an authentic one, but rather is externally inauthentic. Sensible signs (e.g., names) of the representations are combined with sensible signs of acknowledgement and rejection; sentences originate; systematic sequences of sentences symbolize deductions; and the judgment process consists, then, in allowing an external progression along the chain of signs to surrogate for genuine inference. A few examples will illustrate this: a is greater than b, b is greater than c, and c is greater than e, and thus a is greater than e; $\mathrm{a} = \mathrm{b}$, $\mathrm{b} = \mathrm{c}$, $\mathrm{c} = \mathrm{d}$, $\mathrm{d} = \mathrm{e}$, and thus $\mathrm{a} = \mathrm{e}$; all A are B, all B are C, all C are E, and thus all A are E. In all of this we may by the letters understand names for the contents upon which our judgment activity is directed. As a rule we carry out such inferences symbolically. Customarily, even in the single steps, we hold ourselves neither to the authentic and full contents nor to surrogating partial contents, but rather simply to the names or letters; so that then there *eo ipso* can be no talk here of an authentic judging and inference. We progress mechanically along the series, connecting and eliminating terms as the pattern requires, and thus arrive at a symbolic judgment (a sentence) which we hold to be a sign of a truth. More commonly, however, the single steps are realized in genuine judging; but by expressing the results of the judging simultaneously in external signs – for example in sentences – those signs enter into the further course of the process as surrogates for the genuine judgments; and the inferential sequence proceeds, then, in the form of external symbols as above.

Again, we do not inquire into the logical justification of these symbolic modes of procedure. We do not even notice that they are symbolic. We follow them out without reflection, and not perhaps on the basis of a previous induction or some other sort of justificatory considerations. They are not logical methods conforming to technical rules, but rather are indigenous mechanical processes. The question [362] here is quite different, and it is: What is the basis for that status as truths which is enjoyed by the results of these natural mechanisms? The response to this question calls for a few reflections. Here it is first to be observed that such symbolic modes of procedure do not possess the same originality as the genuine modes of procedure correlated with them, and that they are, rather, developed out of these latter as convenient simplifications. The uniform character of inference sequences of a certain type, distinctly marked by uniformities in their external expression, leads us of itself, and naturally, without special reflection, to follow the uniformities of expression even where the founding psychical activities are lacking. Once again, it is the power of the association of ideas which is the imperceptible motive force of the process. But here it certainly functions in a wholly unique fashion. The conclusion sentence is not produced in one stroke, in one act. For that would presuppose that we had already repeatedly drawn exactly the same conclusion from the same premisses, whereas it is precisely the characteristic mark of the procedure that it is applied mechanically and with success to each *new* case. The reproduction comes about indirectly, under the mediation of the *form*. We understand by this something like what formal logic does when it speaks of "forms of inference." But in so doing we certainly do not endorse the *explanation* which it gives of such forms, but rather the actual use which it makes of them. The form of an inference consists in the external manner of linkage and arrangement among the premisses. In this way each premiss – and, accordingly, also each one of the names contained in the premisses – is assigned a determinate, systematic position. Of course there are intrinsic characteristics of the judgment system interwoven into the mental process of inference. They ground the systematic form of the linguistic expression and confer upon it a generality reaching far beyond the concrete case. We need not go into that here. It suffices that there are innumerable conceivable inferences expressed in the same form. If, now, we have repeatedly – and, indeed, *genuinely* – carried out inferences of a determinate form, and if the form's systematic type is one easily grasped, then that type will stamp itself upon the memory. As a later consequence, then, a conformable set of premisses can suffice by itself for the reproduction of the conclusion sentence. [363]

As soon as we, in speaking and judging, move a few stages into the familiar pattern, reproduction leaps ahead to offer us the form of the sentence which is the conclusion.

But this power of reproduction supplies not merely the form, but also the content filling out the form: that is, the names which fill out the form into the full conclusion sentence. In fact, as was mentioned above, since each name has its systematic place, and in the conclusion names are joined in a completely determined characteristic position (as is true in the above examples of the *first* name and the *last*), the place-value therefore serves as reproductive Moment in calling forth the appropriate names, and in this manner makes possible a complete reproduction of the whole conclusion sentence.

Since by these analyses we have attained to a more precise knowl-

edge of the natural psychological mechanism of symbolic inference, it will be possible to construct the parallel logical procedure, which resolves our question and provides us with the explanation of why that mechanical process *must* produce correct results. In order that such a mechanism could arise and function, the relevant inferences and their linguistic correlates would have to fulfill certain conditions. Let us enumerate them in sequence. First of all, it is in the nature of the linguistic designators that come into play that they be univocal – if not under all circumstances, yet within the systematic combinations which show up here. The systematic forms of conjunction for the words must precisely reflect those present in the thoughts; otherwise the former never would be able to become regular surrogates for the latter. The ambiguities would, in spite of all the powers of reproduction, always force us to actualize the genuine representations, judgments, and inferences; and a *mechanism* would be impossible. But there is yet another and more specific characteristic which the sign-system must possess, with respect to the fact that a reproduction of the conclusion sentence is supposed to be possible on the basis of the premiss sentences alone. Obviously the one part of the system which contains the premisses, in the appropriate order and connexion, must determine the other part, which contains the conclusion sentence, in a purely formal (and, of course, univocal) fashion. Only then can reproductive imagination forthwith construct the lacking [364]
second part, the conclusion (in the manner described above), in the cases where the former part alone is given. Given univocity of designation, it follows that the corresponding judgment system also must be of such a character that the totality of the premiss judgments univocally determines the conclusion judgment. All of which leads to an important consequence. If, namely, a determinate form of inference, or else a general class of inferences characterized by it, fulfills all requirements, then the knowledge of that state of affairs puts us in position to replace – purposefully and for logical reasons – genuine inferring with a symbolical inference. As soon, that is, as a premiss system of the relevant class is given *in concreto*, then we are able – on the basis of the linguistic expressions alone, and without reference to the psychic correlates – to construct the conclusion sentence; and we have full logical certainty of possessing in the judgment corresponding to it the intended, correct judgment as conclusion. What we then do in such a manner for reasons of knowledge, is done by

the mechanism of reproduction out of blind causality. In order for that mechanism to develop and function there are in fact, as we saw, just those very characteristics required in inferences which, *if* they were known, would logically justify the mechanical procedure. The univocality of the linguistic expression and the univocal determination of the inference by the premisses, both in the psychical and in the symbolical aspects: – these are the necessary and sufficient conditions for the *blind* mechanical procedure, on the one side, and for the *logical* mechanical procedure, on the other. In this manner is our problem resolved, and the apparent teleology of the natural process is completely illumined. But of special interest here is the fact that the logical procedure which runs parallel is also a mechanical one, only it is one where the mechanism is purposively set up on the basis of logical considerations. The generality of the mechanism is concentrated in a logical rule which, for the respective class of inference forms, shows how authentic inferring is to be replaced by an external operation upon linguistic signs, and how in this way the linguistic expression of the conclusion judgment is to be produced by that external operation itself. But this is what all *formal* inferring amounts to in the legitimate and true sense of the word. However, that such inferring is not perhaps something insignificant (as one might be led to suspect from the above simple examples), but to the contrary [365] forms a tool of the highest importance for scientific progress, will be most convincingly proven by our own theory of arithmetic.

Our investigations up to now concerned symbols and symbolic processes of the lowest level. In the course of natural unreflective thought, these stand in as surrogates for authentic representations, judgments, and inferential processes. They do so in virtue of the lawlike tendency of our nature, without there being present any specific consciousness of this their function – and much less, then, their employment being governed by logical motives (whether previous or accompanying). But besides these *natural* surrogates (as we may concisely designate them) we use, and to the widest of extents, *conventional* surrogates. We invent symbols and symbolic processes, or use those invented by others, as supports and surrogates for representations and judgment processes. And in this we proceed purposively, fully aware that we are dealing in the symbolical.

To the *logic* of symbolic representations and judgments we now turn with a few reflections. The carrying out of such a logic would set two goals for itself: It would seek to get to the bottom of the function of symbolic representations and judgments in the activity of *theoretical* judgment, and above all, to provide *logical* illumination for those algorithmic modes of procedure which have on such a wide scale become the vehicles of progress in the exact sciences. And it would seek to formulate the rules for testing and devising such methods. Investigations of the type that we have undertaken above for the natural practice of judgment would, of course, have to provide a foundation for these higher investigations. For in fact logical procedure is, *vis-à-vis* the corresponding natural one, not something *toto genera* different. Both processes make use of the psychological laws of our nature; and, to a considerable extent, they make use of the same ones. But *only* in part. And precisely therein lies the difference. As a new Moment there enters the influence of the will, guided by motives toward *knowledge*, and also the capability of regulating the process of judgment, through this new element, in conformity with precisely such logical interests. The natural judgment requires such regulation in virtue of the manifold natural sources of error which [366 bring it about that the natural processes – although guiding us rightly on the average – nonetheless in many specific cases lead into error. In this respect the logical procedure therefore serves in the *assuring* of knowledge. In another respect, again, it serves in the *expansion* of knowledge. For the conventional techniques do not merely do the same thing better than the natural ones. Rather they do incomparably more. In every case, the source for the conventional modes of procedure lies in the natural ones. If, through reflection, the power of the latter to produce truth is brought to specific consciousness, then – given the power of the will over the psychological forces that underlie those natural procedures – there can come about a systematic invention and purposeful application of analogous, but now conventional, methods. Hence we find that, also in this regard, our contention that the analysis of the natural modes of procedure must precede that of the conventional is justified.

The conventional surrogates are a special class of conventional signs. For conventional signs were not, in general, invented to take the place of authentic representations and judgments, but rather to serve as marks for remembering, as sensuous supports for psychical activities, as instruments of communication and conversation, and the like. Only in consequence of constant usage, with the associations which develop, and occasionally also through experimentation – or through a mixture of the two – do conventional signs (provided that they are actually suited for it) assume the character of surrogates, as natural signs similarly assume the character of natural surrogates. Hence in them we have a mixture of art and nature. By far the largest part of symbolic representations and judgment processes rests upon language. But linguistic signs quite certainly were not invented for this purpose, but rather to enable people to interact with one another. In the abstract sciences the arithmetical symbols and operations upon them play the most important role. In the place of an actual, immensely complicated deduction of magnitude relations from other magnitude relations, there intervenes the blind mechanism of sensuous symbols. But if one tracks the historical development, then one [367] easily recognizes that it is not this mechanism as a foreseen goal that has conditioned the invention of the symbols. They originally served, as a matter of fact, as mere marks for distinction and re-identification, and, in virtue of that, also as supports for authentic judgment processes resting upon them. We already have a high developmental level of spiritual culture when we invent conventional surrogates with the full consciousness of their function – or even when we just use them with full consciousness once they are present. Of this sort must be the symbols and symbolic processes within a rigorous, intelligible, and logically developed arithmetic, as well as in formal logic in general; and, accordingly, also in the domains of their application, the abstract sciences. However, the conventional surrogates which we use in common life and science do not have this pure character. Artificial signs which have become surrogates, through the efficacy of those psychological forces earlier mentioned, further our judgment activity without there being a correct understanding of what is really going on.

To the distinction just made we will now add some further ones which are of importance for a theory of signs.

Under the title "surrogate" we comprehend two things: signs or sign combinations which surrogate for representations, and those which surrogate for judgments and inferential processes. Deputization can take effect on a broader scale and on a regular basis only where the representations and judgment processes have a systematic character, which is, then, capable of being reflected in a system of signs and the uniform rules for their combination and equivalent substitution. For formal logic, therefore, systems of signs, with the algorithms grounded upon them, are of very special interest. In relation to psychological and historical origination, it is necessary to distinguish two things in each sign system: the origination of the single signs, and the origination of the system as such. Artificial (invented) signs can develop into a sign system by natural thought taking possession of them – and into a system of such a rich and finely articulated structure that subsequent reflection would first hit upon every other idea than that such a system should have arisen through the blind interaction of natural laws. This holds true, for example, [368] of language. The single signs themselves are conventional. Howsoever crude were the first means of designation at the beginning of the development of language, still they certainly had the character of inventions. Their suitability for the expression of external processes or internal states was the motive for the intentional use of them for the purposes of communication. And the novel and ever new signs, continually being introduced, are also inventions. But from the single signs there arose, in the manner of a natural development, the system of language, with its finely articulated grammatical structure. And the purposiveness and elegance of this systematization is so conspicuous, that the thought that *it* could be a product of blind natural laws already presupposes a high development of psychology. It is the same with arithmetic. The single signs are inventions. Indeed, a further point is true here: The single methods also are inventions. And nonetheless the system of arithmetic as a whole, with its wondrous structure, is not a product of intentional foresight, but rather is a natural development.

In every system of signs we distinguish the basic signs from the derived or composite signs. The derivation of the latter out of the basic signs ensues by means of sign operations. These are systematic modes of procedure for symbolic representing, judging, inferring, etc., which follow determinate rules. Thus, for example, the arithmetical operations, insofar as they are formative of numbers, are rule-governed methods for the production of inauthentic representations. But, insofar as they formulate rules for the formation and transformation of equations or inequations, they are methods for the production of symbolic (and, indeed, true) judgments. The basic signs of number theory are the signs 0, 1, . . . , 9. All remaining number signs – and, further, signs such as $2 + 3$, $5 \cdot 6$, $\frac{4}{2}$, and so on –

are derived signs for inauthentically represented numbers. Every calculation (e.g., addition) is a symbolic formation of a truth by means of certain operations involving the basic signs.

Every conventional operation with signs serves, in a certain manner, the purposes of knowledge. But not every one of them actually leads to what counts as knowledge, in the true and pure sense of logical *insights.* It is only in cases where the procedure itself is a logical one, and where we have logical insight into the fact that, such as it [369] is and because it is so, it must lead to truth, that its result becomes, not a mere *de facto* truth, but rather a knowledge of the truth. Only then do we have full assurance of being protected from error, and we judge not out of blind impulse, not out of more-or-less blind instinct, but rather out of luminous insight. In this sense we distinguish: *pre-logical sign operations*, which are oriented toward truth, and which perhaps attain to it, but do so without the employment (certainly also the invention) of these modes of procedure being based upon logical understanding; and *logical sign operations*, which are followed on the basis of knowledge, and therefore do not merely provide us with truth, but rather with assured truth.

This distinction concerns, as one sees, *all* symbolic judgment processes whatsoever: even the natural ones resulting from the efficacy of idea-association alone and excluding all logical motives. These latter belong totally to the pre-logical level.[3]

It is useful to call attention here to the point that a purposeful employment of signs for the goals of knowledge is by no means, because of that simple fact, a logical one. Already upon the pre-logical level a purposeful search for and employment of signs can occur. One can very easily observe that signs further our knowledge, without, however, being in the least clear about the grounds of that furtherance. This will be especially possible when the sentences (symbolic judgments) obtained in the symbolic manner lead, by means of a transition from the signs to the thoughts, to genuine judgments that legitimize themselves in a verification that can be

[3] [Husserl's marginal note] Logical/pre-logical: This distinction means that an analogous investigation is required here, and that similar questions emerge as in the case of the natural mechanisms. How does it happen that we can calculate for centuries without an understanding of the calculatory system? But the explanation on this point simultaneously provides those supplementations which turn the pre-logical into the logical.

carried out for any given case. This is how it is in mathematics. One may assert: General arithmetic, with its negative, irrational, and imaginary ("impossible") numbers, was invented and applied for centuries before it was understood. Concerning the signification of [370] these numbers the most contradictory and incredible theories have been held; but that has not hindered their use. One could quite certainly convince oneself of the correctness of any sentence deduced by means of them through an easy verification. And, after innumerable experiences of this sort, one naturally comes to trust in the unrestricted applicability of these modes of procedure, expanding and refining them more and more – all without the slightest insight into the *logic* of the matter, which in spite of many attempts, from the time of a *Leibniz*, *D'Alembert* and *Carnot* up to today, has made no essential progress.

This is, in fact, how things actually stand with logical methods, e.g., those of induction. The natural scientist makes use of these methods with the greatest success, and without feeling the least bit cramped by his own unclarity, or even by the unclarity of the logicians, concerning the sense, limitations and epistemic value of them. In the case of induction we must again distinguish the pre-logical and the logical inductive procedure. Even where the two lead to the same result (which doesn't even hold true as a rough average), they still do it in a wholly different manner; and only the logical one yields *knowledge*. To ground induction upon the blind psychological mechanism of habit, which works but does not justify: that is to confuse pre-logical with logical induction, or (with Hume) to deny in general the possibility of a rational justification of induction.

This is drastic confirmation of the fact that a utilization of symbols for scientific purposes, and with scientific success, is still not therefore a *logical* utilization. We of course have no thought of totally rejecting the pre-logical use of signs. No doubt it on the average leads to what is correct. But still, *only* on the average. And just because of that we require, for science, the utilization of logically justified signs alone. Here one could cite against us precisely our above example – that of arithmetic. It is in fact correct that arithmetic in a developed state is to a large extent independent of a logical understanding of its technical modes of procedure. However, arithmetic has not sprung as a finished invention from the head of a single individual. It is the product of a development of centuries. It has arisen through a kind [371] of natural selection. In the struggle for existence the truth triumphed over error, which proved itself untenable; and arithmetical methods were developed accordingly, in that they were successively modified to exclude errors which would otherwise still be possible.[4] The correctness of the result could quite well serve as touchstone for the method; since one could check the result (by means of the verification mentioned above) without using the method. But how much mental energy has been wasted on this route, more governed by chance than logic! One thinks of the endless controversies over negative and imaginary numbers, over the infinitely small and the infinitely large, over the paradoxes of divergent series, and so on. The progress of arithmetic would have been quick and secure, instead of slow and wavering, if already upon the development of its methods there had been clarity concerning their logical character. And there likewise can be no doubt that for the future continued development of arithmetic also – so far as an expansion of the arithmetical domain is intended – insight into its logical character must be of decisive influence, promoting progress.

Outside of arithmetic we find yet many more confirmations of the fact that logically untested symbols can lead to error. After all, logicians also have long since called attention to this fact in the case of the most important of sign-systems which we possess, that of language. In what sense language furthers thought – and, on the other hand again, hinders it – this is at present discussed in every system of logic which endeavors to have an effect upon practice. We are warned against trusting in words too much, told to represent their full sense to ourselves in each proof, to guard against equivocations, and the like – rules which, although highly useful, are still limited to an all too narrow sphere. In that for the most part only the symbolic character of the simpler forms of language, that of words and sentences, was attended to, this symbolic character was [372]
overlooked in the mechanical-symbolic modes of procedure which

[4] [Marginal note] Methods which emerged and proved their value within narrower domains were – more through the instincts of genius than through logical reasoning – applied far beyond the limits of those domains, wherever analogous problems invited it. This led to frequent errors, which in turn forced one to reflections upon the methods, and to appropriate reformulations thereof, until – avoiding error now on this side, now on that side – finally the limits of their validity were staked out. In this manner there triumphed, in the struggle for existence, the more exact over the vague, and ultimately the correct over the incorrect, methods.

already offer themselves for use to natural thinking, and which replace a more or less complicated process of inference by complex linguistic devices. I mean here the simple and composite syllogisms. Although they were treated to extravagant lengths in traditional formal logic, their rules nonetheless remained misconstrued. What was held to be rules of genuine inferring were (precisely as *formal* rules) in fact rules of symbolic inferring. This misinterpretation of the true state of affairs so unfavorably influenced the manner in which the subject matter was treated, that the theory of knowledge lost its way, and, on the other hand, *Praxis* was not in the least furthered. Had the symbolic character of the syllogistic (the larger part and the heart of the old formal logic) and of general arithmetic been recognized and brought to strict precision through penetrating investigations, then the theoretical understanding of these "formal" disciplines would have been able to exercise a clarifying and fructifying influence upon philosophy and the special sciences. But as matters stand today, the thickest clouds of confusion mislead and hinder both in the formal disciplines and in philosophy and the special sciences. Characteristic of the unclarity of the logicians is the fact that they either do not trouble themselves at all about theories of algorithms (and that is the rule), or else they do it in that manner, so incidental and superficial, which is the surest symptom of unclarity. With the latter I have in mind the discussions of *Mill* (*Logic*, Book IV, Chapter vi, subsection 6) and *Bain* (*Logic*, First Part, Appendix B). If one takes only the most common and the simplest of algorithms, that in which one counts and calculates with numbers, one will search logical works in vain for light on what really makes such mechanical operations, with mere written characters or word signs, capable of vastly expanding our actual knowledge concerning the number concepts – making possible for us accomplishments which were inconceivable to the greatest thinkers of antiquity. And on the other side again we find, as the characteristic sign of unclarity among the mathematicians, strange theories which they (one in one way, others in another way) have taken to be the philosophy of their discipline, and which often enough have led them – and the most original minds more than all – [373] into sterile detours. A truly fruitful formal logic constitutes itself from the outset as a logic of signs. When sufficiently developed, it will form one of the most important parts of logic in general (as the art of *knowledge*). The task of logic is the same here as elsewhere: to take possession of the natural modes of procedure in the judging mind; to test them, and to provide insight into their value for knowledge; in order finally to be able to determine with precision their limits, extent, and significance, and to be able to formulate relevant general rules on that basis. If it conceives its task correctly, then it will not have to restrict itself to keeping step with the pre-logical ways of using signs. Deeper insight into the essence of signs and sign techniques will, rather, empower it also to devise such symbolic modes of procedure as have not yet occurred to the human mind, and to establish rules for devising them. The relationship of the logic of signs to symbolic operations in the *praxis* of life and of science will thus be analogous, for example, to the relationship of inductive logic to practical inductions. Also here, as has been late enough in its acknowledgement, it would be the task of logic to take possession of these great natural tools for the formation of judgment, and, by means of scientific reflections concerning justification, limits, and import, to make from the natural and logically unjustified procedure one which is conventional and logically justified, and which guarantees not mere persuasion, but rather assured knowledge.

REVIEW OF ERNST SCHRÖDER'S *VORLESUNGEN ÜBER DIE ALGEBRA DER LOGIK*[1]

In the "Foreword" to this work the author describes its goals as follows:

"The reader will gather from the title that only the so-called deductive or formal logic is dealt with. The treatment of formal logic as a calculus would indeed *pre-eminently* deserve the designation 'exact logic.' Through such a treatment formal logic is set free from the fetters in which word-language has bound the human mind by the power of habit. Also, it is such a treatment alone that is capable of giving the laws of valid thought the most rigorous, concise, and clear formulation; and, in consequence of its advantage in this regard, it finds itself in the position to reveal numerous and significant deficiencies – if not errors – in older presentations."

Through recent works by the American C. S. *Peirce* and his followers, as well as by the author himself, this mode of treatment is supposed to have made such significant progress "that a definitive presentation and arrangement of a first and chief part of the whole system already seems attainable." And that is what the author wants to achieve.

As one can see from this, *Schröder*'s book makes the considerable claim to present a completely reformed deductive logic, in the form of an "algebra of logic." The foundational and guiding conceptions of this new logic derive from the British extensional logic, the viewpoint of which the author represents with a certain enthusiasm and seeks

[1] First appearing in *Göttingische gelehrte Anzeigen*, 1891, pp. 243–278. Republished in *Husserliana XXII*, (Nijhoff: The Hague, 1979), pp. 3–43. This piece is closely related in subject matter to the five which follow here, and to Appendices I–III, but especially I. [D.W.]

to support upon new grounds. This first large-scale attempt at an [4] extensional logic upon German soil offers a welcome opportunity to measure the old and the new logic against each other. For purposes of critical discussion we will, therefore, favor those parts of the work which are suited to throw some light upon the relationship between the two logics, as well as upon the essence and value of the deductive calculus.

Volume I, now before us, is prefaced by an extensive *Introduction* (pp. 1–125), "set in general philosophical terms" and divided into three sections.

The *first*, *Section A*, contains, in the words of its title, "preliminary reflections upon the character and delimitation of the problem to be solved, with observations about induction, deduction, contradiction and valid thought, *the* thinking subject, its representations, and the thing." Only a part of these somewhat disparate reflections can be regarded as truly introductory to the calculus, or at least as a constituent of a deductive logic. Many matters probably are taken up only out of the author's felt need to express himself for once concerning certain philosophical problems (such as the unity of consciousness, the origin of our belief in an external world, the freedom or necessitation of the will, and the like). We shall here restrict ourselves to that which is our main point of interest, and which gives clarity concerning the *genuine* "character and delimitation of the problem to be solved." Now, to be sure, we learn from the Foreword "that we are concerned here with deductive or formal logic." But it will nonetheless be well to look more closely at what the author understands by these (not wholly unambiguous) terms, and thus at *which* area he intends to deal with; and then – at which area he really *does* deal with.

"Logic (runs Schröder's definition, in agreement with most of our logicians) . . . concerns itself with all of those rules the following of which furthers knowledge of the truth." (p. 1) But he nonetheless intends to restrict himself to "logic in the narrower sense, . . . in order to work it out more thoroughly." "This *deductive* or formal logic concerns itself with the laws of *valid* thinking." The author places it in contrast with *inductive* logic; and, indeed, he supposes this property of the "validity" of thought to be what determines the division of the [5] whole of logic into these two domains. Then far-ranging reflections and criticisms (pp. 4–15) are devoted to the definition of validity.

Although these remarks surely are, on all essential points, highly questionable, they may be passed over here. It suffices that we know the investigations at hand actually to be oriented toward the delimitation of deductive logic in the *common* sense of the word – and, thus, toward the logic of deduction in its contrast with the logic of induction – and that the presentation of such a logic of deduction is the goal which the author sets for himself.

Or "intends to set," I should more precisely say; for one can hardly deceive himself more about his true goals than the author does here. And I will go into the details of the evidence of this deception, since it is characteristic of the whole of extensional logic. Already upon surveying the author's statements about the essential elements of deductive logic, we notice that a mere partial domain within it is confused with deductive logic itself. Deductive logic includes, we read, "first, the so-called 'analytic truths' or 'tautologies,' which present themselves as 'identity judgments'," (p. 9) and of which "All black cows are black" will serve as an example. "Second, . . . it includes necessary progressions of thought from beliefs already present . . . to new beliefs; this being, precisely, deduction proper." (p. 10) It immediately strikes us that under the first heading is by no means designated the totality of immediately Evident propositions (*Evidenzen*), but rather only a small part thereof: those which belong within the *"formal" theory of inference*. An added remark by the author makes this perfectly clear. "Our reflections [obviously referring to the calculus] even put us in a position to enumerate these identity judgments completely, within set limits." But this assertion has a sense only for those evident propositions of the special domain mentioned. One indeed can very well set the task of giving the totality of analytic judgments which can be purely formally constructed from given terms *S*, *P*, . . . *Z*: that is, constructed in total disregard of the peculiarities of the respective contents, attending only to the formal conditions of judgment which can found them. One obtains, for example, the following: "Provided that something is an *S*, it is not a non-*S*." "Provided that something is an *S*, and simultaneously is a *P*, it is either an *S* or a *P*." And so on. However, there are innumerable Evident propositions which, being dependent upon the peculiar [6]
character of the terms judged of, are not included here. For example, if *S* and *P* are two numbers, say 2 and 3, which can be authentically represented (i.e., they are not numbers defined only through indirect

symbolizations), then we have the directly Evident proposition that $2 < 3$. And this obviously is no "identity judgment" *in the sense* presupposed by the formal theory of inference. And through this restriction it is at the same time clear what is meant by "deduction" under the second heading above. It is *pure* deduction. This is surely made clear by a cursory survey of the work as a whole. In fact, the domain which is uniquely and solely governed by the logical calculus is none other than that of *purely formal deduction.* More precisely stated, it is the domain of pure implications between any judgments whatsoever – in the case of which, *as* pure, there is no consideration of the peculiar characteristics of the contents of the terms judged of. Because of this we have, in the concrete case, the possibility of replacing the terms with general signs, of inferring according to the corresponding general schema, and only then replacing the signs by their specific meanings. Whatever can be deduced from any system of given premisses on the basis of their mere "form" falls into this domain. But nothing more does. Above all, this domain does not take in the total range of deductive logic in general. There also does not fall within it a single one of the deductive sciences. Not one of them is an unalloyed system of pure deductions from a totality of given premisses – although each one often enough makes use of such deductions in particular steps.

Now one might well object that logic, *as* the general theory of deduction, must have no recourse to what is conditioned upon the specific matter of judgment, and that it therefore has precisely to deal only with deduction in general. The first part of this is quite right. Of *course* deductive logic does not work with specific concepts, such as numbers, figures, and the like, but always deals with logical activities which are used in all deductive disciplines and which constitute the essence of their methods. But pure deduction is only one of these activities. The deductive sciences do not just deduce. They also operate, construct, and calculate. Is the arithmetical *operation* in any (genuinely executed, not merely symbolical) addition or subtraction a deduction? Or is the geometrical operation in any construction? And [7]
all highly developed deductive disciplines utilize further symbolic techniques for the derivation of truths: they *calculate* with various algorithms. But is calculation deduction? Not at all. Calculation is a blind procedure with symbols, according to mechanically reiterated rules for the transformation and transposition of the signs in the

respective algorithm. Also, "setting up equivalences" for a problem [within the algorithm] is no deduction, just as little as is the final move in a deduction proceeding symbolically, *viz.*, the interpretation of the resultant end-formula. The entire procedure replaces and thereby spares us many pure deductions, but is *itself* no such deduction. The theory of all those mental activities which, though not themselves deductions, yet serve in the derivation of scientific truths, obviously belongs within deductive logic. But it does *not* belong within the domain of pure deduction. It was a chief error of the old formal logic that, while limiting itself to this narrow domain of pure deduction, it still believed itself to attain the goals of logic. The algorithmic logic, which in fact is the old logic's direct descendant, has made this error its own. And this in addition to another and not less momentous error, as we will immediately find.

When I ask, namely: Does Schröder's logic (or the algorithmic logic of extensions in general) at least set for itself and solve the problems which constitute the essential substrate of this narrow domain of pure deduction? In *its* opinion it certainly does. (See the Schröderian "ideal" on p. 17.) But there is, once again, a great gulf between the actual and the presumed goal. We are not here even presented with a *logic* of pure deduction alone, but rather with no more and no less than a *calculus* of pure deduction. Already in the Scholastic logic we find a degeneration into a mere technique of deduction. It was mainly devoted to the development of rules of technique. By following these rules within the forms which, in general, that logic took into consideration, one could mechanically construct the conclusion for any given set of premisses, and thereby *spare* himself genuine deduction. This primitive and limited technique is the seed from which the proud edifice of the logical calculus has grown. Far removed from being a theory of pure deduction, it is, rather, a device making such deduction *superfluous*. It is nothing other than a technique for [8]
manipulating signs: one which, by its system of rules, makes it possible, given a fitting symbolization of the premisses (their number and complexity may, moreover, be ever so great), to arrive at the totality of pure deductions invested in those premisses, or at their symbolic correlates. But calculation is no deduction. Rather, it is an external (*äusserliches*) surrogate for deduction.

Certainly here, as in the case of any calculatory discipline, the hard questions about the essence and logical justification of the calculative method are unavoidable. All the more so, as it is upon the response to these questions that the epistemic status of the results from such disciplines is above all dependent. If the "algebra of logic" were, instead of being a dexterous technique, at least a theoretically grounded, special method in logic, then it would have to give us some clarification on these questions. But the *logic* of this algebraic calculus does not fall within the mental horizon of the investigators who take the calculus to just *be* deductive logic – especially since, in fact, the mental operations upon which the calculus is based do not themselves belong within that domain of pure deduction which it exclusively governs.

The logical calculus is, thus, a calculus of pure deduction; but it is not *its* logic. In it we have its logic just as little as the *arithmetica universalis*, which spans the whole domain of numbers, is a *logic* of that domain. Of the deductive mental processes involved, we discover just as little in the one case as in the other. Accordingly, the "laws" of the calculus are also nothing less than they are the norms of all "valid thinking," or, more precisely, of inference conforming to pure implications. Those laws are not rules to which everyone does and must conform, so far as they infer correctly. Rather, they are only rules which one *can* follow on any occasion, with full confidence of a correct result. In light of this, they may well be suited for testing the *result* of any pure deduction; but whether the deduction itself was a correct one, that is something which they cannot decide. The mere multiplicity of algorithmic methods – which, materially and formally, forge quite different ways toward the solution of the same problems (cf., here only the methods of *Boole*, *Jevons* and *Schröder*, among others) – teaches us that in the algorithmic formulae we in nowise find mirrored the canon for the deductive activities involved in knowing. Let the practice of deduction draw ever so much use [9]
from such methods – and it is certainly very dubious that it does – the logical *theory* of deduction remains where it was. It is, indeed, not even touched upon. This also explains a peculiar fact which can be confirmed, in a similar manner, in the logical and in the mathematical domains: One can be an outstanding technician in logic, while being a very mediocre philosopher of logic; and, again, one can be an outstanding mathematician, while being a very mediocre philosopher of mathematics. (Boole provides an outstanding example of both.) It almost seems as if the mental activities required in the one area

and in the other were heterogeneous in the extreme, since it is so extraordinarily seldom that they are found united in one person.

Section B of the "Introduction," titled "Concerning Signs and Names," is again of some interest. Here the *semiotic* questions which are so important for a deeper grounding of the calculus are dealt with. This section begins with some well-chosen excerpts from Trendelenburg's fine work, "Über Leibnizens Entwurf einer allgemeinen Charakteristik" (in his *Historische Beiträge zur Philosophie*, vol. III), as well as from the pertinent treatises by Leibniz himself: treatises which contain many striking observations concerning the function and value of signs for knowledge, but which are still far removed from being carried through to a usable theory of symbolic methods. Our author does not provide remedies for these deficiencies. What, beyond those excerpts, he offers as his own exposition is not very gratifying. Let us cite some of his main statements: The exact sciences "have more or less expressed a tendency to shift, as much as is possible, the difficulties of the study of things . . . over to the study of signs." (p. 40) "Depending upon whether or not they have attained to this ideal" they have "passed from their original, inductive stage into the deductive stage, or else find themselves yet in the former. This brings to light the fact that, in the deductive sciences, functions of the utmost importance rest upon the sign; and, in fact, that the sign will ultimately form the sole object of consideration in these sciences." (p. 42)

Virtually nothing is brought forward in elucidation and justification of these grossly exaggerated assertions. Wherein consists this [10]
"study of signs"? What is meant by "shifting the difficulties of the study of things over to the study of signs"? And let us measure the assertions of the author against the case of Euclidean Geometry. The signs which it utilizes are figures, whether drawn upon the blackboard or in phantasy. Now, I ask, is the geometrical theorem ever generated by the mere study of *these* signs? Quite to the contrary! The signs are mere supports for the conceptualization of the genuinely intended concepts; and, likewise, the sensible operations upon the figure are nothing more than sensible crutches for the conceptualization of the genuinely intended operations upon concepts, or upon generally represented conceptual objects. And the judging activity step-by-step bears, not upon the signs, but rather upon the objects themselves, which the signs symbolize. Now since signs, consequently, do not

here form the sole object of consideration, are we to say that even Euclidean Geometry remains still in its inductive stage? Only in disciplines which work by algorithm may one speak, in a certain sense, of a shifting of the difficulties in the study of things over to the study of signs. But even in these disciplines it cannot be justifiably claimed that signs form the sole object of consideration. In the process of calculation there certainly is required no reflection upon the concepts underlying the signs. Rather, one attends exclusively to the rules for the signs, the "laws" of the calculus. But the calculation is not the total deduction. Rather, it is only one element in it. The calculation is preceded by the "fixing of equivalence," i.e., the replacement of the problem by a corresponding algorithmic problem, equivalent to the former by the rules of the symbolic method. On the other hand, the calculation is followed by the "interpretation": the conversion of the end-formula into the corresponding judgment required by rule.

On page 24 there begin those observations which are devoted to the "principles of designation," and the manner in which they are introduced shows that in them the author intends to give the specific logical basis for his method of signs. But principles of naming – and these alone are what is subsequently dealt with – are in nowise yet the principles of those peculiar designations which come together in *algorithms*. If Schröder confuses these two different things, then that [11]
is due to a fundamental misunderstanding of the relationship between language and algorithm, which we will discuss later.

After the author explains the distinctions between word and name, between one-word and many-word names, and between syncategorematic and categorematic expressions (pp. 42–44), he then develops the logical standards which must be met by *adequate signs*. These signs must be, first, suitably short (44); second, they must be rationally built up from simpler and simplest signs (47); and, third and finally, on each repetition they must evoke or be accompanied by the same representation (47). The second requirement is obscure. One assumes that the author has in mind the rational composition which is present in algorithmic sign complexes because they faithfully mirror the conceptual content taken into account in the application of the calculus, and do so unambiguously and in conformity with that content's structure.

The third requirement is that of *univocity*. With the intent of explaining it, Schröder takes pains to bring the distinction between univocal and equivocal names into exact conceptualization. In vain, since there is totally lacking in him the true concept of the signification of a name. This requirement of univocity is also expressed in the form: "The name must be of a ... constant signification." (48) However, in the relevant explanations (on pp. 47–48) the author identifies the signification of the name with the representation of the object named by the name, from which there certainly follows the striking consequence that all common nouns are equivocal. It is not as if the author had overlooked the distinction between equivocal and general names. And who *could* do that! But to see a distinction, and to grasp its essence, are quite certainly two different things. Beyond this he is using the term "signification" itself in an equivocal manner, and this to a quite intolerable degree. In the place cited above, the common sense of the term is had in mind, the conflicting and erroneous explanations notwithstanding. But at other times there really is intended the object named by the name – as when, for example, in literal contradiction to the above requirement, species terms are designated as names to which "several significations belong with equal and full right"! (p. 69) And this not sufficing, the class corresponding to the common noun also is said to be its signification. [12] (p. 69n) Thus it is understandable that the author is not capable of a precise rendering of the essence of equivocation. It is due, he says, "simply to the variable use to which we subject the name." But then wherein does the variation consist? In a certain sense the use of a common noun also is variable. There is further connected with this unclarity about the concept of signification the fact that Schröder classifies names such as "round square" as "meaningless" ("*unsinnige*"), in comparison with names with one or more senses. Obviously he here is confusing two very different things: namely (1) whether a signification (a "sense") belongs to a name, and (2) whether or not there exists an object corresponding to a name. "Meaningless" names in the strict sense are names without a signification – pseudo names, such as "Abracadabra." But "round square" is a univocal common noun to which, however, nothing can in truth correspond.

In the course of the further expositions in Section B, there next comes up for treatment the concept of the proper name (56–62), along with the concepts of the concrete and the abstract, the corresponding names (simultaneously discussed), and the concepts of part and property. Here as elsewhere, I can only emphasize a few points from among many which I cannot concede. Undoubtedly under the influence of *Jevons*, in the *concept of the proper name* there are confused two concepts which it is worthwhile to keep separate: the more general concept of the singular term, and the more specific concept of the proper name in the common sense of the phrase – i.e., the "non-connotative" singular term. This distinction certainly refers back to Mill's classification of names into connotative and non-connotative, which is treated quite badly by Schröder. So far as it does not coincide with the distinction between common noun and proper name, it is said to be "insignificant" and to "dissolve into a frivolous haze." (62) Whatever has led the author, whom we certainly may suppose to have read *Mill*'s *Logic*, to make such strong assertions remains a mystery to me. Indeed, *Mill* expressly gives examples (Book I, chap. 2, subsection 5) of singular names (in Schröder's language: "proper names") which are connotative. Moreover, Schröder imputes to Mill's term "not connotative" the sense of "not designating," as if the non-connotative names were without a sense, and as if, moreover, Mill had never written his excellent, precise elucidations! [13]

Let us now turn to the Schröderian theory of *common nouns* and *collective nouns*. (63–75) The definition of the former is given in a purely nominalistic fashion, in a manner similar to that used by formalist arithmeticians with the basic arithmetical concepts, in that they fix the conventions for the use of certain signs, while concepts themselves remain neatly out of play. Thus there is here formulated, to begin with, a rule which is to "govern the manner in which the common noun is used as a predicate." Namely, the common noun is to designate several things, and, indeed, is to do so in such a way that it belongs to each of those things with an equal right. The totality of the things comprehended by the common noun in this manner is the *class*. (p. 67) Moreover we have a second rule, which governs the use of a common noun as subject: "With the common noun, when something is asserted *of* it, one is always allowed to understand at will this or that individual of the class." (p. 68) To this is attached the definition: "An assertion with a common noun as subject, representing a class, is called a *general* judgment." Thus one takes any group of objects, affixes one and the same sign to each of them, establishes the above two conventions for that sign, and one then has a common noun. At least according to Schröder! For that *this* absolutely is not the linguistic function and sense of the common noun requires no proof. It is in general incorrect to say that each common noun "represents a class." If I speak of "a man," then I mean "something which has the property of rationality, etc." Here there is present no representation of a class, and the name also is no class name. "An *A*" is a class name only when it stands for "one out of the totality of *A*'s." It may be convenient, in the algorithmic treatment of the signifiers of judgments, to replace general names, in each case, with the class names equivalent to them. But does that mean that the two are absolutely identical?

The author continues with a discussion of *the relation of general names to collective names.* Characteristic of the obscurity of his view is the assertion that it is "of very little consequence for logic, to set up a distinction between names which are collective and those which are not." (72) We could, after all, designate any thing in which any parts at all are distinguishable by means of a collective name. [14] Therefore the distinction can "hardly be justified, other than as a grammatical or psychological one." (!) – However, the question is not whether a given thing can be divided up and resolved into a multiplicity, but rather whether in naming we represent and mean that multiplicity. It would be palpably absurd to call an apple a collection. That it is divisible into parts has nothing to do with the matter. And if we actually divide it up, then the totality of the parts is indeed a collective. But are "apple" and "the totality of the parts of the apple" synonymous names? It also shows nothing against the distinction here in question, that many names (e.g., "army") can function on some occasions as a collective term, and at other times as a general term; just as little as anything is shown against the distinction between unity and multiplicity by the fact that any multiplicity can also be counted as a unity, e.g., as when it is counted together with other multiplicities.[2]

In the midst of these discussions, Schröder finds occasion to define the *relationship of mathematics to logic.* All will be eager to learn

[2] It seems to me that precise distinctions between the various types of names here discussed would be the following: A *collective* name is one which is determined in its signification to designate a multiplicity (*Vielheit*), while a *distributive* name is one which is determined in its signification to designate any one thing from a multiplicity. The latter is a *class name*, provided that the multiplicity in question is the entirety of the objects of a concept (or is a conceptual extension). The role of the *common noun*, by contrast, is to designate the object of a concept *as* an object of that concept.

how a renowned mathematician, who also has devoted so much study to logic, delimits the two reciprocal domains. The explanation is brief: The collective mode of unification is said to be "of particular significance for the disciplines that deal with number and magnitude .... The study of the laws of this unification [is], accordingly, to be assigned to arithmetic and the theory of magnitudes, but not to logic (in the narrower sense) .... At this point of divergence a large group of disciplines branches off from logic, in order to contrast themselves ... with logic as independent." (75) This view, which seems to take the mathematical sciences to be partial domains within logic – though independently developed ones – is not exactly one that is clearly grounded, and I do not need to subject it to a precise analysis. A few words will suffice to reduce the author to absurdity – from his own standpoint, *viz.*, that of the logical calculus. For this calculus deals with class relationships. But classes themselves are [15]
nothing other than collectivities, and only insofar as they are collectivities does the calculus consider them. Indeed, this is so far true that the calculus in its *general* conception is to be designated as a general calculus of collectivities, from which the class calculus proceeds only by means of the special interpretation of the collectivities as extensions of concepts. In the sense of the passages quoted, it would therefore follow that the logical calculus – being, in Schröder's view, deductive logic brought to exact formulation – is to be assigned to arithmetic and the theory of magnitudes.

Now the division of names into positive and negative, and into absolute and relative, is still to follow. In regard to the former division we once again encounter a startling fact: "Being positive or negative is ... merely a superficial and, so to speak, grammatical property of the name ... which has no logical import at all, though admittedly it has a psychological import in some circumstances." (!)

Section *C* of the Introduction is of special interest for us, because it contains *discussions of principle* between the new and the old logic, and a vindication of the former.

The theory of general *concepts* is pursuant upon the theory of common *names* developed in Section *B*. This order of things is characteristic. Now for the first time there is raised the question: "Which things are we to call by the same common name?" (80) I can well waive a critique of Schröder's attempt "to analyze genetically the way in which we succeed in forming a concept... of things represented by a common name" (83); especially since this attempt is all too much lacking in clarity and consistency. But for an understanding of the following discussions, one must consider the attendant definitions.

The *content* of the concept "... is formed by the properties common to the things designated with the common name. Specifically, the concept's '*factual*' content consists of properties of things actually reflected upon in the formation of the concept; but its *ideal* content consists of the total set of all common properties whatever..., which, however, it is perhaps never possible to capture in thought." (83)

Now, presupposing this, let us engage with the author in the following line of reflection. Since concepts can be considered from two points of view, the content and the extension, there also are "two possible ways in which a concept could, conceivably, be defined." [16]
It could be defined by *division*, the specification of its extension; or by *definition*, the specification of its content. Each way has its peculiar advantages and disadvantages. "If the extension of a concept encompasses indefinitely many individuals ..., then that extension can never be exhaustively specified by recourse to those individuals. Rather, in that case one is required, in order to specify the extension, also to invoke such subclasses as are themselves in turn open, and, if they are to be specified, can be so only by means of specification of the content of a concept appertaining to them." Now what conclusion will the reader expect to be drawn? Naturally this: that a purely extensional specification is, in general, no way to define a concept; that all definitions of concepts are definitions of contents; and that wherever we hear of the definition of a concept by means of its extension, what is meant, and alone can be meant, is an indirect definition of the conceptual content to be defined, by means of another conceptual content corresponding equivalently to the first in virtue of having the same extension. And this, by itself, suffices to make it known that the ideal of an "extensional logic," i.e., a logic which in principle considers *only* extensions of concepts, is futile (*ein nichtiges*), because it is objectless. But the author merely draws the conclusion that "the specification of extension therefore appears as the more imperfect of the two ways of defining a concept." "Appears," he says. For the imperfections of content specification turn out to be still greater, and even insurmountable. Therefore logic must be based upon specification of extensions alone.

What, now, are those imperfections? We can restrict ourselves to examination of the following: "The properties which are common to the individuals falling under the concept – and which, in their union, constitute the ideal content – in general never can be completely enumerated." (86) But how, then, is a definition to be the determination of the "ideal" content? Where is the logic which ever has posited this utter impossibility as its goal, and has understood "definition" in this manner? And does it not clearly follow from what Schröder himself here has said that the task of such a determination (think only of the infinity of negative and relative properties of a 'thing') is apriori impossible to carry out? However, he calmly proceeds: "Therefore definition gives up all claim to direct statement of the total content [17] of the concept. It rests satisfied to state directly and explicitly only a part of that content; and it gives us the remainder only mediately and implicitly, in that it emphasizes, from among the agreeing properties, a select group which . . . jointly imply the remainder." These are the "characteristic properties." (87) Herein lies an obvious absurdity. No one possesses an ideal conceptual content, unless it is in the symbolic form: "Totality of the properties common to the objects of the concept" (that is to say, its "factual" content). What sense is there now to speak of an "emphasizing" of a group of properties out of this totality, when after all no one possesses the totality in the authentic sense; or moreover to speak of such a group from the totality implying "the remaining" properties? If the "definition" rests satisfied with this, then it is contented in a pure absurdity.

Now we are sufficiently prepared to meet and evaluate Schröder's first and *main argument* for the treatment of logic "*purely as a logic of extensions*," and against the treatment of it as a "logic of conceptual contents." This argument holds that "in all cases . . . the determination of concepts by means of definition . . . is supported . . . upon the laws of necessary inference." Now, Schröder argues, at the outset there are given to us only the concepts which are attached to the ready-made common nouns of language. For the formation of all further concepts "the only means at our disposal is, then, definition, in the use of which, of course, logic must already be presupposed." In the light of this it then appears to him that "the endeavor to present logic as a logic of the conceptual content – and for millennia the effort has predominately been to realize a logic of this sort – commits the fallacy of *Hysteron Proteron*." (89)

Here all is confusion. That the determination of an "ideal" concept would presuppose deductive logic – indeed, even the whole of logic – is beyond doubt. But where, I again ask, is the logic which understands by "definition" such a determination of concept, setting it as its goal? Further, it is incorrect to say that definition is the only means at our disposal for the attainment of new concepts in addition to those given with language – regardless of how we take the equivocal terms "concept" and "definition." This point perhaps requires no further discussion. The blow struck against content logic is, after all of this, a mere beating of the air. And this is already true in virtue of the [18] simple fact that the content logic here combated is only a mirage in the author's mind. How he could identify this fiction of a logic of his "ideal" conceptual contents with the totality of logic up to this point in time remains completely incomprehensible to me. And, what is more, he does not even hesitate (*loc. cit.*) to hurl the reproach of a vicious circle against the old logic, as if it – whether in the theory of the logical concept, or the theory of the logical judgment, or elsewhere – assumed that we actually possess his "ideal" concepts, if only in a concealed manner.

But the following fact constitutes the best refutation of the Schröderian argument: In exactly the same sense in which the so-called extensional logic is a valid theory, a "logic of ideal contents" can also be constructed, and with a technique which is even identical with that of the extensional logic. Indeed, still further, logical problems solved by the two also are identically the same. I have demonstrated this in another place.[3]

But the author advances yet *other arguments* for his view. The next argument consists of an allusion to the difficulties which stand in the way of a general theory of concept formation, and to the *de facto* conflict of philosophical viewpoints about concept formation. It appears to him to be "no scientific procedure . . . to set out to raise upon such uncertain and much-contested foundations the edifice of a science which, like logic, claims to formulate only absolutely certain, because conceptually necessary, . . . truths." I respond to this by a question: What would Schröder the mathematician say should anyone apply this same argument to his discipline (and to what discipline

[3] Cf., the essay, "Der Folgerungskalkül und die Inhaltslogik," *Vierteljahrsschrift für wissenschaftliche Philosophie*, 15, no. 2, 1891, pp. 168–189, with a "Postscript" on pp. 351–356 of the same volume. (Translation of these follow immediately below. [D.W.])

are these words, "uncertain and much-contested foundations," more appropriate)? Berkeley made just such an application, and was much censured. And there also is an error here in another respect. That logic, among other disciplines, deals in a scientific manner with conceptually necessary truths is certain. But is a science *of* conceptually necessary truths to be, in virtue of that simple fact, a conceptually necessary – i.e., a deductive – science? Only of that small area of logic which takes in the pure relations of implication between judgments [19] is it correct to say this, as indubitably follows from the possibility of an algorithmic treatment of that area.

The paradoxic force of a third argument – "that there are many concepts which, as to their content, do not exist at all, and which nevertheless enjoy a . . . sharply delimited extension" – influences the author through confusion of his own terminology with that of other logicians, who, like *Lotze*, do not characterize negative concepts (not-human, for example) as concepts. From which nothing is to be concluded.

The final argument consists of a reference to results. And on this count it appears to the author that "the millennia-long efforts to bring logic through the consideration of conceptual contents into a soundly progressive growth have been frustrated." But will one indeed be inclined, after the proofs given above, to concede to the author – who is so outstanding as a mathematician – the decisive judgment in this matter?

So these are the motives which led Schröder to believe that logic must be constructed as an extensional logic. He also, with such a construction, intends to take the "easiest route," in which "the powers we have can reach the farthest." (89) Moreover, he takes the apparent one-sidedness which follows upon the restriction to conceptual extensions not really to be so serious. For, in virtue of the reciprocal relationship between extension and intension, "each is untroubled by the other, but still is to a great extent fixed simultaneously with it." (100) However it is, by contrast, to be noted that, for the independent formulation of an extensional logic in opposition to an intensional logic, the mere possibility of a study of class relationships without recourse to the relationships of the contents *corresponding to those classes* would in no wise suffice by itself. Rather, what alone would suffice is the utter independence of that study from all contents and content relationships whatsoever. But an extensional logic which is independent in *this* manner is impossible in principle. In fact, every judgment about extensions is truly a judgment of content. If we judge that class *A* is contained in class *B*, then it is affirmed that the object of the concept *class A* is object of the concept *contained in class B*. Now we could also replace these concepts by their extensions, and form the relevant class judgments. But the same point is to be made in turn about these class judgments. And so on *in infinitum*. [20]

Thus, so little is it true that the logic of extension is to be treated independently of the logic of intension, and so little is the former "fixed untroubled by the latter," that when doing extensional logic we yet stand within intensional logic, or subordinate to it. But with this we also see Schröder's supposition, that intensional logic cannot make any reference at all to class relations without incurring the objection of inconsistency (*loc. cit.*), to be completely erroneous. In opposition we can state that, insofar as the logic of extensions has any sense and meaning at all – and thus not as a new *logic*, but rather as a special logical technique – it belongs totally within intensional logic.

In the "Introduction" there also is a large-scale exposition (in essentials, concurring) of a curious theory of the judgment from *Peirce*. In this theory, incredibly but actually, the judgment is explained as a special case of *habit of thought*. (105–118)

More interesting to us are the discussions, closer related to the genuine goals of this work, of the question as to "why [in this work] *logic is presented as an algebra*, and why as so presented, in contrast to other modes of treatment, it seems justified in claiming pre-eminently for itself the title of an 'exact' logic." (118) The essential line of thought here is: The old logic availed itself of *natural* language for the expression of thought – that expression being as good as indispensable for the occurrence of thinking in the form of judgments. The new logic, on the other hand, invents an *artificial* language for this purpose; and for good reasons. For natural language, with its well-known deficiencies, its equivocations and synonymies, inhibits and misleads logical thinking – on specific points as well as in the development of logical theory. The emancipation from natural languages through the creation of an artificial language kept free from such defects implies, accordingly, a step forward for logic and – transforms logic into a calculus.

Whoever might have to form a picture of the new logic merely in terms of this characterization would indeed be not a little astonished upon discovering its actual content. It is not true that this "exact" logic is nothing other than a logic on the basis of a new language. As we have already emphasized, it is not *a logic* at [21] all, but rather is a calculus serving specific logical purposes. And hence the talk of a "presentation of logic *as* an algebra" is wholly inappropriate. Disregarding this, the basic error which one commits with the above characterization of the new logic rests upon confusion about the essential difference between a *language* and an *algorithm.* Schröder constantly expresses himself as if the algebraic procedure were already given by the introduction of a symbolic language in place of natural language consisting of words. This is an old error. We find it already in *Descartes* and *Hobbes.* We even find it in *Leibniz,* who upon the assumption of it set out to invent a symbolic (*charakteristische*) language, thinking therein to have attained simultaneously to a *calculus ratiocinator.* We find it taken to the extreme by *Condillac,* who straightaway identifies language and calculus. But the two concepts are fundamentally different. A language is not a symbolic method for the systematic derivation of conclusions, and a calculus is not a symbolic method for the systematic expression of mental phenomena. The peculiar function of *language* consists in the symbolic expression of mental phenomena, which expression we in part need for communication of those phenomena, and in part need as a sensuous support for our own inner movements of thought. The art corresponding to linguistic designation is grammar. Thus, grammar does not teach us how we should judge, and it also does not give rules concerning how we can derive correct judgments indirectly, through symbolic devices. Rather it only teaches us how we are to express judgments correctly in language. On the other hand, the peculiar function of the *calculus* consists in its being a method for the symbolic derivation of conclusions within a certain sphere of knowledge. Thus it is an art which, through an appropriate symbolization of thoughts, substitutes a calculation process – i.e., a rule-governed process of transposing and replacing signs with signs – for actual inferring; and then by means of the correlation of symbols and thoughts set up at the outset, it derives the desired judgments from the resultant end-formulae. And even this *correlation,* which constitutes a part of the symbolic process of derivation, does not have the character of a linguistic designation. For the function of the sign here absolutely is not to accompany the *thought as its expression.*

For the rest, it is not to be denied that a language expressing judgments might simultaneously be a deductive calculus. But with [22] the one the other is not automatically given. And it is above all certain that the logical calculus is *only* a calculus, and absolutely is not a language. When the author (on page 64) specifies the ideal to be striven for as "the discovery of . . . the most rational system of designation for the naming of all objects and for the expression of all processes of thought" – in fact even "the vindication of that system as a necessary one" – he has once again fundamentally deceived himself about his own goals.

We have, in this manner, totally rejected the principles of "exact logic" presented in the "Introduction." Is it not then superfluous still to continue on with the discussion? Is not the logical calculus itself already rejected? Not at all! It is a characteristic phenomenon which we observe in all calculational disciplines, that not merely the application, but even the invention, of algorithmic methods shows itself to be independent, to the greatest degree, of insight into their essence and into the grounds of their value for knowledge. General arithmetic, the most highly developed of calculative disciplines, flourishes onward and upward even though the most gifted of its representatives are, and always have been, far removed from a deeper grasp of its fundamental principles. In fact, the unclarity is so far-reaching that there is not even minimal agreement upon the true conceptual foundations of this science. Arithmetic has to do with numbers. But what is a number? One takes it to be the cardinal number, another the ordinal number, while yet another holds it to be measurable magnitude; and so on. Thus we have the curious spectacle of a science which does not know what it really is dealing with. This is not the place to track down the deep-lying reasons for this state of affairs. But we do derive from this the rule that we must never estimate the value of an algorithm in terms of what its advocate may present to us as his philosophy. This also holds true for the logical calculus. It is here as in similar cases that *D'Alembert*'s saying is above all applicable: "Carry on, and confidence will come to you!" And indeed, those who have bravely worked on through the calculus, and have familiarized themselves to some degree with its technique, will find it hard still to deny that it performs a function similar to that of the arithmetical calculus in the domain of number – even though one not so grand in scope. One can hardly deny that the calculus puts at our disposal methods which, in the restricted domain accessible [23] to it, spare us extremely complicated and laborious deductions, and achieve their results more quickly, more certainly, and above all more easily – presupposing, of course, a full mastery of the calculus! – because mere calculation is involved. Moreover, the logical justification of this very simple algorithm is not so obscure that it is difficult to arrive at and to grasp the cogency inherent in the procedure.

The critique of the systematic presentation of the calculus, to which I now turn, must attend to two aspects, as we consider both *the logical* and *the technical* sides of our subject matter. With regard to the first I must, unfortunately, continue to decline to share the views of the author; but with regard to the latter, he is the master. Through the invention of new methods, and the improvement of methods previously given, he documents that excellence as a mathematician which science has long recognized in him. And inasmuch as in the calculus we have a substantial enrichment of the old logic, the laudable algorithmic results of the author turn out to the advantage of that discipline itself, and in a manner which I certainly would not like to see underestimated.

In order to begin immediately with what is familiar, I note that the range of questions dealt with in the present volume coincides, at least in the main, with the one which found exposition in the small older work by the author, *Der Operationskreis des Logikkalküls* (1877). But in the *form* of the treatment, a fleeting comparison alone brings a certain divergence to light. In spite of essential changes introduced with the *Boole*an technique, in the smaller work the author after all remained true to that technique in one main point: He too replaced judgments by *equations* between classes; and, accordingly, the steps in the calculus proceeded by equivalence transformations of equations into other equations. By contrast, in the new presentation the author follows *Peirce*: Judgments are now replaced by the *subsumption relation* between classes, and the calculus consists of *equivalence transformations* of subsumptions into other subsumptions. Superficially considered, equations are indeed to be found here also. But they are only intended as abbreviated designations for certain pairs of subsumptions. [24]

Only subsection 1 of the first Lecture (pp. 126–141) deals with the concept of subsumption in its relationship to the concepts of identity and subordination, as well as with the reasons for favoring it in the construction of the calculus. The distinction between subordination and identity is illustrated by the examples: "Gold is yellow" and "Table salt is sodium chloride." But this is not done in a wholly appropriate manner, in that the author, instead of working right from the outset with the extensions of concepts – upon which the point exclusively depends, as is immediately clear – stresses the impossibility or the possibility of conversion as the characteristic distinction. The further discussions of the identity relation (128) are very questionable, and their deficiencies are connected with the mistaken views of the author on the essence and signification of names. If we take an accurate view of the matter, then we see the distinction between the two relations to consist simply in the fact that, in the first case, that of subordination, the subject class is a *part* of the predicate class, whereas in the case of identity the two classes coincide.

After his clarifications on these relationships, the author holds us to be adequately prepared "to grasp the true signification of the copula": – "... The categorical judgment always expresses the fact that the subject ... is either subordinate to, or else is identical with, ... the predicate." Were they taken literally, crude errors would lie in these words. But, to the contrary, the accompanying explanations indicate that "the interpretation mentioned can in every case at least produce unmodified the 'logical content' of the judgments concerned – not infrequently, to be sure, at the expense of a modification in the 'psychological content'." (132) "But if in this way we can only bring it about that all judgments appear in a common form, and so become accessible to a general [*viz.*, the algorithmic] treatment, then by the attainment of such immense benefits the *modus procedendi* mentioned is abundantly justified."

To this latter point we gladly agree. However, we must protest the logically objectionable manner of speaking which is documented in the above passage, as well as upon innumerable later occasions. (On p. 152 we see "That all judging really acts through subsumption." And on p. 156, that judging "turns out to be subsumption," etc.) This manner of speaking leads the author into errors upon several occasions. But, as Schröder himself once pertinently observes, the recasting of categorical judgments into the form of subsumptions of classes is done "merely to favor logical technique." Accordingly, the [25]
Schröderian distinction between the "logical" and the "psychological

content" of a judgment – or, more precisely, of an assertion – also is not to be granted. The true logical content of an assertion is its judgment content, and thus is that which it affirms. But in no wise does the relevant class-judgment express *that* judgment content. Here we need another and more appropriate distinction: namely, that between the *logical* and the *algorithmic content* of an assertion. If it is true that not all judgments, but rather only all judgments about class relations, admit of a direct algorithmic treatment; and if it is further true that all other judgments can be transformed into equivalent judgments about classes: – then in the application, in this way possible, of the class calculus to any arbitrary judgments there arises the task of first transforming all assertions into their equivalent assertions about classes. And of these latter assertions we then can say, that they for the first time bring the algorithmic content of the former assertions to expression. From this there emerges, as the basis for the calculational theory of inference, the proposition: Every assertion which is not itself already an assertion about classes can be equivalently transformed into such an assertion, and can, accordingly, be expressed in the signs of the algorithm and subjected to its rules. Or, more briefly: Every assertion possesses an algorithmic content which is identical to or equivalent with its logical content.

Thus the general applicability of the calculus is based upon the "equivalence" of judgments. This equivalence should rightly have been the very first thing to be analyzed. Two judgments are equivalent if they can be reciprocally deduced from each other. In fact the two may, so far as the interests of knowledge are concerned, be taken as one. Where the one is given, the other may also be regarded as given. What can be demonstrated of the one holds also of the other. In practice, therefore, one will select from the two that judgment with which one can more conveniently operate. And so the procedure, which here is required by the calculus in particular, is completely justified from the logical point of view. But one must, on either side, seek for nothing more behind it all than precisely this simple fact of equivalence.

Now the construction of the calculus here will not rest upon the relationships of subordination and identity, expressed by the signs "(" and "=", but rather upon the relationship of subsumption, which is a [26] composite of the first two, and is expressed by the sign "= (=."[4] The profuse justification given for this preference for subsumption is more than dubious (133–134). Subsumption is said to be the "simpler," the "more primitive" relationship; it is supposed to express only one idea, but the others, two. "a = (= b" expresses namely, the fact that all *a* are *b*. By contrast, for example, "a = b" says that all *a* are *b* and that all *b* are *a*; and so on. But this is all clearly wrong. Schröder constantly confuses the linguistic expression of the class judgment with the linguistic expression of the content judgment equivalent to the class judgment. If one only holds rigorously to the genuine content of the symbols, then for each of the author's assertions precisely the opposite is true. For example, "a = b" states that the class *a* is the same as the class *b*; "a = (= b," on the other hand, states that the class *a* is subordinate to or identical with class *b*. The subsumption relation therefore is, as the sign precisely expresses it, composed of identity *and* subordination; and these latter are, accordingly, the "more primitive." *If* there really is an argument in all of this, then we must rather prefer these relations for the construction of the calculus. But there just is no argument here. In the construction of the calculus it is not asked which relation is the simpler, but rather which is the more useful in an algorithm. This cannot be decided apriori, and hence the whole question is out of place in the introduction of the presentation. Schröder's further reflections on the usefulness of introducing the subsumption sign into *Arithmetic* are highly credible. Through such a sign the common errors which occur in calculating with ambiguous expressions would be completely avoidable. But of course this proves nothing about the suitability of this sign, and the corresponding relation, for a systematic development of the logical calculus.

Now in order to follow out the systematic development, I turn immediately over to subsection 3 on p. 155, which has to do with laying the foundation of the "*identity calculus* upon domains of a manifold (*Mannigfaltigkeit*)" – that calculus including the *class* calculus, originally the goal, as a specific singular case. "Here we think

[4] Schröder uses as the designation of subordination a parabolic curve, open to the right; and for the designation of subsumption he uses the same figure, but overlaid by an identity sign. For lack of the corresponding type pieces, the printer has chosen to use the signs "(" and "= (=", which are to some degree similar. The latter is to be read, "subordinate to or identical with," or, more briefly, "sub."

of a manifold of elements as given .... The nature of this manifold, as well as the kind of elements in it, is totally arbitrary. The considerations here are to be general, and will ... claim validity for any conceivable manifold, of whatever elements. Any combination of [27]
elements of the manifold we call a *domain* of that manifold." (157) The subsumption relation, a = (= b, is taken as the first and fundamental relation which can obtain between two domains. "a = (= b" expresses the fact that the domain *a* is contained in the domain *b*, their identity being regarded as the limiting case of being-contained-in. "The concepts and propositions of the identity calculus will admit of the most diverse of applications, differentiated only by the interpretation assigned to the general symbols .... We can, in particular, understand the letters to refer to: (A) Domains of a manifold of elements; (B) Classes ... of individuals – and also, especially, concepts considered as to their extension; (C) Concepts considered as to their content ...; (D) Judgments ...; (E) Inferences ...; (F) Functional equations, algorithms, ... 'groups' ('*Gruppen*'). And, in short, given a suitable interpretation of the signs, pretty well anything of which one can think." (160)

I must confess that this first introduction and foundation of the new calculus does not appear satisfactory to me. That calculus is supposed to refer to arbitrary "domains" of any manifold whatever – or, to set aside an unnecessary qualification, to arbitrary manifolds. If we replace this sophisticated term with the synonymous common term, "set" ("*Mengen*"), then what we have is a *general calculus of sets*. The first question would be: What are the possible relations presented by two arbitrary sets, *a* and *b*, as such? The response would immediately yield five possible cases: *a* is a partial set of *b*, or *b* is a partial set of *a*, or *a* and *b* are identical, or they intersect, or they mutually exclude each other. Thus we immediately have here the perfect analogues of *Euler*'s five relations between spheres. If, now, one considers an arbitrary number of sets exclusively with regard to such relations, then one can foresee that from the knowledge of some such relations the knowledge of others would also be derivable, and perhaps even derivable by calculation. So mere reflection upon what lies in the concept of a general calculus of sets immediately provides us with a broad basis for the construction of that calculus. But Schröder does not penetrate to what is clearly the heart of the matter, *viz.*, that in the identity calculus we have to do only with a calculus of mere set relations as such. And from this there arise right from the outset, as well as at later points, numerous misunderstandings. [28] Recourse to the concept of set is disdained even in the setting up of the axiomatic principles, and hence the calculus becomes a curious mongrelization of set calculus with a blind play of signs.

Already in the above survey of applications, obscurities of principle show up. Under (A) stands the general interpretation. Under (B), as something new, is the class interpretation – but distinguished from the interpretation in terms of conceptual extensions, which is supposed to be only a special case of it. But what is it that differentiates a class of objects which is not a logical class, not an extension of a concept, from a set in general? I find not a shade of difference. Application (C), to conceptual contents, is justifiable: One can view those contents as totalities of their essential properties (*Merkmale*), and relate them to each other in the manner of sets. But how do things stand with applications (D) and (E)? Are judgments and inferences sets? And things are scarcely otherwise in the case of (F). Considered *merely* as pluralities of formulae, formal systems do fall under the calculus of sets. But as soon as we treat them as something more (as, for example, the author does in Appendix IV), *this* application comes to an end. It is not that I wish to deny the applicability of the calculus, considered purely as a symbolic technique, to all of these cases. But what I contest is Schröder's fundamental error in holding that the calculus *as* "identity" calculus takes in all of those applications. I would deny that the founding concepts of the calculus of sets (and what other than this would the "identity" calculus be?) also found, or could found, all of the applications which the calculus as such admits. In general, I doubt whether there is *any* conceptual foundation of the calculus which really does that.

We have here that same phenomenon which also is responsible for so much trouble and perplexity in the domain of mathematics; namely, we have a calculus (in mathematics, the algebraic) that admits of applications which are wholly heterogeneous and conceptually unrelated to each other. This rests simply upon the fact that in the differing conceptual domains analogous relationships are found, which then permit analogous forms of inference – the simple and, therewith, the more complicated forms as well. But a calculus is nothing other than a system of *formal* inference, i.e., inference grounded purely upon the mode of expression. Wherever the external forms of the primitive relations and relational concatenations are analogous, these forms lead us, given the same symbolization, to identically the same [29] calculus, however much the subject areas may differ from each other.

Now the analogies which make possible the transference of our calculus to different domains are so evident that they even lie within linguistic consciousness. One quite generally speaks of "inclusion," of "exclusion," and of "coincidence," in various extended senses of these terms. And in every case where one so speaks, and can exhibit five relationships corresponding to the five spherical figures [of Euler], there the calculus also is applicable. Thus it is, for example, with judgments in their conditional relationships. Thus it is with concepts, and so on. If one considers the matter, it is not difficult to find applications of the calculus which were never thought of by its inventors. In fact, this is so far true that we can put over against the extensional calculus an algorithm of contents, wholly based upon the foundations of the old logic. This algorithm is even formally identical to the extensional calculus, and constitutes the most striking of refutation of that cherished dogma of extensional logicians, to the effect that *only* recourse to classes makes possible an algorithmic system of deduction. Whoever goes to the bottom of these matters will nowhere have need of such fantastic devices as do *Boole* and Schröder when they come to interpret the logical calculus as a "calculus of statements." I have shown this in another place. (See the essay cited in the footnote above.)

The fundamental objections which I have here brought forward against Schröder's viewpoint are, however, not objections against his further presentation. If it is true that the common ground for the applicability of the calculus does not lie in an involvement with one determinate class of contents (sets, continua, concepts, judgments, etc.), but rather in an exact analogy among a restricted number of relationships which make possible "operations" and symbolic expressions that are everywhere the same, then in the exposition of the calculus one obviously can proceed in two ways: (i) *Either* one can set it up as a mere technique, or as, so to speak, a mere game with signs, without regard to any applications. And in this case one knows that any domain of things whose relationships, given adequate symbolization, yield the necessary and sufficient primitive formulae of the system must, then, also fall under the calculus as a whole. Thus, the mere game structure (*Spielsystem*) becomes a *deductive calculus*. (ii) *Or else* one proceeds at the outset from a determinate [30] domain of application, grounding the primitive formulae – and with them, in turn, the whole calculus – upon the nature of the concepts in *that* domain. The transference to other domains ensues according to analogy, going by way of the primitive formulae. Of course the exposition must in this case be set up on the principle that the least possible is to be derived from the concepts, and the most possible through deductions from propositions already known. All the more so since one cannot know whether or not relationships which are immediately obvious in one domain may be given to insight only indirectly in another. In any case the principle mentioned makes for the greatest ease of transference of the calculus to foreign domains. Now we can quite well capitalize on Schröder's exposition of the logical calculus by taking it to be an exposition of the second sort, although the basic viewpoint upon which his exposition is based is false. The principle of establishing the most possible through proof and the least possible from concepts also guides him, but from quite other motives – *viz.*, in order to satisfy the requirements of a presumed rigor. In this connection Schröder overlooks the fact that, from the standpoint of their value for knowledge, all algorithmic proofs of directly evident propositions (*Evidenzen*) are to be rejected, since they seek to establish what has ideal certainty by means of what is probable – and thus, so to speak, to establish an infinite by means of a finite. However, for the development of a technique this principle is very advantageous.

Now let us follow out that development of the identity calculus which is presented here, always attended by its interpretation in terms of classes. Two axioms are provided right at the outset: "*Principle I*" is "a = (= a" (Lecture II, subsection 4, p. 168): and "*Principle II*" is "If a = (= b and simultaneously b = (= c, then also a = (= c." (p. 170) The first is, precisely, the algorithmic correlate of the law of identity, while the second is the correlate of the law of the "Barbara" syllogism. The author, to be sure, expresses himself as if those laws were themselves the ordinary-language expressions of formulae I and II. This misleads him, for example, into an identification of Principle II with the "*dictum de omni et nullo*." But the dictum does not come into play at all with the inference in II (where the signs signify sets or classes!). Passing over the remaining discussions – in part dubious, and almost always too lengthy – which concern these two principles, only when we come to p. 184 do we find a step forward in the development. It is with the definition of the "identity equation": "If a = (= b and simultaneously b = (= a, then it may be said that a = b." This obviously is a circular definition, since not merely externally, in the subsumption sign, but rather in the concept of the subsumption relation itself, the concept of identity is explicitly contained. Instead [31] of a definition, what we have here is, rather, a thesis that is Evident.

I also am unable to agree with the manner in which the crucial *signs* "0" and "1" are introduced into calculus. *0* is, namely, defined as a domain which stands in the subsumption relation to every domain *a*, while *1* is defined as a domain *to* which every domain has that relation (188). Expressly rejecting any question about existence, we are from here on "to count the *symbols* '0' and '1' . . . as among the domains of the manifold." Precisely stated, this means that from here on we are to calculate with the signs "0" and "1" as if they were signs of actual domains – something which we absolutely cannot know apriori. But, I ask, with what justification do we do this? The author here proves himself to be a member of that school of formalist arithmeticians who suppose that the oft-mentioned freedom of definition gives them the right "creatively to introduce" an arbitrary number *x* of concepts into a deductive discipline, provided only that one takes care that the definitions are free of contradiction (*widerspruchlose*). But besides contradiction there is yet another thing which is to be logically avoided: opposition to subject matter (*Widerstreit*). Now if such definitions, e.g., that of our "0" and "1," contain incompatibilities (*Unverträglichkeiten*) – and that is something which is absolutely impossible to read off of formal definitions – does not the calculus cease to be a calculus of correct inference for classes? Is a geometry still a *geometry* if, after defining "square circle," it makes use of this definition in its derivations? Thus the question of *quid juris* is very much in order.

Similar objections concern the crucial definitions of *sum* and *product* (Lecture III, subsection 5). They read: "If it is true of given domains *a*, *b*, and *c*, that c = (= a and c = (= b [or else that a = (= c and b = (= c], then it is – more concisely – to be said that c = (= ab [or that a + b = (= c]. (196) These definitions wholly conform to the formalistic character of the previous one. The product (or else the sum) of two classes, *a* and *b*, is here defined "as merely one element of a certain manner of speaking," or as a certain abbreviatory style of writing. But with what justification is this style of writing specified – a style which after all has the form of a subsumption and, as such, presupposes a conceptual sense? Such "styles of writing" are [32] in fact not without consequences. They imply that one then immediately applies to them all of the propositions which have been proven for *conceptually founded* styles of writing of analogous form (here, proven for subsumption). And in fact, in the expositions which follow, Schröder makes immediate application of Principle I to such subsumptions containing those "mere signs," "ab" and "a + b." (Cf. the proof for theorem VI, p. 199.) But then how is it that one is permitted to take the principle a = (= a, which had a sense only for "domains," and use it without further ado on those "senseless signs"?

Schröder comes to his own aid here by explaining that he claims the principle of identity to hold, not just for manifolds alone, but rather "for all possible objects of thought" whatsoever, and thus for mere "names as well, without regard to whether those names have a sense or not." Quite correct. But a = (= a, as we have previously stressed, certainly is not the principle of identity, but rather is only a certain principle that applies to sets. "a = (= a" does not express the fact that *a* is *a*, but rather the fact that the set *a* is subordinate to or identical with the set *a*. What intelligible sense does it then make to say that a sign "ab" is subordinate to or identical – as sets are identical – with the sign "ab"? Furthermore, these signs are also made subject to the content-laden principle II, the correlate of the form Barbara (see pp. 202, 204, 207, etc.). And this is done without it ever being shown that such apparent subsumptions, involving signs such as "a • b" and "a + b," are rightly subsumed under the principle in question. Here is an essential gap in the chain of demonstrations.

In truth, through these definitions – similarly as in the case of the definition of identity – one defines away the propositions which presuppose authentic definitions of sum and product, to be given on the basis of the concept of set. Hence, we cannot accept the "convention" which Schröder proposes to us; namely, "that we in any case assign the symbols '0,' '1,' 'ab,' and 'a + b' to the domains of our manifold, if need be downgrading them to the status of 'inauthentic domains' ("senseless names, if you wish") of this manifold. There is lacking any demonstration of the permissibility of such a convention in case the calculus should become – as is intended – a calculus for actual manifolds, and not for "inauthentic" ones, i.e., for senseless names.

The question concerning the "interpretation" of all of these symbols – i.e., that concerning the (possible) existence of objects which [33] fall under the conceptual determinations expressed in the definitions of the symbols – only comes up for further discussion in subsection 7, on pp. 211ff. The existence proof leads to the familiar interpretations. It is successful throughout, but not in the case of *0*. There is no domain *0* which, according to the definition, must be contained in every other domain. There are, in fact, disjunct domains which have no domain in common. The *0* can, thus be added on only as an "imaginary" domain.

The addition of the *0* as an imaginary domain – an addition so crucial for the calculus – appears to me to be in nowise so clear a matter as it here and elsewhere is treated. With what justification may the *0* nonetheless be added, since in the system of the calculus as presented above, the "creative" definition of *0* yields no legitimation of existence, and since further, as is conceded, a domain does not correspond to the concept of *0* – and, in virtue of the incompatibility of its determinations, even *cannot* correspond? For this addition I find not even the shadow of a proof. The *0* of the identity calculus presents the same problem as the –1 of the arithmetical calculus. In the one and in the other case, the relevant proof is adduced only from considerations of algorithmic technique. Such proof would show that any subsumption conclusion derived utilizing the *0*, where that conclusion contains only signs founded on reality, is of necessity a valid conclusion, given the sense of these "real" signs and of the laws exclusively concerning them. With "creative" definitions nothing is accomplished, even though they be formed in such a way that they retain the consistency of the calculus. The question is not whether the calculus is consistent, but whether it still is a calculus *of sets*.

I also cannot agree with the "positive postulate" concerning *1* (i.e., the universe class, which contains all "domains"). I hold the requirement that the elements of the manifold be compatible with each other to be superfluous. The calculus applies to arbitrary manifolds. Where we have obtained the elements – whether in the domain of reality, fiction, or absurdity – does not matter to it. It is also applicable to sets of impossible and incompatible objects, so far as these may be of interest. One must take into consideration the fact that that unification of incompatibles which would be necessary in a set is merely collective unification; and one must not, as the author does, confuse [34] this unification with the unification found in the judging act of belief. Even explicit contradictions are unifiable in the sense of a collection. They are non-unifiable only in the sense of truth and persuasion.

Lecture IV (subsections 8 and 9) is concerned with transferring what until now was established in general terms to the calculus of classes, and with the ordinary-language "*interpretations*" of that calculus preliminary to its application. These interpretations are, in essence, a matter of showing how those linguistic forms which correspond to the class signs "0," "1," "ab," and "a + b," fit into the transformation of content judgments into the equivalent class judgments. They have nothing to do with the "linguistic expression" of these signs themselves. Upon close examination, the application of the class calculus to content judgments requires two steps: (a) The translation of assertions into equivalent assertions about classes, and (b) of these latter in turn into formulae of the calculus (or conversely, with "interpretation" of the formulae). This twofold translation can be avoided only by *immediately* setting the class symbols directly into a correspondence with the concepts in the original assertions – the symbols corresponding to those concepts indirectly when we come to them by way of the equivalent extensional assertions. And the direct route is also the sense and aim of "interpretation" here. Wherever, for example, we find a substantive "*a*" attributively combined with an adjective "*b*," there we can immediately write down the equivalent product "ab," without first forming the concept of the "class which takes in the totality of the objects had in common by the class *a* and the class *b*" – although this *is* precisely the sense of "ab." One must keep this in mind in order to avoid taking offense at the facts themselves (*an der Sache selbst*) in virtue of offense at the author's inexact mode of speaking.

Among the interpretations given here, only those of the formulae 0 = (= a and a = (= 1, which hold true for any *a*, occasion some more substantive remarks. The following are, respectively, designated by the author as the "proper results of their translation": Nothing is *a* (e.g., gold), and *a* (gold) is everything. That these assertions are false, he then very well observes. But instead of seeking the error in the translation, he takes pains to find it in a deficiency of "word language." We can leave untouched the philosophical observations about language (241–242) connected with this, since it can precisely be shown that the misunderstanding lies in the interpretation – and at a point, moreover, upon which all extensional logicians since *Boole* [35] have run aground. The obvious difficulty with this interpretation is closely tied to the one which – as in the identity calculus, so also here – infects the concept of the null. If the distinction between real and imaginary classes (and to the latter the class *0* certainly is to belong) were a sound one, then there would be no way to escape the most tangible of absurdities, to which the formulae 0 = (= a and a = (= 1 – and, for special cases of *a*, the formula a = (= 0 – lead, as simple translation of them shows. The extensional logicians have brought this difficulty upon themselves. They have overlooked the fact that in a formal theory of deduction only a purely hypothetical inferring comes into question, and that, consequently, all of the categorical judgments which show up in such a theory are only apparently categorical. Accordingly, in the extensional logic which constructs an algorithm for the formal theory of deduction, the corresponding subsumption judgments, of the form "the class *a* is subsumed under class *b*," also are merely *hypothetical*. The existence of the subject class is not asserted; but the meaning is, rather, that if there is an extension belonging to the concept *a*, then one belongs to the concept *b* as well, and one under which the former extension is subsumed. Hence it is idle to speak of real and imaginary classes in reference to our domain. In precisely the same sense in which the classes gold, metal, etc., enter into the pure theory of inference, the classes of round squares, non-existent entities, and so on, also do. We can, for example, quite well state that the class of round squares belongs in the class of non-existent entities. With this we immediately attain to the correct interpretation of the symbols "0" and "1." "1" is the sign for the extension of the concept of existence, and "0" is the sign for the extension of the concept of non-existence. The formula 0 = (= a expresses, consequently, the algorithmic content of the judgment: If something is non-existent, then it also possesses any arbitrary property *a*. And, in fact, if *n* is non-existent, then that same *n* is a non-existent as a red, sweet, or non-red (etc.) thing; and to the non-existent as such, therefore, all properties actually belong. On the other hand, the formula a = (= 1 expresses the algorithmic content of the judgment: If something is an *a*, then it also possesses the property of existence. And this, again, is self-evident, since the categorical judgment, "Something is an *a*," includes the existence of the subject.

Thus, careful consideration of the purely hypothetical character of [36] all deductions in our domain suffices to set aside all difficulties with the concepts *0* and *1*, as well as with the interpretation of the above formulae.

The author introduces an important revision into the foundations of the calculus upon the occasion of this discussion. He rejects *the Boolean concept of the 1*, as the class of all possible objects of thought, on the basis of the following line of argument: If we are permitted to bring all possible objects of thought into the manifold *1*, then by *a* we are allowed to understand the class of the classes of the manifold which are = *1*. This *a* would, as a particular real (*reelles*) object, include *1* itself; and, further, it would include *0*, since *0* is to be included in every other class. Since, thus, *1* and *0* are those objects which have to be regarded as = *1*, then not only must 1 = 1, but rather it must also be true that 0 = 1. But this latter equation is an absurdity. One could with its help, as the author shows, deduce the absurd formula a = b, for any *a* and *b*. (p. 245)

This argument, which at first glance is quite impressive, is sophistical. Let us form the class *K*, whose *elements* are themselves classes, and indeed are those which together = *1*. Then *K* obviously would contain the class *1* as an element, since 1 = 1. But does this *K* also then contain the class *0* as an *element*? Absolutely not! Certainly it contains *0*, as does every other class. But it contains it as a subordinate class, not as an element. It is by definition true of all *elements* of the class *K* that they = *1*; but this is not true of its subordinate *classes*. It thus is not true that 0 = 1. The elements of the null class certainly are also elements of *K*; and of them it holds true, once again, that they = 1. But this also agrees with their signification as objects from the class of the non-existent.

Thus, the objection to the Boolean concept of the unity class (*Einheitsbegriff*) miscarries, and therewith also the Schröderian restriction upon that concept (which, as is easily seen, also would concern my concept of the unity class, presented above). Only this remains correct in these discussions, which are carried farther in this passage (*loc. cit.*): that in the cases where we simultaneously have, besides certain classes, also classes *of* those classes, the calculus may not be blindly applied. In the sense of the calculus of sets as such, any set ceases to have the status of a set as soon as it is considered as [37] an element of another set; and this latter in turn has the status of a set only in relation to its primary and authentic elements, but not in relation to whatever elements *of* those elements there may be. If one does not keep this in mind, then actual errors in inference can arise, as the author has strikingly shown – but also, unfortunately, shown by committing the errors himself. But it is going too far to claim that the calculus is not actually applicable to such "mixed" manifolds. (248)

In the Vth Lecture (subsections 10 and 11), now, the algorithmic consequences of the definitions and propositions previously given are drawn; and in a fine exposition a long series of propositions are proven, which leads up to a fundamental theorem of the calculus – *viz.*, the "*distributive law*," $a(b + c) = (ab + ac)$. However, only the proof of half of this law is successful: $(ab + ac) = (= a(b + c)$. Lecture VI (subsection 12) deals with the "unprovability" of the other half: $a(b + c) = (= (ab + ac)$. The author succeeds in rigorously demonstrating that this latter subsumption is not deducible by means of the concepts and principles introduced up to this point. He produces, namely, a calculus of "groups" for which the basic formulae all hold true, whereas that subsumption certainly does not. Therefore it cannot be a mere algorithmic consequence of the basic formulae presupposed.

All of this I admit, but find nothing so impressive about it. For what inducement would we have, then, for the supposition that those basic formulae mentioned are the formulae which are necessary and sufficient for the calculus? But if they are not, then obviously not everything can be done with them. In saying this I in nowise wish to place a low estimate upon what is gained from this and similar proofs. Only I find the gain in a wholly different direction than does the author. It is of value to know what is necessarily conditioned upon restricted parts of the foundations of the calculus taken by themselves. In this way one gains the means of transferring at least a relevant partial algorithm over to foreign domains, where the whole algorithm would be no longer transferable. An example of such a transference is the "group" calculus mentioned above, as treated in Appendix IV.

For the further development of the calculus, the author adopts the route of adding, as a new axiom ("Principle III," p. 293), the "unprovable" subsumption in the distributive law (slightly restricted). [38]
In this I find a minor deficiency of the system. The proposition in

question is (for classes) no immediate axiom, and therefore a proof of it must be given – after expanding the "foundations."

From the so-called unprovability of the full distributive law the author draws a peculiar conclusion: namely, that instead of there being one calculus for manifolds in general, there are really two. And these are such that in the one both parts of the distributive law obtain unconditionally, while in the other only the one part of it does. For the former he retains the name "identity" calculus, while for the latter he introduces the new name, "logical" calculus. – I also cannot find this move plausible. If the "identity" calculus really is a general calculus of the manifold, then the full distributive law holds true for it, being a necessity grounded in the concept of the set. Thus there is no bifurcation in the identity calculus. The "group calculus," which the author brings forth as an example of a "logical" calculus, does not fall at all within the domain of the original identity calculus, since it is not a calculus of pure manifold relations. That a calculus deals with manifolds, and that it deals with these manifolds as such and only as such, are two different things, which are confused by Schröder.

In Lecture VII (subsection 13) the third and last basic operation is introduced – that of negation. And this, indeed, is done with the following definition: "The *negation* of a domain *a* is what we call domain a_1, which stands to it in such a relation that both $aa_1 = (= 0$, and $1 = (= a + a_1$." To these formulae the equations $aa_1 = 0$ and $1 = a + a_1$ are algorithmically equivalent; but, in spite of the fact that they are derivable from the above subsumptions, *logical* priority indeed belongs to them – in conformity with our reflections above, concerning the relationship between identity and subsumption. For this definition, therefore, the equations would have been preferable.

With reference to subsection 14, I bring up a general reflection concerning the *duality* discovered by the author to permeate the whole of the logical calculus. He formulates this quite peculiar fact as follows: "In each proposition, and in every general formula of the identity calculus of domains, we may simultaneously interchange [39]
throughout . . . the signs of subordination and superordination, the "0" and the "1," as well as the signs of product and sum; and, in so doing, one of necessity always obtains again . . . a valid proposition . . . ." (316) Now the "deeper grounds" for this phenomenon is seen by the author in the fact that "the subordination of conceptual extensions runs parallel with a superordination of the corresponding conceptual

contents, and, in particular, that the multiplication of extensions can at the same time also be regarded as an addition of the contents." (318) In saying this Schröder totally forgets that the identity calculus as such has to do neither with conceptual contents nor with conceptual extensions, but rather has to do exclusively with manifolds as such. But even if it had to do with the former, his proof would remain a merely apparent proof, as can be easily shown.

Subsection 15, which is very comprehensive, gives us a general *theory of negation*, as a *function of judgment.* For a *logic*, such psychological theories may be of fundamental significance. For a logical *calculus* they are totally insignificant. We therefore abstain from a large-scale presentation and critique of the author's views here. What is essential for the logical calculus could be dealt with in a few words. For its general application to arbitrary judgments, it needs only the uncontestable fact that any negative judgment admits of transformation into an equivalent affirmative judgment with negative subject matter. But that, in the negative judgment, the negation really does belong to the subject matter, which the author would like to prove – this is unnecessary. Moreover, in its most proper domain, the class calculus knows nothing at all of negation in the literal sense of the term.

Subsection 16 (Lecture VIII) deals with the *interpretation* and *application* of negation for *extensions of concepts*. Since *1* enters into the definition of negation, the restriction which the author held to be indispensable for the concept of the unity class wins a significance for that definition. If the *1* must not, as Boole has it, signify the totality concept for all that can be thought of, then, in virtue of the formula $a + a_1 = 1$, a_1 (i.e., "the negation" of *a*) cannot be the extension of the concept non-a. But in conformity with this, the formulae $aa_1 = 0$ and $a + a_1 = 1$ also cannot, in the Schröderian calculus, be regarded as the class equivalents of the principles of *contradiction* and *excluded middle*; whereas the author sees in these formulae even the "most concise expressions" for those principles. However, he [40] does suppose, with the restriction of the concept of the *1*, also to have discovered an essential limit for the latter principle itself. It would, on his view, lead to what is false under certain circumstances, if one were rigorously to maintain that each and every thing must be *a* or non-a. "It may be true that every sheep is either black or not black, while for all that it is not valid to say that every breed of sheep is either black or not black, in that one breed may contain, besides white sheep, black ones as well." One only need inquire into the precise sense of the misconstrued expression, "a breed of sheep is black," and the fallacy here is obvious. So, until further notice, the old principle will retain its validity in its old extension. But, as we have shown above, the stated restriction upon the class *1* is not required for the logical calculus (although a certain modification of the calculus is required. Cf., p. 271 above). Consequently, the definition of negation also is, in this respect, unrestricted. Hence *Boole* also remains in the right on this point; and those definitional formulae for negation are, then, really the full class equivalents of the logical principles.

For the rest, discussions of principle become more and more infrequent from this point of the book on, and the development of the calculus takes a more rapid and more interesting course. The technical presentation – or, as we also could say, the mathematical – is exemplary; and in relation to the algorithmic theories presented to date, it is rich in fruitful simplifications and in elegant developments. For those who are familiar with the Boolean calculus, especially in *Venn*'s excellent presentation, I call attention to Schröder's defense of his own concept of *identity addition* over against the Boolean concept of *exclusive addition* (368). The objections by *Venn* are convincingly refuted. It certainly is true that, for *Boole*, exclusive addition was an essential constituent of the system. It alone made possible, it seems to me, the ingenious transference of the arithmetical algorithm to the domain of classes, through which at one stroke the logical calculus stood forth, almost as if grounded upon a miracle.

The development of *Boole*'s methods, modified by Schröder, reaches up to Lecture XII (subsection 23). In this subsection itself, the author gives us a presentation of his interesting investigations [41]
into the *inverse operations* of the calculus: the "identity" subtraction and division. Both of these are operations which are to be carried out only under certain conditions, and which, in general, yield results that are infinitely equivocal. Negation appears as an ordinary and univocal special case of both. By using negation alone the calculus fully manages the solution of its essential problems. Thus the superfluity of the inverse operation is shown.

But might there not also be other special cases of these operations which could be of use for the technique of calculation and which could occasionally be the basis of considerable simplifications? One must indeed admit this possibility, even if one gladly concedes that a calculation with the *total* domain of inversion operations – with their complicated formulae and their ballast of "valence conditions" – would be of such inconvenience and difficulty as to curtail any advantage. Certainly in the case of exclusive addition, at least, we find many simple, special forms (cf., for example, p. 486); and so, then, it is actually *Boole* and *Venn*, alone making use of this addition, who employ the inverse profitably. The author himself provides striking testimony of the algorithmic usefulness of even ambiguous and senseless expressions from the inverse domain. Following the common rules for function development – which are proven only for direct operations, and thus also univocal ones, he develops (495) this infinitely ambiguous difference:

$$a \div b = (1 \div 1)ab + (1 \div 0)ab_1 + (0 \div 1)a_1b + (0 \div 0)a_1b_1.$$

And then he continues: "Now the coefficient $0 \div 1$ is senseless .... In order that it may be dropped out, the corresponding coefficient a_1b will have to be $= 0$, which provides the valence condition." Logically considered, this is a fallacious inference, even if we disregard the fact that the admissibility of such a development is arbitrarily postulated. If $0 \div 1$ is senseless, then we also must not, as the author here does, write that $((0 \div 1).0) = 0$. For this equation (as well as the one above) is then itself a senseless one in relation to the conceptual foundation of the calculus. And yet the technical usefulness of the procedure leaps out at us. We find precise analogues in arithmetic. One calculates with senseless signs such as $\surd - 1$, and for example formulates the equation $(\surd - 1)^4 = 1$, although such equations are without sense in relation to arithmetical concepts. It thus does not seem to me [42] to have been at all demonstrated that our calculus cannot also, like the arithmetical, gain considerable utility through the introduction of such signs, proceeding out of inverse operations – however it may stand with their conceptual foundation. Certainly on both sides – the arithmetical as well as in the logical calculus – there is required in each case proof of the admissibility of methods which operate with signs that, in *Cauchy*'s much decried phrase, "are without sense when taken literally." I hope to supply this proof in another place, and rigorously to prove my own view, which runs directly contrary to the usual views, on these questions.

In Lecture XIII (subsection 25), the reader finds a plentitude of problems for the application of the *Boole-Schröder* methods, whereby he can be convinced of what the calculus accomplishes. Only the scantiest familiarity with the rules suffices to make evident the superiority of calculational methods in contrast to those involving actual inferring; and this, indeed, with regard to the saving of time as well as with regard to gains in certitude. However, this fact in nowise by itself suffices to confer upon the calculus significant practical value. If the calculus really took in the whole of deduction, then certainly every philosopher, mathematician, and natural scientist would have to make great haste to master it. Instead of this, the domain which it actually governs to advantage is a very paltry one. Indeed, pure deductions do show up all throughout life and the sciences. But as a rule they are of such a simple type that to solve them by means of the calculus would be the most laughable of detours. To draw pure consequences from complicated systems of premises is, up to now, still a very rare thing, and natural procedures have always sufficed in order to evaluate such consequences. So what scientific researcher will agree to spend the considerable time and effort requisite to learn the calculus and maintain a subsequent mastery of it, in order to spare at best a few small hours in the solution of such highly infrequent problems. Perhaps the future will bring forth areas of fruitful application. As matters stand at present, the representatives of the new discipline struggle with the singular difficulty of now discovering the problems which are to be solved by their fine methods; and on no point do they show greater acumen than in the construction of [43]
illustrative examples – which, therefore, appear quite artificial.

However, these objections concern only the logical calculus of conceptual extensions, not the "identity" calculus of mere sets. This latter, which coincides formally with the former, can indeed be profitably applied in many particular fields of mathematics – for example, in the theory of functions, where manifolds of values of arguments frequently come into consideration. Likewise in the calculation of probabilities, where sets of chances make the application possible. Beginnings have already been made in these matters, but here too we do not have sufficient results definitively to decide the question about *practical* value. But I would in no case wish to cast doubt upon the extraordinary *theoretical* interest that belongs to the algorithmic treatment of the theory of pure deductions, as well as of pure set theory.

The last Lecture (XIV) contains discussions with regard to other methods. First, there is an excellent statement and critique of *Jevons'* method (560–568), which, in comparison with those of *Boole* (which they are supposed to replace), must be designated as a great leap backward, and much less still would bear comparison with the newer methods of Schröder and C. S. Peirce. Then there is a presentation of the ingenious graphic methods of Venn, which suffice for problems involving no more than five class symbols (up to page 573). A special subsection (23) is devoted to the methods of *McColl* and C. S. *Peirce*, the latter of which is fundamentally different from the *Boole-Schröder* methods, and impressive in its exceptional originality, simplicity, and elegance. It appears eminently usable, especially after the simplification which Schröder introduces into parts of it; and it would accordingly have been desirable to have had a greater quantity of exercises attached to it and worked out. The methods of *McColl* are passed off by the author as only a special manner of applying the Boolean methods, but one which falls far behind them in convenience of application.

The concluding part of the volume consists of six appendices, which are more of mathematical than philosophical interest.

– Halle A. S.

THE DEDUCTIVE CALCULUS AND THE LOGIC OF CONTENTS[1]

Among the representatives of the algorithmic theory of deduction, it is commonly maintained that only the 'interpretation' of all judgments as class judgments can lead to a genuine calculus. It is, they suppose, essential to the logical calculus that it be a calculus of classes. Now the efforts of *Jevons* and *Wundt* certainly stand in some opposition to such views. In the construction of their algorithms, the viewpoint of content is taken to be preferable to that of extension. But a more precise analysis of these algorithms shows that, in spite of the opposite intention, they are in essentials extensional algorithms, and that a certain wavering between the two viewpoints hinders the clarity and consistency of their development. In the case of *Jevons*, this is clearly manifested as soon as one follows out the various turns in his concept of identity – and he does reduce all judgments to identities. For one then sees that each assertion that contains any universal term at all leads to an identity only by way of an equivalent assertion about classes. When *Jevons*[2] sets up the equation, "circle = curve of least perimeter," then the identity sign surely cannot refer to the two conceptual contents; and it likewise cannot refer to the identity of a single object which falls under the two contents, since the equation is, after all, supposed to represent a universal judgment. Thus it only remains for the identity to hold true of the two classes involved. And with that we stand, again, within the logic of extension.

And similarly for *Wundt*'s presentation. He sets out with reflections [45] drawn from the logic of contents; but slips unnoticed into

[1] First published in *Vierteljahrsschrift für wissenschaftliche Philosophie*, Vol. 15, #2, 1891, pp. 168–189. Republished in *Husserliana XXII*, pp. 44–66. Husserl's working notes for this piece are in Appendix III below. [D.W.]

[2] W. St. Jevons, *The Principles of Science*, London, 1883, p. 38.

the logic of extensions. For example, when he in words defines a "summation of concepts," but in fact defines a summation of their classes.[3] It does not correspond with my purposes to go more deeply into criticism of these *apparent* efforts toward a logic of contents. Let it suffice that I concede to convinced extensionalist logicians that *up to now* consistent developments of algebraic methods of deduction have been attained only in their way.[4] Certainly they will be little pacified by this concession. They have not merely asserted, but rather also have argued. They have thought they could show, on apriori grounds, that a calculus of *contents* is, as such, impossible, and that the detour by way of classes is therefore unavoidable. This is found for example with *Venn*, in his *Symbolic Logic*. And recently it has turned up again in E. *Schröder*, whom one can probably call the most extreme representative of extensionalist logic – to whom that logic is, when all is said, deductive logic itself, instead of merely a method of deduction.

Now the best counterargument is a *fait accompli*. I shall show that a calculus of pure deduction can be constructed upon the basis of considerations rigorously adhering to content logic; and this not in one way, but even in a number of ways. And it is a most striking fact to start from, that such a calculus, in any given case, is to be set up in such a way that it follows identically the same formal course as any of the extensional calculi previously existing (whether that of [46]

---

[3] W. Wundt, *Logik. Eine Untersuchung der Principien der Erkenntniss und der Methoden wissenschaftlicher Forschung*, Vol. I, *Erkenntnisslehre*, Stuttgart, 1880, p. 234. Cf. the examples.

[4] I say "consistent" developments, but not "logically incontestable." In fact, the logic of the logical calculus is still very much in a bad way. Among its representatives there is not the slightest clarity either about the boundaries of this discipline, or about its relationship to deductive logic, on the one hand, and to arithmetic on the other. The *logical* considerations upon which the techniques are erected are mostly of a kind that do not stand up to the most cursory examination. If then the calculus steps forward with the grand claim of being a wholly reformed *logic*: one which is – for the first time – "exact," it is quite understandable that precisely the more scientific logicians tend, for the most part, to reject it. However, the logical foundation of arithmetic is just as frail. But does that suffice for its rejection? I believe that the logical 'algebra' also, in spite of its so very limited practical utility, must not be underestimated; and that, for the sake of its positive results, it must be of great interest to the logician. And the possibility of such results is here, as in arithmetic, a serious problem that is worth taking in hand. In a review (in the *Göttingische gelehrte Anzeigen*) of a work by E. Schröder on the calculus – a work to be mentioned several times herein – I, by the way, have occasion to discuss a series of more general

Boole, *Schröder*, *Peirce*, or others). To this circumstance I here owe a brevity of presentation which otherwise would be unobtainable. And to that end I shall rely upon the pre-eminent extensional calculus of our time; the one which *Schröder* has developed in his *Vorlesungen über die Algebra der Logik*.

Schröder considers the class calculus as a special application of a more general calculus; namely, of the "identity calculus" – or, as I would prefer to call it, the pure calculus of sets (*Mengenkalkul*). If *A* and *B* are any two arbitrary sets, then it is obvious that they can, as such, stand in only one of five possible relationships. Namely: (1) *A* includes *B*; (2) *B* includes *A*; (3) they are identically the same set (they "coincide"); (4) they stand in the relation of "intersection"; (5) they mutually exclude each other. These are the perfect analogues of the five relationships of *Euler*'s spheres. If, now, one considers an arbitrary number of sets exclusively with regard to such pure set relations, it is foreseeable that from the knowledge of certain relationships the knowledge of others can be derived. Precisely this derivation is what is provided for, in the symbolic manner, by calculation within the "identity calculus" referred to above. Classes, then, are special cases of sets, so that class relations are special cases of set relations; and only insofar as they are that does the class calculus – which thus proves itself actually to be a special application of the general calculus of sets – consider them.

But besides this application of the calculus of sets to classes, *Schröder* further specifies a whole series of other applications, among which is also an application to conceptual contents.[5] Conceptual contents (taken in the common sense) really can, in fact, be regarded as totalities of properties, and can, accordingly, be set into relation in the manner of classes. The whole totality of properties of one conceptual content can actually be included in that of a second, or else excluded from it, and so on. In this way application of the identity calculus becomes possible, and there is originated an algorithm of content relationships. But it surely is a rather useless one. It would neither directly nor indirectly constitute that which one calls a *logical* [47]
calculus – i.e., a calculus which encompasses and governs the total domain of pure deduction. In its domain would fall only analytical

---

questions which concern the logic of the calculus.

[5] *Vorlesungen über die Algebra der Logik (Exakte Logik)*, Vol. I, Leipzig, 1890, p. 160, γ.

judgments in *Kant*'s sense: thus, categorical judgments in which the predicate concept is actually contained in the subject concept. It could account for assertions such as "Gold is a metal," as being of the form: "The properties of the concept metal are included in those of the concept gold." But it could not do so for synthetic assertions, such as "All equal-sided triangles are equal-angled," since the properties of being equiangular are, precisely, not actually contained in those of the equilateral triangle.

But it could be that, by an application of the identity calculus to *conceptual contents* (of which he omitted clarification), *Schröder* had something else in mind. He indeed distinguishes the "factual content" of a concept (which is nothing other than content in the sense common in logic) from the "ideal" content. This latter takes in the *totality* of valid properties which obtain of a conceptual object as such – as, for example, the ideal content of the concept of gold takes in the totality of properties which can be validly asserted of gold as such. One sees immediately that ideal contents, in their turn, admit of being placed into relation in the manner of sets; and so it seems that nothing stands in the way of an algorithmic treatment of them by means of the identity calculus. Now certainly it is true that Schröder has sought to show that a "logic of ideal contents" would involve a circularity. And he supposes this even to concern *all* of content logic up to now, from *Aristotle* to *Sigwart*, since he identifies *that* logic with *his* fiction of a logic of ideal contents. This latter point does not concern us here. As to the former point, the heart of *Schröder*'s argument consists in holding such a content logic to be impossible because of the necessity, which he imputes to it, of operating with *actually given* ideal contents; whereas, of the infinity of properties of a conceptul object, surely only an infinitesimally small number can be genuinely given. This line of argument is wholly in error. And how little it is capable of justifying an exclusive preference for extensional logic, and an apriori rejection of content logic, is shown by the fact, presently to be proven, that exactly the same things which can be accomplished by an extensional logic can also be accomplished by the logic of ideal contents: – that all problems solved by the first, the [48]
other logic is also in a position to solve, and this utilizing even the exact same rules, formulae, and calculations. Such will be the first content calculus that I here set up.

I can be brief, especially since this calculus is of no great value.

I first lay down the thesis that the ideal content of any concept is given or is not given to us in exactly the same sense as is the related conceptual extension. In fact it also is (almost always) true of the latter that only a very small part of it, if any part at all, is genuinely given to us. But for such an actual givenness, in the sense of a genuine grasp (*Kenntnis*) of all of the conceptual objects, there is never any need whatsoever. It is sufficient that we have the univocally determinate *symbolic* representation of the "totality of objects which possess the properties of the concept." Wherever we operate with classes, whether within extensional logic or without, we work exclusively upon a basis of such symbolic representations. But exactly the same holds true of the ideal contents of a concept. That they never are *actually given* to us in nowise grounds, as *Schröder* supposes, an objection against the possibility of using them in logic: – e.g., in the construction of a "logic of ideal contents." Similarly as with classes, we possess them precisely in the form of univocally determinate symbolic representations. Their content in each case is: "The totality of properties which belong to the conceptual objects as such (i.e., to any and every thing which falls under the concept)." Thus, if we insist upon operating with ideal contents, as for the moment we shall, then these symbolic representations are completely sufficient for us; and, as is seen, we actually possess them in precisely the same sense as we do the corresponding class representations.

We need no more than this in order to set up the calculus of ideal contents. If we consider such contents as totalities of their properties, then they offer occasion for set-like comparison, in terms of the relationships of inclusion (or of being included in), of exclusion, of intersection, and of coincidence. We can, thus, apply the identity calculus directly, by making its signs stand for any ideal contents whatever (instead of for any classes whatever, as is done by the class calculus). But the calculus resulting from this has, unlike the previously mentioned content calculus, the character and powers of [49] a genuine logical calculus: it takes in the entire province of pure deduction. In fact, every judgment can be equivalently rewritten as a judgment about the relationships between ideal contents (thus, to a certain extent, as an analytic judgment), and thereby each judgment becomes subject to the calculus. For the judgment, "All men are mortal," we substitute the other, "The ideal concept of man includes that of mortality" – which, although of course not identical with the former, is nonetheless equivalent with it. In general, in any categorical judgment the ideal concept of the predicate is subsumed in that of the subject (or else is excluded from it); and the requisite transformation is to be carried out accordingly. In the process, negative judgments are replaced by the affirmative judgments with negative matter that are equivalent to them, and particular judgments by certain universal judgments equivalent to them, exactly as in the extensional calculus.

Already from this, one perceives that the new calculus is in a position to solve identically the same problems – and, indeed, by means of identically the same formulae – as the old, and is distinguished from the old only by the fact that it never at any place speaks of conceptual extensions, but always of ideal contents instead. The same calculation obtains a different sense in the solving of the same problem. This one fact alone suffices to show that the class calculus is in nowise *the* logical calculus in which salvation alone resides.

A more detailed exposition of the ideas just indicated does not seem to me to be required. And all the less, since I think I can show that this content calculus, as well as the old extensional calculus, demands – with regard to the logical goals toward which it strives – wholly superfluous rephrasings, and thus takes the long way around. I will seek to show that exactly the same algebraic technique which is set up as a calculus of sets, and which serves (in its special forms as a calculus of classes or one of ideal contents) in the indirect solution of the problems of a pure theory of inference, also can serve these purposes directly and without any detours through extensions or ideal contents. I will, in a word, show that this technique can be regarded *directly* as a calculus of content logic (i.e., of ordinary deductions in the logic of content), and that it can be set up as such right from the start.

The formula *SaP* of formal logic, commonly expressed as "All *S* [50]
are *P*," has undergone a whole series of different "interpretations," for the purposes of deductive technique – whether the archaic technique of [traditional] 'formal' logicians, or the more ingenious one of the calculus logicians. Taking in several forms which show up in the scientific logic of contents of our times, I will divide these interpretations into three classes; namely, in the following manner:

I. Interpretations Purely in Terms of Contents and indeed:

A. Without Quantification of Contents

(a) The hypothetical interpretation: Provided something is an *S* (has the property *S*), it is a *P* (has also the property *P*).

(b) The negative interpretation: It is not true that there is something which is both an *S* and not a *P*; or, what is the same: There is no *S* which is not *P*

These two elementary and fundamental forms of the universal judgment have until now received no algorithmic utilization.

B. With Quantification of Contents

"The ideal content of *S* includes the ideal content of *P*." That is, the totality of properties belonging to an *S* as such includes the totality of properties belonging to a *P* as such. This is the fundamental form for the calculus of "ideal" contents. Now one could set beside this fully quantified form two incompletely quantified forms also, by quantifying either *S* alone or *P* alone, in the manner here considered. I omit to carry this out, since it can lay claim to no significant interest.

II. Interpretations Purely in Terms of Extensions

These originate through quantification of both the subject concept and the predicate concept by means of the corresponding classes. Here belong:

A. The Subsumption Form

The class of *S* is subsumed in the class of *P*. This is the fundamental form for the extensional calculus in C. S. *Peirce* and *Schröder*. [51]

B. The Identity Forms

(a) The class of *S* is identical with that of the *S* which are *P*. *Jevons*' calculus is founded upon this.

(b) The class of *S* which are not *P* is an "empty" class. This is the formula used by Boole.[6]

III. Mixed Interpretations

These originate in that now the subject, now the predicate – and, indeed, each alone in the given case – are quantified in the class sense. Here belongs the form:

A. All *S* are *P*. It is the form of traditional logic. A few words of interpretation may be in place here.

By using this formula, we first form the concept of the class of *S* (all *S*). But it is not this class which is the subject of the assertion, but rather an object of that class, as such. We proceed, e.g., from the representation of the totality of men, and affix thereto the judgment: Provided something belongs to this totality, it also has the property of mortality (or else the negative: Of this totality, none are non-mortal). The corresponding form with quantified predicate is:

B. Provided something is an *S* (an *S* as such), it belongs to the class of *P*; e.g., Gold belongs to the metals.

I need not say that it is in nowise the same judgment that underlies this long series of assertion forms. It is clear that, partly in affirmative form, and partly in negative, different matters are judged (*beurteilt*). However, the essential fact is that all of these assertions are equivalent to one another; i.e., can be reciprocally deduced from each other.

Obviously our list can still be expanded in various ways. The forms from I.B. on are all affirmative in appearance. But, when carefully considered, they are not truly affirmative assertions. Rather they are hypothetical; for in the sense of a *pure* theory of deduction, the existence of the subject cannot at any point be presupposed. Accord- [52]
ingly, then, we also would – similarly as under the heading I.A. –

6 The Hamiltonian forms, although they are completely quantified, do not belong here, since none of them offers us a full equivalent of the universal judgment. Such an equivalent arises only through disjunctive combination of the two Hamiltonian forms, "All *A* are all *B*" and "All *A* are some *B*." But by this we obtain precisely the subsumption form *A* above.

place beside each hypothetical form an equivalent negative form, whereby the table would be enlarged by a number of new (though not precisely simple) forms. We could undertake further transformations analogous to the kind presupposed by *Jevons* for class judgments, only deploying in place of his identity concept that of conceptual equivalence, which includes the former. We would obtain, for example, the form $S = PS$; i.e., provided something possesses the property S, it also possesses the property PS, and conversely. All judgments can be transformed in such a manner into conceptually equivalent judgments – whereby, of course, considerable complications ensue.

And there also are equivalent forms of various other sorts, to enumerate which no one would seriously care to attempt. The limitations which I have imposed upon myself in the above classification follow from the specific interest which is here decisive for me – *viz.*, that of constructing the most adequate *logical* algorithms possible. Disregarding forms which have already been used in a logical calculus, we should, therefore, bring into consideration only such as are especially recommended for algorithmic treatment by their logical character, their frequent employment in natural thinking, or their superior fitness for use in a technique. The following general reflection, which we also need for developments further on, will serve to justify our limitation.

If one enters into the various attempts to bring the art of pure deductions up to the level of a calculative technique, one then notices essential distinctions over against the mode of procedure in the old formal logic. The latter, guided more by theoretical than by technical interests, considered each form of judgment and inference for its own sake. By contrast, calculus logic seeks to diminish the number of forms to be considered by as much as is possible, and to compress them all into one uniform pattern. It succeeds in doing this by means of transformations of judgments into equivalent judgments. All negative judgments can be transformed into equivalent affirmative judgments with negative matter. So why develop a threefold technique: one each for the pure affirmative, the pure negative, and mixed inferences? It is most convenient to draw our inferences with affirmative judgments. And so all judgments are, on principle, [53]
converted into affirmative judgments. The same fate of elimination befalls particular judgments, in their turn; and for the same reasons. There are certain universal judgments which are equivalent to them,

and since purely universal inference seems most convenient, all judgments are transformed into just such judgments. And, once more, the same ensues with singular judgments, which in fact already are customarily subsumbed under the universal in formal logic – although of a true subsumption one cannot speak, in this case. Not to be satisfied with this, class logic goes yet considerably further. Only *via* the path through classes does it believe that a calculus can be founded. Hence it transforms all universal judgments into class judgments, and operates exclusively with these.

So, wherever it is convenient, calculus logic makes ample use of the device of equivalence transformations of judgments. It is not in the least disconcerted by the fact that in this way it removes itself from the movement of the original thinking – and often very considerably so. And it is completely justified in this, provided only that it no longer makes a claim to stand for a *logic* of deduction, instead of a mere *technique* thereof. Nothing has so much contributed to the widespread disdain of the calculus as the vague confusion of these two objectives, so essentially different, on the part of those who have invented and advocated the calculus. Now the aspiration to construct a deductive calculus is, in itself, quite legitimate, and in relation thereto we find deviations from the natural movement of logical thinking to be admissible. But we do not, after all, extend this to cover *every* deviation whatever, so long as it merely falls within limits set by the equivalence of judgments. It is clear that we may accept only such deviations as do not result in an unnecessary complication, either in the construction of the calculus, or in its application. The principle operative here is that of *the least possible means*. And this surely entails that we should strive for a deductive calculus that has the closest possible connection with the natural movement of thought. The closer we hold ourselves to natural thinking, the easier becomes the conversion of deductive problems into formulae of the calculus, and, in turn, the conversion of the derived formulae into the judgment which is the conclusion. Now our principle excludes from the outset most from among the great number of equivalent assertion forms. All of the more complicated ones indicated above [54]
(without entering them in our table) are to be avoided. It is certain that they are not in a position to serve as the basis of a simpler calculus, while in the application of the calculus they are certain to presuppose disagreeable and superfluous complications. Obviously

the ideal would be to ground the deductive calculus from the outset upon the forms which are most primitive and the most at home in natural thinking. Now the forms which have been chosen by *Boole*, *Jevons*, C. S. *Peirce*, *Schröder*, and others, certainly are not of this sort. But the selection of *those* forms, after all, rested more upon contingencies of the historical development of logic in Britain than it did upon the facts themselves (*an der Sache selbst*). In truth, the ideal just referred to actually can be attained; and older German logicians – such as, above all, *Leibniz* and *Lambert* – were on the way to attaining it. It is not correct to say that these logicians did not and could not reach their goal because they avoided the detour through class judgments; for this detour actually is totally superfluous.

If, now, we pick out from our list the forms which have not yet been utilized in algorithms, we find, first and foremost, under I.A. the two fundamental logical forms (a) and (b): "Provided something is an *S*, it is a *P*," and "There is no *S* which is not-*P*." These represent the precise, typical expressions of all universal judgments. Insofar as the remaining forms do present universal judgments, they must also, therefore, be subsumed under these. And that easily results from only attending to the true subjects and predicates of those other forms. But for the purposes of the calculus this is not of importance. Each form expresses a type of relation between the terms *S* and *P*, and it is just these relations which differ in each case. In this regard, thus, the two forms yet remaining – (the forms under III), "All *A* are *B*" and "Provided something is an *A* it belongs to the *B*'s" – are also novel and well-distinguished. The first of these is especially common to natural thinking, though in such occurrences it is often enough understood in the sense of (a) under I.A. – i.e., as the primitive hypothetical form.

In conformity with the above considerations, we will exclude the basic negative form, as well as the negative interpretation of III.A. – where "All *S* are *P*" = "Among the *S*'s there is none which is non-*P*." We do this since for a calculus the preference for affirmative or, more precisely stated, hypothetical inference offers undoubted advantages. [55]
The simplest of the three other forms is, obviously, the hypothetical elementary form. In relation to it the others, no doubt, can be regarded as complications. Thus, for example, the signification of the form, "All *S* are *P*" (i.e., "Provided something belongs to the totality of the *S*'s, it is a *P*"), is conceptually more complex than that of the form, "Provided something is an *S*, it is a *P*." The attempt to exploit this

most primitive of forms in a deductive calculus claims, therefore, exceptional logical interest. We will make the attempt.

The calculus which we now wish to set up shall be a calculus of conceptual objects, and precisely in virtue of that fact will it prove itself to be a truly *logical* calculus. For all judgments bear, not upon classes, nor upon conceptual contents, but rather purely and simply upon conceptual objects, as has already been strongly emphasized by *Mill*.[7] As a general calculus built upon conceptual objects, it must, of course, disregard the peculiarities of such objects, or the concepts characterizing them as such, and base itself purely and simply upon the possible relationships that can be presented by objects of *any* given concept, as such. In this respect the mode of consideration is a purely "formal" one, and all deduction is *eo ipso* only hypothetical.

THE FIVE RELATIONSHIPS. Let us turn the discussion to the relationships which can obtain between objects of any two concepts A and B. Five possible cases present themselves, which form an exhaustive disjunction:

1. An object of the concept A, as such, is also object of the concept B; i.e., provided something has the property A, it also has the property B – but the converse relationship is not to obtain.
2. An object of the concept B, as such, is also object of the concept A; but not conversely.
3. An object of the concept A is, as such, an object of the concept B; and conversely.
4. An object of the concept A is, as such, *not* object of the concept B; and conversely. [56]
5. An object of the concept A is also (or is not) object of the concept B; but it is not precisely *as* object of the concept A that it also is (or is not) the object of B; so that none of the cases 1–4 occur.

These five relationships exactly correspond to *Euler*'s five well-known spherical models for conceptual extensions.

---

[7] This assertion may appear objectionable because it is so strongly worded. One will object that we can occasionally make judgments also about classes and about conceptual contents. However, wherever this occurs, classes and contents are themselves judged of *as* objects of certain concepts – although, of course, not as objects of the concepts belonging to them in the original sense.

I further stress the fact, of which use shall subsequently be made, that just as we can speak of relationships between the objects of the concepts *A* and *B*, we quite as well can also speak of relationships of these concepts themselves, with reference to their objects. In this mode of consideration, language – no doubt sensitive to the analogy – allows exactly the same expressions as for the relationships of classes or of spatial areas. One metaphorically says of two concepts that the one includes the other, that they mutually exclude each other, or possibly that they coincide. And only in the case of simple isolation, where all conditioning relationship between the concepts is in fact absent, does the physical model not apply. Also, such expressions as these are admissible: The concept *A* 'implies' ('*bedingt*') the concept *B*, or is the 'mark' of it. Precisely in this sense, for example, Kant's principle of inference speaks of "*nota.*" But the exact interpretation of such expressions always lies in primitive hypothetical judgments of the type which we here presuppose.

I have not needed the above table of possible relationships in its entirety for my further presentation. However, that table makes the analogy with Euler's table of spheres stand out in such a manner that, right from the outset, one sees that to the calculatory technique with extensions (to which the sphere relationships are so closely connected) there must correspond, in our domain, a parallel and formally identical technique.

In order to make the relation of our calculus to the extensional calculus clear right from the start, I select for symbolization of certain relationships of objects (or, otherwise considered: certain relationships between properties) exactly the same signs which, in *Schröder*'s presentation of the class calculus, are used for certain relationships of extensions. Also, in the ordering of definitions, axioms, and theorems, that same presentation is to guide us, so far as is practicable. Any calculus is based upon a certain number of fundamental formulae, all calculation in it being grounded in them *via* their iterated [57]
application and combination. Thus, if we succeed in showing that all of the *fundamental* formulae which have been proven sufficient for the construction of the class calculus have validity for our symbols – which have a wholly different conceptual foundation – then we have a rigorous proof that the whole of the remaining calculus also, considered as pure technique, is valid for our domain, in spite of the completely changed intellectual substance.

SYMBOLS. For symbolization of relationships 1, 2, and 3 we must, following our model, use the signs "(", ")", and "=", respectively. The essential sign for the fundamental formulae is one composed of the first and third signs; namely, "= (=", which by its constitution *could* point to a certain compositeness in the relationship to be designated by it – as it in fact does in the class calculus. But this is not correct in our domain, and therefore the designation seems less fitting here. In fact we need two judgments for the exact expression of relationship 1, for example – the availability of the simple sign "(" notwithstanding. That relationship is, thus, a composite relationship. By contrast, the relationship to be symbolized by the sign "= (=" is a simple one. We *define*, namely, the inscription, "A = (= B," by means of the linguistic expression: "Provided something is an *A*, it is a *B*.

If this judgment holds true, then either relationship 1 or 3 obtains – as the composite sign (superfluously) indicates.

TWO AXIOMS

I. A = (= A.
Provided something is an *A*, it is an *A*:
The principle of identity.

II. If A = (= B and B = (= C, then also A = (= C.

This is the familiar *modus Barbara*. Therewith the principle of deduction is the Kantian principle, *nota notae est etiam nota rei ipsius* (or, more precisely: *notionis ipsius*).[8]

These two axioms are the equivalents of Schröder's first two 'principles.'[9]

THE DEFINITION OF EQUIVALENCE. If A = (= B and simultaneously B = (= A, then we write: A = B.

This definition symbolically expresses the fact that between the objects of concepts *A* and *B* relationship 3 obtains. Here the concepts are "equivalent" in relation to their objects. To this relationship there

---

[8] As Höfler, *Philosophische Propaedeutik*, in co-operation with A. Meinong, Part I, Prague/Leipzig, 1890, p. 168, correctly observes.

[9] *Op. cit.*, p. 168 and p. 170.

corresponds the identity of classes in the class calculus.[10]

DEFINITION OF PRODUCT $A \cdot B$. Under $A \cdot B$ we understand the object of that concept which simultaneously takes in the properties of *A* and those of *B*, and these alone (thus, the total of the properties of the two concepts). $A \cdot B$ is, thus, "Something which is simultaneously an *A* and a *B*."

There then holds true the *Axiom*:

III. If C = (= A and simultaneously C = (= B,
then also C = (= A · B.

E.g., a perfect man (*as such*, that is) is virtuous. A perfect man is wise. Thus, a perfect man is simultaneously virtuous and wise. This axiom corresponds to Schröder's (alleged) definition of class product.[11]

DEFINITION OF THE SUM $A + B$. Under $A + B$ we understand the object of the concept "either *A* or *B*"; or, if one wishes, the object of any genus concept which contains the concepts *A* and *B* as its species. The disjunction here is not to be understood in the exclusive sense, but in the inclusive.

Now there obtains the *Axiom*:

IV. If A = (= C and simultaneously B = (= C,
then also A + B = (= C holds true. [59]

E.g., every virtuous man is praiseworthy; and every wise man is praiseworthy. Thus, everyone who is either virtuous or wise is praiseworthy. This axiom corresponds to *Schröder*'s quasi-definition of the class sum (*loc cit.*).

DEFINITION OF *1*. "1" is to be the sign for the object of the concept of existence, of being in the sense of truth. Accordingly, this axiom holds true:

---

[10] Cf. Schröder, *op. cit.*, p. 184.

[11] Schröder, *op. cit.*, p. 196. Cf. my critique, soon to appear in the *Göttingische gelehrte Anzeigen*. (Translated immediately preceeding, this volume. [D.W.])

V. Whatever *A* signifies, it is true that A = (= 1.

One must not, perchance, read this formula: *A* (*e.g.*, gold, a round square, etc.) is an existent. Rather it is: Provided something is an object of concept *A*, it also is an object of the concept of existence. And this is quite evident. The judgment, "Something is an *A*," includes, like any categorical judgment, the existence of the subject.

DEFINITION OF *0*. "0" is to be the sign for the object of the concept of non-existence. It, thus, signifies "Something of which it is true that it is not, in the sense of truth." And, accordingly, this axiom holds true:

VI. Whatever *A* signifies, it is true that 0 = (= A.

That is, provided something possesses the property of non-existence, it also possesses every other property. In fact: If *n* is any non-existent thing, then also *nA* is a non-existent, whatever concept *A* may be. Consequently, any object of the concept *0* simultaneously falls under any other concept *A*.

At this point I also wish to discuss the sense of formulae such as A = (= 0, which occasionally show up – since at first glance they are quite paradoxical, and seem directly to contradict the universally valid formula V, A = (= 1. Now the first formula really says: Provided something is an object of the concept *A*, it also is object of the concept of non-existence. But the second says: Provided something is object of the concept *A*, it also is object of the concept of existence. The contradiction is obvious. But it also is legitimate, as soon as we assume that an *A* as such cannot exist. If there is no *A*, then it is true both that *if* an *A* were, it would have to fall under the concept of the existent, and, on the other hand, that it must fall under the concept of the non-existent. That is all quite as it should be. The hypostatization [60] of the existence of a non-existent implies, *eo ipso*, a contradiction. For this reason, then, it becomes understandable that the formula A = (= 0 in the calculus can be considered as the representative of its equivalent judgment, "*A* does not exist."

THEOREMS. The formal principles developed to this point suffice – as *Schröder* (following C. S. *Peirce*) has shown – to demonstrate, within the calculus, a long sequence of theorems: e.g.,

$$A \cdot A = A, \quad \text{and} \quad A + A = A;$$

and, further, the laws of "commutation" and "association," respectively, for multiplication and addition. I.e., we have:

$$A \cdot B = B \cdot A, \quad \text{and} \quad A + B = B + A; \quad \text{as well as}$$

$$A \cdot (B \cdot C) = (A \cdot B) \cdot C, \quad \text{and} \quad A + (B + C) = (A + B) + C.$$

And still other theorems can be demonstrated by calculation. We claim all of these for ourselves, since such calculative deductions never have anything to do with the original concepts, but rather deal exclusively with the rules for the signs, which are expressed in the basic formulae. Many of these propositions, moreover, are also obvious as immediate axioms, as soon as one has recourse to the meaning of the signs. To the propositions provable by means of the principles given up to now also belongs the theorem,

$$(AB + AC) = (= A(B + C),$$

which is "the provable subsumption in the law of distribution," according to *Schröder.*[12] But not so for the corresponding theorem,

$$\text{VII.} \quad A\,(B + C) = (= (AB + AC),$$

which is Schröder's "unprovable" subsumption in this same law.[13] This latter theorem is proven by an obvious argument from Jevons. The left side of the expression speaks of something simultaneously being both *A* and either *B* or *C*. This same "something" is then, obviously, either simultaneously *A* and *B*, or simultaneously *A* and *C*. But that is precisely what the right side of the above expression says. – According to the definition of equivalence, the two distribution theorems then supplement each other to form the "full" distributive law: [61]

VII^{a} $\quad A(B + C) = (AB + AC)$, which is a major formula of the calculus.

[12] Schröder, *op. cit.*, p. 280.

[13] Schröder, *op. cit.*, pp. 282ff.

DEFINITION OF NEGATION. Under A_1 we understand the non-*A*, in the sense customary in logic: i.e., something to which the property of not-being-*A* belongs, and of which that negative property can be asserted. With the aid of this concept there result the axiomatic formulae:

VIII. $\mathrm{A} \cdot \mathrm{A}_1 = (= 0$, and $1 = (= \mathrm{A} + \mathrm{A}_1$.

The first of these, conforming to a previous observation, may be read in the form: What simultaneously is an *A* and a non-*A* does not exist. The second one states: Provided something falls under the concept of existence (or, what is the same: All that exists ...), it is either an *A* or a non-*A*.

The calculus does not deal directly with negative judgments. They are replaced by affirmative (rightly: hypothetical) judgments with negative matter.

So with this we have attained our goal. We have shown that the *total* formal basis upon which the class calculus rests is valid for the relationships between conceptual objects. Thus with one stroke we have made the technique of that calculus over into a directly and authentically logical calculus. Any given formulae which *Schröder* or others have derived, and any procedures which they have devised, we may lay claim to for ourselves. Accordingly, we can solve any logical problem which in general falls within our range, directly and without the detour through classes – and with identically the same calculations as those required by the class calculus. With this the logic of extensions is definitively proven to be superfluous.

We have based our calculus upon the form of the pure content judgment, "Provided something is an *A*, it is a *B*." Could we not also utilize one of the other forms – e.g., "All *A* are *B*" – for the same purpose, and upon it too ground a logical algorithm? Certainly, I reply. And nothing is easier than to construct that algorithm, since it in turn is formally identical with the calculi worked out up to now, provided only that we make a suitable choice of definitions and symbols.

I return to an earlier observation, that the judgment form utilized in the previous calculus also can be regarded as a relation between conceptual contents. The concept *A* stands to the concept *B* precisely [62]
in such a relation that, inasmuch as something falls under the first, it

also falls under the latter. Accordingly, the whole calculus also can be regarded as a calculus for these relationships of concepts. But on this interpretation the definitions refer, not to the objects, but rather to the concepts themselves. "$A \cdot B$" would then signify the totality of the properties of *A* and of *B*; "$A + B$" the disjunctive conceptual content *A* or *B*. "1" would signify the concept of existence; and so on. Only the formulae, such as "A = (= B," would bring the *relation* to expression – by means of the sign combining the terms.

Now this conceptualization can be immediately carried over. We once more consider the form "All *A* are *B*" as a relation between the conceptual contents *A* and *B*. The content *A* stands in a relation to the content *B*, such that the totality of objects which falls under the first also must fall under the latter. We then retain all definitions (referring them to concepts, as just stipulated), but now assign to the symbol "= (=" the capability of expressing the new relation. Nothing further is necessary; for now all axioms, I–VIII, hold true of the new relations also. And, indeed, they do so for the simple reason that they are logically equivalent with the corresponding axioms for the previous relations and signs. But with this the entire calculus is transferred, as the necessary consequence of the definitions and axioms.

Now precisely the same also holds for the third formula taken into consideration above. And we can at once add: It holds true for any equivalent affirmative form with the same terms. One of these forms is, for example, the extensional form. It too can be regarded as a relation between conceptual contents: the concept *A* stands in a relation to the concept *B*, such that the class of objects which possess the properties of the former is subordinate to the class of objects which possess the properties of the latter. If we had developed the calculus as, for example, a calculus of [conceptual] objects, we now could instantly, in a manner wholly analogous to the foregoing, transfer it over to the relations between extensions. Another form is that of the "ideal contents." We could proceed in a similar manner here also. And so on.

In short, whichever equivalent forms with the same terms we may take up, it is certain apriori that the calculus must be identically the same. In fact, since all of these various relational forms present [63]
logically equivalent judgments, we obviously can choose the steps which found the calculus in such a way that they are exactly parallel at all points; and, indeed, are so in such a manner that the correspond-

ing steps are either identical *or* equivalent. Given suitable choice, the definitions are *identical*; for they refer to operations exclusively upon the conceptual contents, into which nothing of the difference of the relations enters. But the axioms, which derive from the relations, with the aid of concepts of operations, are throughout *equivalent* judgments that correspond point by point. These, regarded as relations of their terms, are of analogous form; and therefore, given suitable choice of signs for relations and operations, they must pass over into formulae which are identically the same. Since such a type of *modus procedendi* can constantly result in routes that are exactly parallel; and since, further, the corresponding steps are either identical, or at least are expressible in identical form (given appropriate choice of symbolism) – what wonder then, that with a formal basis at all points the same, there also results a calculus which is identically the same? Thus there is explained this phenomenon – so very surprising at first glance – of the sameness of the algorithm in so many areas that are conceptually diverse.

Now from this there follows, further, an important result. Whichever one of such equivalent forms of the universal judgment may be present in a given problem, we never need first to carry out a transformation into a standard form, but rather may begin directly by the introduction of formulae. If by chance the judgments are given as class judgments, then we apply the calculus as, so to speak, a class calculus. If they are given as judgments about conceptual objects, then we calculate in the object calculus. And so forth. The underlying principles of inference may be quite different *logically*. Now it is the *dictum*, now the *nota*, now the corresponding principle for classes, or whatever other principles naturally equivalent to each other. In calculating we notice nothing of all these differences. The formulae remain the same.

The algorithms which we have considered up to this point sought and attained what is, in essentials, the same goal. The one which was most worthy of attention for logic was the calculus of conceptual objects. Now we enter into another domain, in which *judgments* take the place of concepts. Of course relationships wholly analogous to [64]
those which concepts present in relation to their objects also show up between judgments in relation to their matters (*Materien*). The analogy is such a complete one that it, once again, founds identically the same algorithmic technique. Already in *Boole* we find this

acknowledged. But also here, as elsewhere, his strength lies more in ingenious invention than in providing logical foundations. Through the most violent of reinterpretations he forces the subsumption of the new calculus under the calculus of classes. The basic judgment of the new domain, "If the judgment *A* is true, then the judgment *B* also is true," he replaces by the bizarre form, "For the time in which the judgment *A* is true, the judgment *B* also is true" – the conceptual content of which is not unessentially altered. To this relationship of times within times, then, the class calculus – or, rather, the general calculus of sets – is directly applicable. *Schröder*, the latest one to revise the calculus, follows *Boole* in this matter; while already in Venn's presentation, to be sure, the superfluity of such a move had been correctly acknowledged.[14] However, *Venn*'s presentation also[15] leaves much to be desired in rigor and in persuasive force. He does not here get wholly free of the standpoint of classes, and goes wrong (as, by the way, he also does in the class calculus) on an essential point: namely, the interpretation of the signs "0" and "1." Without justifying this criticism in detail, I will summarize, in a few words, all that is necessary for an independent founding of this "calculus of assertions."

The analogues of the five sphere relationships of *Euler* are:

1. If judgment *A* is true, then judgment *B* is true, but not conversely. The first implies the latter, or "includes" it; and, indeed, does so unilaterally.
2. If judgment *B* is true, then judgment *A* is true, but not conversely.
3. The case of mutual implication or logical equivalence of judgments, of "coincidence" with respect to their logical value.
4. The case of mutual exclusion: If *A* is true, then *B* is not; and conversely.
5. Two judgments between which there is no implicational relationship. [65]

SYMBOLS. The inscription, "A = (= B," is to be read: "If the judgment *A* is true, then the judgment *B* is true"; or, more briefly: The judgment *A* implies judgment *B*."

---

[14] *Symbolic Logic*, London, 1881, p. 372.

[15] *Op. cit.*, chap. XVIII.

AXIOMS. Again, the two following axioms hold true:

I. A = (= A, and
II. If A = (= B and simultaneously B = (= C, then also A = (= C.

THE DEFINITION OF EQUIVALENCE. If A = (= B and simultaneously B = (= A, then we write: A = B.

THE DEFINITION OF THE PRODUCT *A · B*. Under "A · B" we understand the simultaneous validity of both of the judgments *A* and *B*. Then, again, we have the axiom:

III. If C = (= A and simultaneously C = (= B, then also C = (= A · B.

THE DEFINITION OF THE SUM *A + B*. The sign "A + B" signifies the validity of either the judgment *A* or the judgment *B*; and perhaps of both together. And there follows upon this the axiom:

IV. If A = (= C and simultaneously B = (= C, then also A + B = (= C.

THE DEFINITION OF *1*. The sign "1" signifies the validity of the proposition: "Of two contradictory judgments, one is true and one is false" – thus, the validity of the combined principles of contradiction and excluded third. Then we have the axiom:

V. A = (= 1, whatever *A* signifies.

In fact, in the truth of any judgment in general we already have presupposed that of the principle just mentioned. Without that principle [66] there would be no distinction between truth and error. If, therefore, any judgment is true, that principle also is true.

THE DEFINITION OF 0. The sign "0" signifies, accordingly, the validity of the negation of just that very principle, and hence the invalidity of it. And then we have the axiom that:

VI. 0 = (= A, whatever A signifies.

And, actually, if the opposite of the fundamental principle of logic holds true, then there exists no distinction between truth and error, and, accordingly, any arbitrary assertion may be made. (189)

I doubt that any other interpretation of the symbols "0" and "1" would be capable of grounding these formulae V and VI, which are necessary for the calculus.

There now obtain, once again, all of the theorems mentioned on p. 108, and in particular the full distributive law,

VII[a] A(B + C) = (AB + AC),

of the correctness of which one is immediately convinced by putting it into linguistic expression.

THE DEFINITION OF NEGATION. "A_1" designates the invalidity of judgment A. Obviously, then these axioms also hold true:

VIII. AA_1 = (= 0, and 1 = (= $A + A_1$.

With this we have proven a totality of basic formulae sufficient for the construction of the *entire* symbolic technique of the calculus, and have, consequently, grounded the "calculus of assertions" entirely out of its own resources.

ADDENDA TO: THE DEDUCTIVE CALCULUS AND THE LOGIC OF CONTENTS[1]

1. A certain analogy of the Boolean formula, $AB_1 = 0$ – i.e., "The class of *A* which are non-*B* is an empty class" – to the basic negative form of the universal judgment, "There is no *A* which is non-*B*," misled me into the assumption that a calculus which wished to proceed from this latter form would have to follow the route of the Boolean or related technique. How one does sometimes overlook the obvious! A glance at formulae I–VIII (pp. 105–109) would have sufficed to show *that they all can be interpreted in the sense of the negative basic form*, so that the whole of *Schröder*'s calculus also is adapted to that form. The inscription, "A = (= B," is then always to be read precisely as "There is no *A* which is non-*B*." Or: "An *A* which is not a *B* does not exist."

This same result also follows directly from the considerations on pp. 109–111, which provide the proof that, in general, a calculus grounded for *one* form of the universal judgment can be transferred, identically the same, to *any arbitrary equivalent form*; i.e., can be interpreted in the sense of any such form, *salva veritate*. The restriction to equivalent *affirmative* forms (*loc. cit.*) was not objectively justified, and was occasioned only through the prejudiced assumption mentioned above.[2]

I would in addition expressly point out that, for interpretations of A = (= B, there are admissible even those forms of assertion which (indifferently of whether or not they occur in a unified expression) [68] coordinate to the universal judgment not, in truth, *one* judgment, but rather a *complex* of judgments, as being equivalent. Of this

[1] First appearing in *Vierteljahrsschrift für wissenschaftliche Philosophie*, Vol. 15, 1891, pp. 351–356. Republished in *Husserliana XXII*, pp. 67–72.

[2] Accordingly, on p. 110, line 23, the word "affirmative" is to be deleted.

type are, for example, the equivalence forms indicated on p. 99, any one of which contains a *pair* of judgments that is equivalent to the universal judgment. We perhaps could understand A = (= B in the sense of the assertion: "The concepts of *A* and of A-which-are-B are equivalent." I.e., "Provided something is an *A*, it also is an *A* which is B; and provided something is an *A* which is B, it also is an *A*" – certainly a ridiculous, because completely useless, complication. Nevertheless, we could take *all* formulae of the calculus in the sense of *this* form; as we can do with *any* form, howsoever complicated, if only it is *equivalent.* This circumstance makes it especially clear that the only suitable route to the establishment of the calculus consists in selecting one of the most primitive and most convenient of forms. For all other [equivalent] forms, the calculus is then *eo ipso* conjointly founded. And the cause for the exclusion of the more complicated and less customary forms is not that they would ground calculi which are merely less complete. To the contrary, the calculus is – given suitable choice of algorithmic concepts, signs, etc. – everywhere identically the same. But rather it is that they allow us to ground the same calculus with less convenience.

2. In setting up the calculus, the symbols "0" and "1" (and the formulae V and VI which depend upon them) are the sole points to offer substantial difficulties. But I would like to stress, first and foremost, that even if our efforts in relation to them were to fail, and also would continue to fail from here on, the content calculus would then be in no worse position than the extensional calculus. For exactly the same difficulties which the symbols in question offer in the former, they also offer in the latter. Any correct interpretation of them in the *one* calculus necessarily provides, *via* the route of an easy transformation to equivalents (in terms of extensions – of contents, or else of conceptual objects), a correct interpretation in the *other.* And exactly the same is true of the grounding of formulae V and VI. If it actually has been carried out for the extensional calculus, then we did not need also to attempt it in undertaking a calculus of contents. And I frankly admit that my own efforts with regard to "0" and "1" do indeed admit of many serious doubts: doubts which had already arisen for me in the composition of the essay; whereas I, unfortunately, sensed their full weight only as it came from the [69]
press. Happily, I am in a position to provide new proofs, and of a kind which appears to remove all doubt.

In the calculus of conceptual objects (or of implicational relationships between properties) I maintain the same interpretations of the symbols "0" and "1," but nonetheless give new proofs of the formulae V and VI: proofs which are not merely more satisfactory than the earlier ones, but also set the true sense of the remarkably paradoxical formula VI in a correct light.

First, an observation concerning the sense of the inscription, "A = (= B." It signifies: "If something is an *A*, then it is a *B*." I.e., *not*: If something is *represented* with the property *A*, then it is or must be represented with the property *B*; but rather: If something is *recognized* (*anerkannt*) as possessing the property *A*, then it must also be recognized as possessing the property *B*. Or, what is equivalent: "If something *exists* which possesses the property *A*, then it also possesses the property *B*."

Now if this is correct, then the following formula,

V. A = (= 1,

valid for any *A*, states that "If something exists which possesses the property *A*, then it exists." And this is a proposition which is indubitable for any *A*.

The matter is not so simple for the formula,

VI. 0 = (= A,

which likewise is to be proven as valid for any arbitrary *A*. It signifies: "If something exists which possesses the property of non-existence, then it possesses the property A," whatever *A* may be. In other words, it then possesses any and every property.

Proof. (1) If something exists, then it is either an *A* or a non-*A*, whatever *A* may signify. On the other hand, if something does *not* exist, then it also is not an *A*, whatever *A* may signify. We can, now, precisely because of the arbitrariness of *A*, replace it with non-*A*. [70] There follows, accordingly: (2) If something does not exist, then it is not a non-*A*.

If now we advance the hypothesis that something exists which does not exist (or: that a non-existent exists), then we know, according to (1), that it must be either an *A* or a non-*A*. But since it *as* non-existent is not a non-*A* (as we just saw under (2)), then it must be an *A*. I.e., 0 = (= A.

One also can carry out the proof in this way: If there is no *N*, then there is no *N* which is *A*, whatever *A* may signify. But if *A* is arbitrary, then we can substitute non-*A* for it, and thus judge: There is no *N* which is non-*A*. This latter judgment is equivalent with: If there is an *N*, then it is *A*.

Thus: If there is no *N*, then the simultaneous assumption that there is an *N* draws after it the consequence that the same *N* would have to be an *A*, whatever *A* signifies. Or: If a non-existent existed, then it would have any and all properties.

If we read the inscription, "A = (= B," in the sense of the negative basic form of the universal judgment ("There is no *A* which is non-*B*"), then the proof of our formula VI becomes most simple. It then signifies: "There is no non-existent which is non-A." In fact, if there is no *N*, then there is no *N* which is *A*, whatever property *A* signifies. And thus there also is no *N* which is non-A. With that, the proof is already given.

The validity of formula VI for any *A* implies that for any arbitrary but determinate *A* the propositions,

0 = (= A, and 0 = (= A_1,

simultaneously hold true, although they seem directly to contradict each other. The explanation of this paradox lies in the absurdity of the hypostatization of the existence of a non-existent, which violates the fundamental principles of logic. If we hold fast to these principles, and nevertheless make that hypothesis, then it is no wonder that, in a [71]
logically correct manner, we attain opposed consequences.

Formula VI expresses a true proposition (not an *Axiom*, as p. 107 says) concerning the relationship between two absurdities. The antecedent, which makes the presupposition of the existence of a non-existent, is absurd; and also the consequent is absurd, holding that this non-existent itself must possess every property (thus, contradictorily opposed properties also!). But the consequence is logically correct which holds that, if the one proposition were true, the other would have to be also.

3. I would still add to this that the symbols "0" and "1" can also be interpreted in another way, which many will perhaps prefer. The

most convenient mode of expression is gained when we regard the letter-symbols in the calculus as signs for properties, and then give these definitions:

> "1" signifies the property of *possessing some property or other*; and
> "0" signifies the property of *possessing no property*.

The formula,

V. $A = (= 1$,

then states the immediate truth: "If something possesses the property A, then it possesses the property of possessing some property or other (i.e., then it possesses some property)."

The formula,

VI. $0 = (= A$,

on the other hand, states the paradoxical proposition: "If something possesses the property of possessing no property, then it possesses the property A," (i.e., because A is arbitrary, it possesses *any and every property*).

Proof. If something has the property of possessing no property, it also does not possess the property non-A, where A signifies any arbitrary property. But everything which possesses a property exists, and what exists is either A or non-A. Consequently: If something has the property of having no property, then it possesses the property A – and then it possesses any and all properties.

The elucidation of the paradox of proposition VI proceeds exactly as with the foregoing interpretation of the symbols "0" and "1." [72]

The expression "$A = (= 0$" also is now, once again (cf., p. 107 above), suited to represent the proposition: "A does not exist." For what possesses no properties at all, also does not exist. Did it exist, then to it there would indeed belong, in each case, one of every pair of contradictory properties, A and A_1.

4. I withdraw my interpretation of the symbols "0" and "1" for the calculus of assertions, and replace it with the following new one:

"1" signifies the judgment, "*There is a valid judgment.*"
"0" signifies the judgment, "*There is no valid judgment.*"

The formula,

A = (= 1,

then states; "If the judgment *A* is valid, then there is a valid judgment." This requires no proof.

The formula,

0 = (= A, valid for any A,

contains, once again, a paradoxical proposition: "If the judgment, 'There is *no* valid judgment,' is valid, then the judgment *A* is valid," and thus – since *A* is arbitrary – *any arbitrary* judgment is valid.

Proof. If there were no valid judgment, then the judgment A_1 (the contradictory opposite of *A*) would also be invalid. But if A_1 were invalid, then *A* would have to be valid. Since *A* can signify any arbitrary judgment, the hypothesis advanced therefore leads to the consequence that any arbitrary judgment would have to be valid. Here also proposition VI expresses a true judgment about the relationship between two absurdities.

That in this proof, as in the previous analogous one, we are operating with absurdities is, I think, no objection. We do the same in any indirect proof. We consider the absurdity of a hypothesis to be proven by the correct derivation of an absurd consequence. Therewith it is conceded that even absurdities may function as hypotheses in a purely hypothetical line of reasoning.

A. VOIGT'S "ELEMENTAL LOGIC,"IN RELATION TO MY STATEMENTS ON THE LOGIC OF THE LOGICAL CALCULUS[1]

As is well known, the "algebra of logic" was founded and developed as an "extensional" or "class" logic. In fact, it can well be designated as the generally dominant view, that an algebraic formalism can be constructed for logic *only* by recourse to classes and class relationships. This view is so commonly expressed and argued for that references to literature are hardly necessary.[2] Also, the most recent and most extensive work on the calculus stands totally within this point of view.[3] "To set logic up from the outset as a logic of conceptual contents," *Schröder* says (I, p. 99), "might well match the attempt to build the roof before the house." And this remark is fitting here, since for *Schröder* the concept of (a scientifically exact, deductive) logic coincides with that of the algebra of logic. At no time, so far as I know, has there been made a serious and consistent attempt to found calculative methods of deduction directly upon the judgment forms of "content logic," without an admixture of class

[1] First appearing in *Vierteljahrsschrift für wissenschaftliche Philosophie*, Vol. 17, 1893, pp. 111–120. Republished in *Husserliana XXII*, pp. 73–82.

[2] The guiding theoretical conceptualization no doubt originally was the following: Classes are sets (*Mengen*). Sets are magnitudes. But magnitudes do not admit merely of the algebraic mode of treatment (in this case the "analytical," in the normal sense of the arithmetician), approaching them in terms of measure and number. Rather, they also can be dealt with directly, without such – often superfluous and inapplicable – mediation: as one can see in many geometrical analyses. (It seems that for Wundt also (*Logik*, Vol. I, Stuttgart 1880, 231ff.) this line of thought was decisive.) In this possibility of taking classes to be magnitudes – but as not being numerically determinable – one saw the basis for the applicability of algebraic methods.

[3] E. Schröder, *Vorlesungen über die Algebra der Logik: (Exakte Logik)*, Vol. I, Leipzig 1890, Vol. II, section 1, Leipzig, 1891.

relationships artificially introduced.[4] [74]

If, contrary to my knowledge (for a considerable part of the English and American literature is inaccessible to me, scattered in periodicals and in journals of learned societies), endeavors along these lines have emerged, then they certainly have not been generally noted and convincing. How otherwise are we to explain the singular artificialities to which *Boole* and his followers, up to today, have seen themselves forced, solely in order to make the transformation of judgments into class judgments possible in every case? Thus, for example, the judgment, "If God is just, then the evil are punished," must be thought of as replaced by this other judgment: The class of temporal points at which the judgment "God is just" is true, is subsumed under (i.e., is identical with or included in) the class of temporal points at which the judgment "The evil are punished" is true. Surely one would have spared himself such things as this, if one had understood that implications between properties and between judged states of affairs (*beurteilbaren Sachverhalten*) can, just as they are, be subjected to the calculational mode of treatment. And, further, how is it to be explained that for so many logicians (among whom, F. A. *Lange*) the existence of a logical calculus is regarded as the main argument for the claim that the essence of the simple judgment is exhausted by its extensional form? Still further, why is it that the terms "extensional logic" and "algebraic logic" are on all sides held to be synonyms? And, finally, there is the fact that in the first volume of his work – intended to be, in essentials, of encyclopedic completeness – *Schröder* gives not the least intimation of the existence of such attempts, although in the same volume the question about the possibility of a "logic of contents" is discussed most thoroughly.

Please excuse the copiousness with which I support my assertion of the (absolute, or nearly absolute) generality of the prejudice in question. This is required, as will immediately be clear, by compelling motives.

To break through this prejudice was the purpose of my essay, "The Deductive Calculus and the Logic of Contents," in volume XV

[4] It is indeed customary, in logico-algebraic investigations, to speak simply of concepts and judgments, not of classes and class relations; but still the latter is, again and again, right beneath the surface, as a more exact analysis shows. Cf., for example, my discussions, in this Journal for 1891 (p. 168 [p. [44–45]]), with reference to Jevons and Wundt.

(1891) of this Journal (pp. 168–189). By means of a few extremely [75] simple – and, I daresay, completely intelligible – explanations, I believe myself to have shown that, for judgments which do not to begin with refer to relations of classes, the detour through classes is a totally superfluous one; since any arbitrary form of judgment can be taken as the starting point for calculational methods of deduction. My procedure consisted simply of this: with reference to the best present form of the extensional calculus, which we owe to Schröder, to carry out a proof that the basic formulae of *that* calculus – from which all formulae and calculations derive in a purely analytical fashion – can from the start be given the interpretation of "content logic." Thereby it was shown, with mathematical rigor, that the entire calculus, purely formally considered – thus, disregarding its original interpretation in terms of classes – is valid for the domain delimited by the new interpretations, and that, therefore, it also could be grounded apriori in that domain. Thus, however many systems of such parallel interpretations as there may be, there are that many content calculi; of which, however, as is easily seen, a whole series stands in the relation of logical equivalence to each other – as well as to the extension calculus.[5]

I was, of course, very eager to see what reception my theoretical and critical expositions[6] would find from extensionalist logicians. I could not hope to convince them on all essential points. But, on the other hand, I could not doubt that the refutation of the old prejudice [76]

[5] In my demonstrations I did, indeed, neglect the domain of particular judgments, in that – misled by the mode of their formal statement in older investigators – I supposed that one could, for algorithmic purposes, replace particular judgments with certain universal judgments. When I noticed this obvious blunder later, I nevertheless thought it superfluous to add the needed reinterpretations of formulae for the domain of particular judgments in a separate publication. That such a reinterpretation can be carried out is settled, in fact, following obvious points made in my essay; and the *de facto* execution, conforming to the pattern given, would be a quite insignificant task. – And yet a second blunder occurs in my essay, of which I first became aware through Voigt's reference in the article "*Was ist Logik?*" (this Journal, Vol. XVI, 1892, p. 304). A calculus of "ideal contents," the possibility of which I had discussed with fleeting references, would not, as I maintained, possess the character of an "identity" calculus, but rather that of a "group calculus" (in E. Schröder's terminology) – which, however, is a matter of complete indifference: both in general, and also for the essential content and the results of my work.

[6] The latter in a review, thoroughly going into questions of principle, of Vol. I of Schröder's *Vorlesungen*, in the *Göttingische gelehrte Anzeigen*, 1891, No. 7, pp. 343–378.

which identifies the concept of extensional logic with that of algebraic logic, and the new foundation of the syllogistic algebra which that refutation brought forth, would have to find agreement, given the Evidence of my demonstrations. More quickly than I expected, the latter has occurred. A representative of the school contested – Mr. Andreas *Voigt*, in an article entitled "*Was ist Logik?*" (this Journal, volume XVI, 1892, pp. 289–332) – has abandoned the viewpoint of extensional logic, which he hitherto had accepted.[7] He has without further ado adopted my ideas, and has constructed an 'elemental logic' on the basis of them. And the critical statements of my review, cited above, also have not failed to impress Mr. *Voigt.* Where he fully grasps their substance, he agrees with them too.

But as happy as all this seems, it has its dark side. Painfully disturbed by my decisive rejection of the inappropriate pretensions raised by the "algebra of logic," Mr. Voigt clothes the adoption or approval of my expositions in a most paradoxical form. That is, in the form of blunt rejection! Now the manner in which Mr. Voigt goes about this is such that I find it necessary to provide a fair-minded reader of this Journal with materials with which to decide whether or not Mr. Voigt has always adhered to the duties of the conscientious literary man. Whoever reads Mr. Voigt's essay without the most careful comparison to my works, obtains throughout the impression that Mr. Voigt owes nothing to me – absolutely nothing at all: as if, rather, all I have stated and demonstrated were either falsehoods or well-known trivialities. Mr. Voigt, further, thinks it well, in his criticisms of me, to convert my most unambiguously clarified intentions, meanings, and theories into their opposite. And, in fact, he constantly ascribes to me the strict opposite of what I have expressly stated. Moreover, he takes pleasure in imputing to me "manners of combat" ("*Kampfesweisen*") of which I have never approved, much less practiced.

1. I said that Mr. Voigt has simply accepted my ideas and statements concerning the logical foundations of the calculus, shown above to be completely novel, and that upon them he bases his
presentation of "elemental logic" – in which we also have, in fact, [77]
nothing other than a direct grounding of the calculus for the judgment forms of content logic. Mr. Voigt measures out my service in this

[7] Cf. A. Voigt, *Die Auflösung von Urtheilssystemen*, Leipzig, 1890, subsections 1–3.

matter solely and only with the following remark (*loc. cit.*, p. 313): "We must acknowledge as Husserl's service that he has adopted and disseminated *Schröder*'s symbolism ... " But no less characteristic are other passages. E.g., on p. 310 we find the words: "... Now that *we* have given a positive statement of elemental logic, as it *may have dimly presented itself* to Husserl." Or on p. 312, and frequently, as Mr. *Voigt* continuously asserts the failures of my efforts (and of my interpretations of the symbols "0" and "1") – but without undertaking, even with *apparent* reasons, to *prove* that I fail. There is not a word to indicate that I have brought forth a totally new conceptualization for the foundations of algebraic logic, and that in this "elemental logic" of Mr. Voigt's we only have an essentially unmodified extension of that conceptualization of mine. If Mr. Voigt changes the order of the definitions and axioms, incidentally demonstrates one of my axioms (and they truly *are* axioms, not at all postulates), inserts here and there a calculation, or continues my Idea also over into the domain of problematical or particular judgments – then these are unessential matters, in comparison with the main point. Similarly with the analysis of A = (= B into two relations: $x \prec A$, $x \prec B$. This analysis was placed in his hands by my interpretation of A = (= B (*viz.*, if something is an *A*, it also is a *B*); and its applicability had not escaped me. However, its introduction was incompatible with my well-calculated purpose of strictly adhering to *Schröder*'s symbolism, definitions, and axioms.

But now the reader will object: How about those "failing" interpretations of the symbols "0" and "1," repeatedly urged by Mr. Voigt? Perhaps the essential difference [between our positions] lies in his emendation of them? Not in this either! For, we hear with astonishment, my supposedly *false* interpretations are partly identical, and partly equivalent – and thus, in any case, are algorithmically identical – with Mr. Voigt's *correct* interpretations! Since there can be doubt only about the "0" and the "1" of the "calculus of assertions" (Mr. Voigt's logic of whole propositions), and the objection of falsehood [78] is expressly referred precisely to them (313), the demonstration may also follow through for them alone. Mr. Voigt writes for "0" and "1," respectively, "F" and "T;" and he understands under "T" an unconditionally true proposition, and under "F" an unconditionally false one. Then A = (= T (which is formula 1 on his p. 313), and F = (= A (which is formula 1′ on p. 314), hold true, whatever *A* may be. The latter formula means: If a false proposition were true, then any arbitrary proposition would be true. For me, "1" signifies the proposition, "There is a valid proposition," and "0" signifies the proposition, "There is no valid proposition" (this Journal, volume XV, p. 355 of my "Addenda" – only here I write "proposition" ("*Satz*"), while there "judgment"). Now, as is easily shown, these equations hold true:

1 = T, and 0 = F;

for if in the above formula we replace *A* with *1* (in formula 1), and *A* with *0* (in formula 1′), there then results:

1 = (= T, and F = (= 0.

But, on the other hand, the relationships,

T = (= 1, and 0 = (= F,

also hold good. And the first of these in fact states that 1, i.e., the proposition, "There is a valid proposition," is true; whereas the latter states that 0, i.e., the proposition, "There is no valid proposition," is false. All of which is evidently correct, with regard to any axiom you may choose – and, moreover, requires no calculation. For since one can admit, for *T* and *F*, any arbitrary unconditionally true or false proposition, then one also can admit the ones which I marked (*loc. cit.*) by "1" and "0." And this no less holds true even of the interpretation which I at first put forth (p. 188 of my essay, *loc. cit.*), and which I indeed later withdrew: but not, as Mr. *Voigt* misinterprets me, because I found them false. They were only inept.[8] Now after this proof of the absolute equivalence of my "failings" with Mr. *Voigt*'s correct interpretations of "0" and "1," I ask the reader to check the relevant passages in Voigt's essay (310 and bottom 313).

Whereas Mr. Voigt's assertions of the above sort give the impression that he subscribes to the ideas which he has taken over from me, [79] he speaks in other passages as if, in those ideas, we have to do with pure trivialities or things known to everyone: from which he also

[8] I withdrew them because I was not satisfied with the derivation of the basic formulae (V and VI of my essay) related to them. But these, also, are not false.

quite certainly sets out. "If the one logic is provided," p. 319 tells us, "then the creation of the other is merely a matter of reinterpretation: of translation, as it were, of one language into the other. So the fact that the formalism of the one logic completely coincides with that of the other is not, as Husserl supposes, 'a most striking fact to start from.' Rather, it is *a fact which is settled from the outset on, and is easily confirmable.*" After the fact, as after the exhibition of any obvious truth, such things are easily said. Mr. *Voigt* also forgets to add that still *one year* before the appearance of my essay he had not even a suspicion of that "fact which is settled from the outset" – as one can tell from subsections 1–3 of his monograph of 1890, cited above.

In several passages Mr. *Voigt* frankly enjoys himself with my translations of *Schröder*'s formulae, and presents matters as if my work stood to *Schröder*'s in the mere relation of "translation" to "original." But he is completely silent about the fact that the idea of the possibility of just *such* a translation, and the actual carrying through with it, is something novel and of fundamental importance for the logic of the calculus.

2. I have stated above that Mr. Voigt attributes to me beliefs and assertions the exact opposite to what I have expressed. I will here give a few illustrations of this.

A. Mr. Voigt says (p. 296 of "*Was ist Logik?*") "His [Husserl's] critique is a *deterrent to the study of algebraic logic*, a warning about this trend in logic which *can teach philosophers nothing*, and has the highest interest and attraction for the inveterate mathematician." (Cf. also p. 290, p. 332, and frequently elsewhere.) – Mr. Voigt probably has in mind the following passage from a note on p. 169 of my essay: "I believe that the logical 'algebra' also, in spite of its so very limited practical utility, must not be underestimated; and that, for the sake of its positive results, it must be *of great interest to the logician.*" And [80] so on. (Cf. also my review, pp. 260 and 277–8.)

B. On p. 329 we read: "Husserl appears to hold it to be a peculiarity of extensional logic, that it does not presuppose the existence of concepts, whereas the logic of content has to do only with real concepts .... As proceeds from the overall manner in which his definitions are set up, Husserl holds the symbolic language of logic to be a mere translation from word-language. And one therefore can, without further ado, assign to the judgment A = (= B the same signification as is expressed in words with the judgment, 'The *A*'s are *B*'s,' which presupposes the reality of the concepts *A* and *B*." To what extent these sentences correspond to the truth one perceives from, among other things, the fact that, on p. 257 of my review in the *Göttingische gelehrte Anzeigen* (1891), I designated extensional logic as one distinct, special method belonging entirely within content logic.[9] It may also be seen from the fact that the explanation of the sign "A = (= B" which I *really* gave is supplied by the sentence: "Provided something is an *A*, it is a *B*," for which an equivalent and only an equivalent sentence may be substituted. (This Journal, p. 180 and p. 185, volume XV.) Among such equivalents are found, of course, no truly categorical propositions, i.e., those involving the existence of the subject (Cf. *Gött. gel. Anz.*, *loc. cit.*, p. 271); much less, then, the one foisted upon me above. And as to the further imputation, that I hold the symbolic language of logic to be a mere translation of word-language, it, again, is contradicted by everything I have written: among other things, by the "manner" in which my definitions are set up. For example, "By 'B · A' (or 'A + B,' or "0," "1," etc.) we understand the *object of the concept* which ...." (*loc. cit.*, pp. 187ff.)

3. How easily Mr. *Voigt* brings himself to ascribe to me a lack of conscience as a critic is shown by the following example: He – in order to characterize "the manner of combat (*Kampfesweise*) of this critic" (p. 321) – makes the capital objection to me (p. 322) that I have [81]
kept silent about one of Schröder's "most important" arguments for an extensional logic: *viz.*, "... That the logic of content is absolutely incapable of dealing with particular judgments." But if one follows up on the passage referred to in *Schröder* (I, p. 100), what is found? A *side comment*, set out of the text by square brackets. Without any discussion, be it ever so brief, it contains the claim – as enigmatic as it is brief – that "the logic of content, consequently, would be *incapable of forming* particular judgments, unless it be as tautological or 'vacuous' judgments." This comes together with a reference to

[9] I have frequently and strongly emphasized the view that in algebraic logic, which claims to be logic itself, we have merely a subordinate fragment of logic. But this wording – "That the algebra of logic ultimately can be conceived of as one subordinate part of logic in the widest sense, is something which Husserl too will concede" (see bottom p. 291 of "*Was ist Logik*?") – is an odd way of expressing agreement with this point.

"the explanations" in subsection 44 of the second volume, which at that time *had not yet appeared*!

4. In order to make ridiculous the critique which I had brought to bear upon *Schröder*'s *Algebra der Logik*, Mr. *Voigt* also ascribes to Schröder views opposed to those which he has actually advanced. Thus, for example (cf. *op. cit.*, pp. 307–310, and especially 310 and 331), he ascribes to him the view that algebraic logic adopts the "formalistic standpoint." This means, when exactly stated, that that logic, with no regard to a particular domain, selects its definitions and basic principles arbitrarily and purely in terms of the symbolism itself. Hence, the concepts or conceptual objects to which the calculus relates are not given at the outset, but rather are delimited purely and simply by the requirement of satisfying the formal demands of the symbolism. The opposite route consists in starting out from a given conceptual domain, and determining the (then absolutely *not* arbitrary) definitions and basic principles (then axioms) in conformity with the requirements of that domain. As to *Schröder*, now, he proceeds at the outset from "manifolds" ("*Mannigfaltigkeiten*"); and hence he chose the *latter* route. To be sure, he mingles into his expositions – what is permitted on the first route, but is logically inadmissible on the second – arbitrary, formal definitions. And for these, then, only subsequently an interpretation (not to be regarded as a justification!) is sought. In the case of "0," even the possibility of any interpretation is lacking. And yet the definition specifying an impossibility is supposed to justify the use of "0!" (Cf. *Gött. gel. Anz.*, pp. 267–272.) All of my objections *would*, of course, have been ridiculous, had Schröder constructed a purely formal calculus instead of a calculus of manifolds – which he certainly did not do.[10] [82]

[10] When, in his statements on pp. 131–132, Mr. Voigt imputes to me the denial of scientific importance to purely formal developments (i.e., the construction of algorithms independently of any concrete conceptual basis), as well as to formal definitions within such formalisms, this is one of the cases where his reporting is partly in direct contradiction, and partly in easily perceivable opposition, to my presentations. See my review, pp. 266ff. Moreover, Mr. Voigt shows – by his note on p. 331, and many other passages – that he is so far removed from any deeper understanding of the logic of symbolic methods that he has not yet even noticed its essential difficulties and problems. That is, in fact, what I have to charge the algebraic logicians with: – In an "algebra of logic" they see the salvation of logic, if not logic itself. But into the *logic of* this algebra, or of any algebra in general, they have so little insight that they offend against that logic as soon as it comes into issue. It certainly also is true that the logic of the algebra is to be derived through

5. Mr. *Voigt* repeatedly speaks of my "manner of combat" in a tone which presupposes that, in my review, I treated E. *Schröder* with insufficient respect. The reader of that review will, however, recall that the rejection of all of Schröder's principal assertions has not prevented me from emphasizing, in the most laudatory of manners, his great service to the calculus. Cf. my review, p. 260 and elsewhere.

But enough of such disagreeable matters. That Mr. Voigt should have failed to give adequate expression of his agreement with my views, where they were decisive for him, would not have moved me to reply. But I *was* so moved by apprehension that the less informed reader might – trusting in the reliability of his reports – attribute to me views and tendencies wholly removed from me. I gladly assume all responsibility for whatever I have written, but in nowise for that which I am *supposed* to have written, according to Mr. Voigt.

---

no species of algebra of logic. Cf. the objections of principle in my review, to which Mr. Voigt has, unfortunately, given too little attention: too little, even, merely correctly to report them, much less to be able to refute.

CONCERNING THE CALCULUS OF THE LOGIC OF CONTENTS: REJOINDER TO MR. HUSSERL'S ARTICLE[1]

by: Andreas Voigt

In his response, Mr. *Husserl* proceeds from the presupposition that the idea of rendering the logical calculus independent of relations between classes is a novel idea, first brought to realization by him. Only in one place (p. 112) does he concede that there could have been attempts at this, contrary to his knowledge, but that, in such a case, they certainly have not been attempts that are generally known and convincing. Accordingly, Mr. Husserl should at least have been somewhat more cautious with his reproach that I would have had to borrow this idea from him, and it just as little fits in with that concession for him to speak in his response (p. 114) of these "... entirely novel ideas and theses, *as demonstrated above*." But I was most of all dumbfounded by the claim (p. 117) that I had "not even a suspicion" – as allegedly could be seen from subsections 1–3 of my dissertation of 1890 – of the coincidence between the formalism of a logic of contents and that of the logic of classes.

It will suffice for me to put the main points of those sections in print here:

From subsection 1: "As in analysis arithmetic displays the science of the numbers, so logic is the science of *concepts*."

From subsection 2: "The most general property of concepts is their *relation to objects*. It consists in the natural interconnection between concepts and objects that is established through our cognizing the objects by means of the concepts ...."

[1] First appearing in *Vierteljahrsschrift für wissenschaftliche Philosophie*, Vol. 17, 1893, pp. 504–507. The article to which Voigt refers is the immediately preceding one in the present volume. Republished in *Husserliana XXII*, pp. 83–86. [D.W.]

"... The totality of those objects which are coordinated with a concept are said to be a *class*. Thus, to every concept there belongs [84] a class ...."

"Considered from the subjective point of view, a concept consists solely in the ability to cognize objects. In the *definition* the concept finds an expression that is universally valid and communicable ...."

"In order for logic to attain the greatest generality possible, it is not to be presupposed in all circumstances that concepts really do embrace objects and that classes contain objects, but rather we also admit to classes which contain no objects at all .... The relation of concepts to objects then certainly becomes a merely potential one; that is, one *would* cognize objects with the aid of the concept *if* there were such objects. This generalization is necessary because there commonly occur in practice concepts of which we do not know whether objects are coordinated with them or not. Indeed, any concept is of this type so long as it is given to us only by means of a definition. Whether and how many objects it embraces is something which we can know only from other sources, and which therefore must be specifically added on *in cases where that must be taken into account in logical inferences*."

"*In spite of this presuppositionlessness with respect to the objects coordinated with the concepts*, and thus given utter indetermination of the class belonging to the concept, a large number of logical operations and inferences is yet possible. The propositions validating them, which consequently hold true of all concepts without exception, we group together as general logic. We then obtain the more specific parts of logic by introducing the distinguishing characteristics of the classes into our logical analysis. It is precisely the consistent and prolific recourse to classes by which the newer algebraic logic has distinguished itself from the older mode of treatment in logic, and has so essentially heightened its applicability and significance."

I had purposely put this Introduction at the opening of my dissertation, devoted chiefly to the study of individual judgments, in order to define my position, according to which class logic *is not logic in general*, but rather is an extension of the older logic, which dealt in concepts alone. And I believe I made this clear enough, for all who *wished* to understand. Had not Mr. *Husserl* explicitly cited my dissertation, I would have believed that it too belonged with the "... attempts that have been published without his knowledge."

How, in spite of his acquaintance with it, he has been able to bring it forth as proof that I have borrowed from him the idea of a logic independent of class relationships is incomprehensible to me. [85]

But in order to remove all doubt, I add the following certified quotation from my original manuscript, in which precisely the relationship between content logic and class logic is discussed in detail. For the understanding of this quotation it should be noted that, in the terminology chosen by me at that time, "genus" was equivalent with a concept given through a general definition alone, in contrast to the class, and that I had originally divided my whole treatise into the two sections: "Logic of Genera" and "Logic of Classes." On the counsel of my doctoral committee I later expurgated all philosophically debatable discussions.

"The division of logic into the logic of genera and the logic of classes, carried out in this treatise, closely corresponds with the distinction between the logic of content and the logic of extension, as drawn by F. A. *Lange* in his *Logische Studien*[2]; but at the same time it refutes the view that these two are only different ways of considering one and the same thing, and therefore have an almost equal justification. The logic of extension corresponds better, according to *Lange*, with the modes of thought and the needs of the modern period, whereas the logic of content is better suited to the outlook of the ancients. Thus it would be almost a matter of taste which of the two modes of analysis one would choose."

"However, it is not mere modes of analysis, but rather quite different parts of logic, which are designated by means of the two phrases above. The logic of genera (of content) does not suffice to solve certain logical problems, and, on the other hand, it also is impossible to reduce the logic of genera to that of classes – since there are concepts which never can be conceived of as classes. There are, above all, nonreal concepts to which there belong no objects whatever, but which nevertheless can be determinately defined and brought under general laws as well as other concepts. Then there are also the concepts of future events, in the case of which one can never know whether none or infinitely many objects (cases) fall under them. Such concepts can be defined only as genera, through the statement of their general distinguishing marks; and there is therefore unconditionally neces-

[2] F. A. Lange, *Logische Studien. Ein Beitrag zur Neubegründung der formalen Logik und der Erkenntnisstheorie*, Iserlohn, 1877.

sary a logic of genera, which leaves wholly aside any possibility of conceiving of concepts as summations of individuals. Therefore, the principles of logic must also be established for genera."

"Expositors of algebraic logic commonly proceed from the view [86] that all concepts can be regarded as summations of individuals, i.e., as classes; and in consequence of this it is therefore frequently held that that way of looking at matters is essential for an algebraic logic, which it also can never get beyond. That this is not at all the case, that the logic of classes as well as the old logic can also be a logic of content, has, so far as I know, first been shown by Mr. *Frege* (*Begriffsschrift*, Halle a. S., 1879), and then, independently, by Mr. *Peirce* in 1880.[3] And even though in *Peirce* there still is a slight hiatus in the proof of certain principles, yet there is no doubt that those principles can provisionally be allowed to stand as axiomatically valid." Etc.[4]

---

[3] C. S. Peirce, "On the Algebra of Logic," *American Journal of Mathematics*, Vol. 3, 1880, pp. 15–57.

[4] I hereby certify, at the request of Dr. Andreas Voigt, that the portion printed above, from the words "The division of logic . . . " up to the words ". . . as axiomatically valid," is taken exactly as worded from a manuscript which he submitted in 1890 in order to attain the degree of doctor in the faculty of philosophy here. In the dissertation as printed, the foregoing is omitted because Professor *Schröder* (in Karlsruhe) and I, to whom the manuscript was given by the Faculty for judgment, had suggested to Mr. Voigt a partial reworking of his treatise.

Dr. J. Lüroth, Full Professor
University of Freiburg i. Br.

Already in this portion of my dissertation are mentioned the two authors who have set up a logical calculus independently of class relations, and of whom Mr. Husserl should at least have been able to know *Frege*, even if the major work by Peirce (in the *American Journal of Mathematics*, Vol. 3, 1880) were not accessible. *Frege* has created a calculus, which is unfortunately very awkward in its form, but in essence coincides with Schröder's and every other calculus. In general, it is quite certain apriori that *any* logical calculus, in whatever way it may be established, must necessarily coincide in essentials with the existing calculi.

HUSSERL'S REPLY TO THE FOREGOING "REJOINDER" BY MR. VOIGT[1]

Mr. *Voigt* is not satisfied with the corrections and denials that, unfortunately, I have been obliged to make in response to his attacks, which are quite inept both in form and in content. He attempts to reply. What I have here before me is already the *second* draft of a reply. The first draft, which went into the contested points in detail, was withdrawn by Mr. Voigt after my rejoinder had been delivered to him by the esteemed Editor. The new draft is silent on almost the whole series of points that I have found it necessary to press, and limits itself to contesting a single one: Namely, the *novelty* of my expositions in volume XV of this Journal. Mr. Voigt believes that he can maintain the following:

(1) That he himself already had knowledge "of the coincidence of the formalism of a content logic with the logic of classes" at the time of the writing of his dissertation in 1890. Or, in order to eliminate a manner of speaking of which I completely disapprove: He wishes to claim that at that time he knew the calculus of implications between [conceptual] objects, the calculus of judged states of affairs, and the like, to be formally identical with that class calculus propagated exclusively by extensionalist logicians.

(2) Others also, such as C. S. *Peirce* and *Frege*, are supposed to have possessed this same knowledge long before me. In general, all of the allegedly new material contained in my work, and which I claim to have been used by Mr. Voigt, is supposed to have been developed in the writings of these investigators.

Response to (1). The first assertion is supposed to be proven by

[1] First appearing in *Vierteljahrsschrift für wissenschaftliche Philosophie*, Vol. 17, 1893, pp. 508–511. Republished in *Husserliana XXII*, pp. 87–91.

bringing together some quotations drawn partly from the dissertation as printed, and partly from the unpublished draft. I can only be [88] grateful to Mr. *Voigt* for this summary. For anyone who even is only superficially familiar with these matters will immediately, and with astonishment, notice that these quotations say nothing – absolutely nothing at all – *to favor* Mr. Voigt, but very much *against* him.

Now first, as to the printed dissertation by Mr. Voigt, I do not (even after recently inspecting it again) find one sentence that would even remotely indicate the knowledge claimed by its author. Here is lacking any beginning point for discussion. Where nothing is expressed, there is nothing to refute. I must yet go much further, and say that in the writing mentioned (and hence also in the quotations from it) there is not even one sentence which an extensional logician of strictest observance could not have stated from time to time. And from this it *a fortiori* follows that Mr. Voigt was greatly deceived concerning his earlier knowledge and its expression in the dissertation.

Now which quotations are supposed to make us think otherwise? Perhaps the assertion that logic, "... in order to attain the greatest possible generality," even admits classes "... which contain no objects"? Or is it, what comes to the same, the "... presuppositionlessness with regard to the objects coordinate to the concept," which is stressed so much by Mr. Voigt? Thus, the "... complete indetermination of the class belonging to the concept"? In that Mr. Voigt brings this forward as an argument, he begs us to accept an obvious fallacy. Is a line of logical analysis which operates with such "indeterminate" classes to be excluded, *in virtue of that fact* alone, from class logic? On the basis of such grounds one could infer, quite analogously, that general arithmetic is not a science of numbers, because it does not presuppose existence of numbers corresponding to its symbols. Even the most extreme of extensionalist logicians can (and even must) maintain the possibility and necessity of operating with empty classes, and with classes for which it is not apriori determined whether they have members or are empty. – Moreover, does it perchance support Mr. Voigt's claim when he proceeded to say: "We obtain specific parts of logic when we draw the intrinsic character of the classes into the logical analysis"? This all the more makes the subsequent sentences witness against him. I suggest that they be re-read. An extensionalist [89] logician could express himself in exactly this same manner.

But, "in order to remove all doubt," Mr. *Voigt* brings certified quotations from the dissertation manuscript which he submitted to the philosophical Faculty in Freiburg. The word-for-word agreement of these quotations with the original is something which that scholarly body might well certify for him, but not their actual agreement with the thesis which was here to be proven. These quotations also simply prove nothing at all. They at most show that Mr. Voigt permits the viewpoints of content and extension logic to run obscurely together, and that he commits also here the error censured above, of denying the right of the class calculus to operate with "indeterminate" or even empty classes. Certainly Messrs. *Schröder* and *Lüroth* have with good reason objected at this point, and required that the entire passage be stricken. Given the manner in which Mr. Voigt divides between considerations of content logic and those of extension logic, it signifies little when he explains the latter to be unessential to an algebraic logic. In any case, we find in all of these sentences not even the slightest intimation of the knowledge claimed at the outset, which they were supposed to prove present.

Here I must add that I so little presented as novel the idea of introducing into the theory of deduction algorithms other than those grounded on class logic, that I even began my essay with criticism of attempts to do just that. But I certainly did suppose that pure and consistently executed attempts of this kind were lacking. And above all, I held and still do hold to be novel the essential thought that identically *the same formulae* which serve the extension calculus can *also* directly serve as formulae in a calculus of implications between properties or between [conceptual] objects, or a calculus of implications between judged states of affairs, and so forth. *This* thought cuts the root of a logic that is extensional as a matter of principle, and thus extensionalist logic loses all of its support. Henceforth one can only speak of a calculational formalism or calculus, which, depending upon the circumstances, serves in the calculational execution of class inferences, or inferences with regard to implications between properties, or with regard to implications between judged states of affairs, and so forth – without it being permissible to regard any one of these "interpretations" as *the* fundamental one to which all others [90]
must in principle be reduced. I can only rejoice that my explanation of this simple state of affairs, which nonetheless remained unnoticed, has succeeded so well that it could seem to be something self-evident to Mr. Voigt, or something which he himself had long known. Yet

Mr. Voigt also thinks – and this belongs in our

Response to (2) – That the supposedly new is already to be found in the writings of Messrs. *Frege* and *Peirce.* I find totally absent here any quotations to prove this; and, beyond that, I notice that Mr. *Voigt*, in the draft of his dissertation, yet did not go so far as to ascribe to these investigators the knowledge here in question. The writings of the profound Jena scholar, whom I esteem very highly, lie before me. I also find in him not one allusion to the ideas in dispute. It moreover is significant that Mr. Frege cannot decisively enough defend himself against confusion of efforts toward *a calculus*, as in a *Boole*, a *Schröder*, or others, with his own efforts toward a *conceptual notation* or *Begriffsschrift* – which intends above all to be a *characteristica universalis.*[2] Frege's conceptual notation is, in general, no calculus, in the genuine sense of the word; even if among its formulae (or better: its 'types') there may be some which parallel certain formulae of the calculus. That the formal parallelism of the conceptual notation with the calculus is complete, is something denied, in the same manner, by Mr. Schröder (in his review in the *Zeitschr. f. Math. u. Phys.*, Vol XXV, 1880, pp. 81–94) and by Mr. *Frege* himself (*op. cit.*, pp. 8ff.). The former even stresses the almost total absence of analogy between calculation and "concepts." I think that this is decisive. No better justified is the invocation of the writings of Mr. *Peirce*. That already in 1880 he possessed the idea of algorithms of "content logic," like others before him, I do not doubt. But is *that* the point in question?

Mr. *Voigt* concludes by saying: "In general, it is quite certain from the outset that any logical calculus, however it may be grounded, must necessarily coincide in essentials with existing calculi." But this is a claim of the same worth as, perhaps, a statement that any geometrical calculus must coincide in essentials with every other one, e.g., the [91] theory of extension (*Ausdehnungslehre*), the quaternion calculus, and analytical geometry. Coincidence in the results is certain, but not in the concepts and methods. And this alone is at issue.

[2] Cf. the treatise, "Über den Zweck der Begriffsschrift," in *Sitzungsberichte der Jenaischen Gesellschaft für Medicin und Naturwissenschaft für das Jahr 1882*, Jena 1883, pp. 1–10.

PSYCHOLOGICAL STUDIES IN THE ELEMENTS OF LOGIC[1]

I. Concerning the Distinction between Abstract and Concrete

Section 1: Independent and Dependent Contents

Any consciousness is, in its totality, a unity within which every element is combined with every other element. However, in the manner of the unification, as well as in the degree of its stability and in its immediacy or mediacy, significant differences are to be found. It is to one such difference that the division of contents here to be considered has reference: – that division, namely, into the *independent*[2] ("separable," "capable of being represented by themselves") and the *dependent* ("inseparable," "incapable of being represented by themselves"). There are certain complex contents which become objects of discrete acts of noticing with relatively greater ease. They press themselves upon our attention as, so to speak, 'natural' unities; and they exhibit *vis-à-vis* the observable contents that show up in union with them a characteristic independence, such as is not found in the members of other types of connections. This is true of the intuitional content of perceptible *things*, in comparison to the less complex, and yet just as intuitable contents which we call the intrinsic qualities (color, form, etc.) of such things.

But it is not easy to state precisely wherein this independence consists. One could perhaps invoke the following as a characterization of it: Objective things reciprocally act upon each other, and there-

[1] First appearing in *Philosophische Monatshefte*, Vol. 30, 1894, pp. 159–191. Republished in *Husserliana XXII*, pp. 92–123.

[2] Cf. Carl Stumpf, *Über den psychologischen Ursprung der Raumvorstellung*, Leipzig, 1873, p. 109.

by they condition relations of dependence between the perceptual appearances, and hence, also, between the intuitive contents of those [93] appearances. We also note, on many occasions, that modification of one appearance (or intuition) will bring with it a corresponding modification in the appearance (or intuition) coordinated with it, while it nonetheless does not lie in the *nature* of the intuited contents themselves that they stand in such relationships of dependence. We find nothing in them which makes the necessity of the connection Evident. The connection in such a case does not itself appertain to the contents as such, but only to the objective things involved. In fact, we can through phantasy wholly suppress the causal connections without finding any change in the intuitive content of the perception.

Now the contents which are to be considered here exhibit just such a separability from contents conjoined with them, but in a much wider range of cases – not merely with regard to the causally conjoined, but also with regard to *all* simultaneously noticeable contents with which these contents are linked, regardless of how. (We disregard, of course, the proper *parts* of the contents, with reference to which one probably would not speak of a "linkage" anyway.) The contents here in question seem to remain what they are, whether the contents linked to them are represented as disappearing or as changing in whatever other way. In phantasy I can retain the head of a horse by itself, and cause the remaining parts of the horse, as well as the whole of its intuitive context, to disappear[3] or change at will – the head all the while remaining intuitively unmodified. More exactly stated, either there are no intuitive modifications of the head, or they are too small to be noticed, or – and this may best correspond to the psychical state of affairs – such modifications are unattended to, as insignificant. In any case, the possible correlated modifications of contents, linked in this manner, which do *not* escape us when appropriate attention is given to what is actually lived through, appear as *merely factual*. There is nothing in the contents involved which would require a functional dependence of their changes as an *Evident necessity*.

[3] Of course such a total disappearance never literally occurs. Rather, if we suppress the context through a total transformation of it in phantasy, then over and over again *some* context or other certainly can slip into place – as occurs, for example, with visual objects. However, the context which thus enters the scene can always be so formed that, while exclusive attention is directed upon the focal content, it is completely retractable. It is simply unnoticed, as if it were not there at all.

Here we have set out from those phenomena which are the first to present themselves to reflection – that is, from *things* appearing to us. In the intuitive content of such things, we have shown the characteristic marks of independence. But we also find these marks [94] in objects which are not things. For example, consider the phenomena of sounds, thought of quite free from any relationship to objects – e.g., as occurs with tones or intuitively closed configurations of tones. In similar conditions this independence also shows up with odors and tastes. But less frequently, and with more difficulty, do intuitions involving the sense of touch admit of the type of separation here at issue.

Dependent contents behave quite otherwise on the points under consideration. The intensity of a tone is not something indifferent to the quality; not something which is, so to speak, external to it. And likewise conversely. We cannot retain the intensity itself as what it is, while arbitrarily changing or wholly suppressing the quality. A change in the quality conditions of necessity a change in the character of the intensity. And this is so whether or not a Moment annexed to both, which we call the pitch, remains unchanged. The intensity simply is nothing off to itself, but rather is something only with and affixed to the quality. If we represent the quality to ourselves as wholly suppressed, then the intensity is also suppressed. And this is not a mere fact, but rather is an Evident necessity. Likewise in the converse case. A change of intensity ineluctably signifies a certain modification of quality, even if the species of the quality remains the same. Here we simply are not dealing with a summation, in which the one term can be varied while the other remains identically (instead of, as here, merely specifically) the same. The two contents interpenetrate. They exist within each other, not outside of each other. Again, the cessation of the intensity conditions a total annihilation of the quality; and this not merely as a matter of fact, but as an Evident necessity. Also a continuous approximation of the intensity toward the null point is sensed by us as a gradual diminishment of the qualitative impression – which, moreover, makes especially clear that variation of quality *with* intensity which was asserted above.[4]

A further example of dependent contents is extension in relation to quality, or shape in relation to both. An immense plentitude of

[4] Cf. Stumpf, *ibid.*, 112ff. For our further discussions see, in general, subsection 5 of that work, to which I here, as elsewhere, owe so much.

yet further examples is offered by the quasi-qualitative Moments of intuition (*von Ehrenfels' Gestalt* qualities, for which *Meinong* has [95] suggested the name "founded contents"[5]), by the quasi-quantitative Moment in the domain of tones, and so on.

These reflections may yield the following definition of concepts: We call *dependent* any content with regard to which we have the Evidence that change or suppression of at least one among the contents given with (but not included in) it *must* yield a change or suppression of that content itself. Any content of which this is not true is *independent*. In this latter case, the thought of the content itself remaining intact while all simultaneous contents are suppressed contains no absurdity. With contents of the former sort we have, one can also say, the Evidence that they, being such as they are, are conceivable only as parts of more inclusive wholes; whereas with contents of the second type this Evidence is lacking.

There is yet another way in which one can attempt to state characterizing marks for the distinction between independent and dependent contents. We are unable to notice, much less to attend to, a figure or color by itself, without having first perceived the object which has the color or figure. Occasionally a 'striking' color or form seems, indeed, to force itself immediately upon us. Yet when we bring the process involved in such cases clearly to mind, it seems likely that here also it is the whole object which first 'strikes' us, but that it does so just in virtue of that peculiar feature upon which, then, our interest (without hesitation) exclusively comes to focus. Thus does the noticing of one content sometimes serve as the foundation for the noticing of another one intimately connected with it. This could lead to the following definition: A content which is noticeable without necessarily presupposing a prior noticing of some other content is *independent*. In the opposite case it is *dependent*. But any such definition would be at a disadvantage, when compared with the one given above, insofar as the Evidence for the state of affairs upon which it is grounded is lacking. But it is still of some interest. [96]

Now we still have to take into consideration those contents which

[5] Cf. the essay "Ueber Gestaltqualitäten," by Ch. von Ehrenfels, in *Vierteljahrsschrift für wissenschaftliche Philosophie*, Vol. XIV, 1890, pp. 249–292. See also my *Philosophie der Arithmetik*, Vol. I, Halle a. S., 1891, chap. 12, as well as Meinong, "Zur Psychologie der Komplexionen und Relationen," in *Zeitschrift für Psychologie und Physiologie der Sinnesorgane*, Vol. II, 1891, p. 253.

occupy a sort of middle position, inasmuch as by their intimate linkage with others they evoke the semblance of dependence, although lacking it in the strict sense of the term. It is the association of ideas which often creates extremely strong unions of this sort. This is the case with that unification of qualities belonging to the different senses which is found in the representation of an intuited *thing*. Those with sight from birth believe that they see the thing with all of its qualities; and each quality appears to them in this visual experience as equally inseparable from the rest. This illusion is even powerfully resistant to reflection, especially in cases where a non-visual quality is strongly associated with a characteristic of that visual image which constitutes the intuitional kernel of the experience in question. A (specifically determinate) tactual roughness commonly coalesces with certain characteristic *visual* unevennesses. And since these latter really are seen and are dependent, there is evoked the illusion that the quality of roughness – now as a total associative complex – is seen and is dependent. Should we wish to comprehend these contents under the term "dependent," then we would have to drop the mark of Evidently necessary functional dependence between the linked contents, and consequently adjoin to the *strict* concept of dependence a *weak* one, or to the narrower one a *wider* one.

Up to now we have regarded independence as an absolute, as a characteristic independence from *all* adjoined contents. Dependence was regarded as the contradictory opposite: as the corresponding dependence upon at least *one* adjoined content. But if our theoretical interest is restricted to some one unification of contents, that still leaves us free (provided that those contents do possess, in relation to each other, the marks of independence) to speak of *relative independence*, in contrast to the *absolute* independence defined up to now. The physical parts of a blackboard are absolutely independent. But the parts of the *surface* of the board are independent only in the relative sense: namely in relation to the other parts of the surface. It is noteworthy that, when one content is inseparable from another, the latter nonetheless can be separable from the former. A line which, taken with certain others, founds a configuration is an independent content. But the configuration itself is dependent relative to the line. For if, for example, the quality of the line is suppressed in thought, then the configuration is also modified, although the line's quality [97]
makes no positive contribution to the configuration. Again, a repre-

sented content is independent relative to an associated judgment, but the judgment is dependent in relation to the represented.

Section 2: Abstract and Concrete Contents

Let us now begin with any unified content, and consider its parts in the light of these distinctions between types of independence just discussed. By disjunct parts we understand, as is common, parts which have no identical parts, including themselves, in common. Then two cases are possible: Either a part considered is independent in relation to all disjunct parts coordinate with it in the whole, or it is not. In the former case we speak of *fragments* (*Stücken*), in the latter of *abstract parts* (*abstractis*), of the given whole. If a whole is decomposed into a number of disjunct fragments, then we say that it is *divided up* (*zerstückt*). Whether the whole of whose division we here speak is or is not itself only an abstract part of another whole is, thus far, left indeterminate. In actual fact, both sorts of cases do occur. Abstract parts can be "divided up," while fragments or absolutely independent contents can themselves be abstractively partitioned. If an abstract whole admits of being divided up in such a way that the resultant fragments are *abstracta* of the same lowest species as that which is determined by the undivided whole, then we call the whole a *physical whole*, and its fragments *physical parts*. Here, as examples, we have division of spatial expanse into spatial expanses, of tone formation into tone formations, etc.

Parts can be either *immediate* or *mediate* parts of the whole considered. Mediate parts are parts of parts. Only exceptionally are parts of parts *also* immediate parts of the whole in question. This is found, for example, with physical parts of physical parts, when both are referred to the same lowest species. By contrast, in a visual intuition extension and color are immediate *abstracta*, while the particular shade and the brightness of the color, as well as the volume and shape of the extension, are mediate *abstracta*. As we brought out above, the noticing of the mediate *abstractum* presupposes noticing the immediate *abstractum*. The complex content which has an *abstractum* as its immediate part is called, in relation to that part, its proper *direct concretum*. (The term "indirect *concretum*" is here to [98]
be used accordingly, needing no further explication.) A *concretum* which is not itself abstract we call an *absolute concretum*. Since

each absolutely independent content possesses abstract parts, each one can also be regarded and designated as an absolute *concretum*. The two concepts are, therefore, coextensive. It thus also becomes clear why the independent content which provides the foundation for processes of abstraction that either are coordinate or are hierarchically arranged is also said to be the (absolute) *concretum* relative to all of the *abstracta* thus obtained. – We call "relative" all of those *concreta* which are not absolute; but where the word "*concretum*" is used without qualification, the absolute *concretum* is, as a rule, intended.

Beyond the authentic uses of the terms "abstract" and "concrete" discussed up to this point, their inauthentic uses still must be brought under review. The relationship of concrete to abstract grounded in intuition is extended to cases where there is no question of any intuition. Any *concept* is commonly said to be abstract, while every *object* of a concept is said to be concrete. That no adequate intuitions (if any at all) correspond to most concepts in living thought, and that many conceptual contents – in virtue of the weakness of our psychic powers, or in virtue of obvious inconsistencies – can never be intuitively perceived as *abstracta* in the appropriate *concreta*, does not trouble us. And the basis of this extension of terminology is clear. Were the conceptual contents and objects genuinely given, then the relevant relationships of concrete to abstract *would* also be there to be intuitively grasped. And since we as a rule have no explicit awareness of the incompleteness involved where something is presented by means of a Representative (*repräsentativen Vorstellen*), that there in fact *is* here an *extension* of terminology is not even noticed.

A partial content which, through its relation to a whole, serves or is fit to serve in the definition of the whole, is called its "mark" [*Merkmal*]. There are abstract as well as concrete marks; and there are even concrete marks in the sense of an absolute concretum (e.g., a well-known landmark of a city).

Section 3: Critical Observations

The distinction between concrete and abstract representations is sometimes fixed by saying that the former can be represented by themselves (*für sich vorgestellt*), while the latter can only be noticed by themselves (*für sich bemerkt*). From *Berkeley*'s polemic against

Locke one can derive something like the following as *Berkeley's* own [99] view: Concrete representations are representations of *things* or parts of things. They are characterized by the fact that they can be represented cut free from the things or parts with which they are united. For example, a head or nose can be represented by itself, separated from the man to which it was originally represented as belonging. Such objects or parts also could *exist* in separation from whatever is united with them. Abstract representations (though, of course, not those which *Berkeley* so zealously contests under that name) are representations of properties. They cannot be represented cut free from their bearers, and they also cannot exist without those bearers. They can only be *attended to* by themselves. In their *Logik Höfler* and *Meinong* further add that this incapacity for separate existence is a matter of Evidence.

Now it seems to me that these and similar definitions suffer from the handicap of operating with such vague and misleading expressions as "to represent by itself" and "to represent separately," without sharply delimiting their signification and clearly explicating the criteria of their application. So far as these definitions are on the right track, they obviously can mean nothing but what we sought to make precise above. Also, I must object to *Berkeley*'s recourse to existing things as an unpsychological move. I have moreover with good reason avoided speaking of abstract and concrete *representations*. I think that it is a good principle to avoid such an equivocal word as "representation" so far as possible. It has been objected that "abstract" and "concrete" are terms which could find application *only* to representations, but never to the things represented.[6] To *things*, certainly not. By why not to the contents? The things are not the actual contents of our representations, but objective unities, and thus are presumed, merely intended, contents.

The basis of the distinction between the abstract and the concrete is often, if not predominantly, sought in the mode of the act of representing. More precisely, it is sought in the mode of the psychic activity through which the one and the other comes to distinct notice. It is on the side of the "abstract representation" that the positive criterion [100 of the distinction is found. The contents of such representations are, namely, said not merely to be represented, but beyond this to

[6] A. Meinong, "Phantasie-Vorstellung und Phantasie," *Zeitschrift für Philosophie und philosophische Kritik*, Vol. 95, 1889, p. 202.

be set off from the *concretum* through the peculiar psychic activity of 'abstracting.' But for my part I have endeavored in vain to find the slightest distinction between consciousness of the abstract and consciousness of the concrete. To abstract, it is said, is to attend to something by itself. But in order to segregate an absolute *concretum* from its encompassing background and make it into an object of special engagement is there not required precisely such an 'abstracting' as described? If I consider this box, I attend to it in particular; and only in virtue of the fact that I do so does it come to special consciousness for me and become an object of intuition. Is *it* therefore to be called an *abstractum*? No one will agree to this. It may indeed be replied that the box could stand out from its background, and accordingly be noticed, without our attending *exclusively* to it, while its shape or color is able to stand forth *only* by means of an exclusive attention, such as we ourselves emphasized above. But the division of attention into exclusive and non-exclusive, thus invoked, cannot ground the distinction with which we are concerned, since the *concretum* can also be attended to by itself, and, wherever it is an object of psychic engagement, must be so attended to. Hence we conclude that it is not permissible to define a concrete representation as one which is not abstracted. And it is just as little permissible to define it as a representation in which – or, rather, in whose content – no abstraction at all has been carried out. For the 'abstraction' which I do in fact bring to bear upon an independent content, e.g., upon the box here intuited, far from allowing it to appear as abstract, gives to it, rather, the stamp of the concrete.[7]

[7] Compare, on the other hand, Meinong's "Hume-Studien I," in *Sitzungsberichte der philosophisch-historischen Klasse der Kaiserlichen Akademie der Wissenschaften in Wien*, Vol. 87, Vienna 1877, p. 200. Likewise see the *Logik* of Meinong and Höfler, p. 23; and, with reference to the whole of the present treatise, see also pp. 21f. of that same valuable work.

II. Concerning Intuitions and Repräsentationen[8]

Section 1: Introductory Analyses of Examples

In order to attain to an adequate delimitation of the concepts of intuition and Representation it will be well to pursue at the outset reflections which are suited to illustrate the range of the actual use of the word "intuition," and the appreciable nuances of its signification.

Intuition (*Anschauung*) in the original sense is just *seeing*, and thus the perceiving of visible objects. Thus people are heard to say: "I want to look at (*anschauen*) that house, that painting, etc." An extension of this use, founded in the dominating role of visual representations in all of external perception, is the identification of intuition with *external* perception. And to this extension there is added, further, the identification of intuition and perception in general. Thus, for example, *Wundt* uses the two terms as synonyms.[9] In order to serve certain psychological interests, *Meinong* wishes to define as an intuition the perceptual representation underlying an actual or possible perceptual judgment.[10] In this way hallucinations also – not just those which actually deceive, but also those which have been found out – would be included under the concept of intuition.

If we wish to adhere to the concept of intuition thus specified, we then are reminded by the psychologists not to understand "perception" in this popular sense. They tell us that not everything is actually perceived which we in non-reflective experience take to be perceived. This is clearly seen in the case of certain judgments of comparison – those, namely, which 'represent' equality or inequality in their objects [*Materie*], while there is lacking a corresponding basis in the actu- [102]
ally present content of the perceptual representation involved. We think that we *see* the equality of the side areas of a perceived cube, although the areas *really* seen are unequal. The perceptual judgment does not elevate the factual content of the perceptual representation to the level of objective consciousness. Rather, it finds in that content

[8] Cf. "Essay A" below, this volume, for related discussions. "*Repräsentation*" and its cognates will be rendered by "Representation" and its cognates, with capital "R," except where the German is used. With a lower case "r," "representation", etc., translates "*Vorstellung*", etc. [D.W.]

[9] *Grundzüge der physiologische Psychologie*, 4th revised edition, Vol. II, Leipzig, 1893, p. 1.

[10] "Phantasie-Vorstellung und Phantasie," *loc. cit.*.

– more precisely, in certain of its Moments – its occasioning cause. Now, however, we have the inconvenience of denying its name to the "perceptual representation" as it naturally presents itself, for what we in fact have in such cases is a holding-to-be-so [*Fürwahrnehmen*] of what is represented (even if only inauthentically) in the perceptual 'representation.' And in the normal case it is even an unmediated holding-to-be-so. (The mediation involved in the procurement of an object for the judgment to be about, the transition from sign to the designated – even in that case where this is a process fully carried out – still does not turn the judgment into a mediated *judgment*.) In the light of all this one feels pressed to distinguish the intuitive content of the perceptual 'representation' from its intended, complete content.[11]

The idea which grounds this distinction is clear. The perceptual representation, one says, passed as an intuition for the ordinary consciousness because in such a representation that consciousness presumed to possess as immanent object what it, rather, merely intended, and what it thus 'represented' in a wholly different sense. I have an intuition of our theater whenever I see it; for in that case I believe myself to have an actual representation of that which the phrase "our theater" means at any time – and of that which, if I do not *see* it, can only be signified, defined, or represented in surrogate *via* a conceptual representation or an image. The ordinary consciousness thinks that it grasps the objective thing itself – that unified complexity, as what it is and is intended as – in *one* glance, in one simple act of intuition. But we know that this is mere illusion. Only a small part of that which we here *presume* to intuit is really intuited. Only a few aspects of [103]
the factual content are present in this act of intuition just as they are intended by the thing-representation which that content mediates, or just as they really coexist in the 'thing itself.' The complete content of the thing-representation only comes to intuitive presence through a continuous flow of contents, in connection with which there are certain *psychic acts*. These acts accompany sequences of obtrusive

[11] Frequently one does hear the misleading terminology which would distinguish the "genuinely perceived" from the "merely represented" that is added in thought. Here there easily blends in a distinction which must not be confused with the one validly drawn in the text above: namely, the distinction between that in the perceptual representation which is conditioned by peripheral stimuli, and that in it which is conditioned by central stimuli. Otherwise expressed, it is the distinction between that in such a representation which originates from the excitation of innate dispositions and that which originates from excitation of the acquired ones.

partial intuitions, identify the ones reciprocally referring to each other, and – running their course within a *single* continuous act – spell out the respective objective unity. By thus calling into question the status of most perceptual representations as full and authentic intuitions we are only following the living intention of the term "intuition," which also validates itself in the case of those false presumptions mentioned above, and which perhaps is to be elucidated as follows: A "representation" in that inauthentic sense where we have a mere deputization by parts, images, signs, and the like, or a mere definition by characteristic marks, is *not* an intuition. By such means the represented is in truth not *set here before us* at all. Rather, an intuition is a "setting before" (*Vorstellung*) in a more authentic sense, where the object is actually put before us in such a manner that it is itself the substrate of the psychical activity. Accordingly, it also is clear that ordinary perception, although it is no intuition of *the thing*, yet may also be characterized as an intuition from another standpoint; that is, from another direction of interest. If a special interest, e.g., a psychological interest, is turned upon the content present at a given moment, as upon the appearance of one side of a thing, just as it is, then we have an intuition *in relation to that momentarily given content.*

Now we are far from having exhausted the range of the actual use of our term "intuition." It is not merely perceptual representations which are called intuitions, but also *phantasy representations*, in the widest sense of that phrase. There is nothing out of the ordinary when one says that he still has an intuition of the swordsman in the Borghese Gallery, instead of saying that he has a vivid memory representation thereof. And of artistic intuition, of the mythic intuitions of a people, and the like, we hear often enough. However, many a psychologist will take exception here also, and on grounds similar to those cited above. He will object that an intuition of the swordsman is something we have, in our example, only in an inauthentic sense. We have a [104]
representation which, in virtue of a similarity of content that perhaps is very far reaching, acts as a surrogate for the intuition that is not present and is merely intended, but a representation which is not the intuition itself. Thus our representation has, once again, an intention that reaches out beyond the immanent content of the act. But only if the fulfilment of the representation is to some extent realized, only if intended content becomes immanent content, would

we be fully justified in speaking of an intuition.[12] And we likewise would be justified in case an intention directed upon *this* content given *hic et nunc* found its satisfaction in that very content itself, just as it is: – in the case, accordingly, where a special interest (e.g., the psychological) were related to the immanent content of the phantasy representation itself. Of course in the case of phantasy representations of external things there comes into consideration, besides the inauthenticity just mentioned, yet that other inauthenticity which we became acquainted with in the corresponding perceptual representation. For also in phantasy a *thing* is, as a rule, replaced by inadequate Representants (*Repräsentanten*) – by a partial and more or less fragmentary "aspect."

Many even more striking examples of the opposition between the intuitional and the non-intuitional are supplied to us by the broad domain of *conceptual* acts of representation. The correct appraisal of these examples does not require preliminary critical reflections. We often are capable of gaining an intuition of that which we represent conceptually. But also we often cannot. In this latter case the lack is sometimes grounded in *de facto* inability, and sometimes in an Evident impossibility. All conceptual representations which involve Evident inconsistencies – such as the well-known round square, wooden iron, and so on – are necessarily non-intuitional. They are not, because of that, meaningless. They are representations exactly in the sense of any other Representations. They have a wholly determinate and well-understood intention, but one which is directed upon something impossible.

A doubt may press forward at this point. The first intention of a conceptual representation frequently is directed upon a corresponding phantasy representation; and when this latter is given, we generally [105] say that we *then* have the intuition intended. If it is pointed out that phantasy representations have themselves the character of those inauthentic representations which in this treatise we are calling "Representations," that will not make us hesitate in holding this manner of speaking to be correct. And we will shortly (section 5 below) find occasion to obviate this doubt. Here we only mention that we did

[12] Perhaps there lies in what has just been discussed the motives which have led many investigators (such as Meinong, *op. cit.*) to prefer, in reference to representations of the imagination, to speak of "intuitive *representations*" instead of simply of "intuitions." This also agrees better with the prevalent manner of speaking.

not deny intuitive character to the phantasy representation in every sense or every case; and so one might already perceive that, if the representations here in question are called "intuitions," they stand under the peculiar conditions which govern the correct application of that term.

A justified use of the term "intuition" is possible, not merely with reference to concrete contents (of which we have mainly spoken up to now), but rather with reference to abstract contents also. We say that one born blind has no intuition of red. To him red is "that which those who see call red." He has only this inauthentic representation of red. He cannot represent red itself as immanent object, as the abstract element in a *concretum*, as does the one who has sight and who, insofar as he can and does represent red, has the intuition of red.

Many investigators will not concede this extension of the concept of intuition to representations of abstract contents. They object that the expression "intuition of red" refers directly and authentically not to the *abstractum*, but rather to the thing which is red, the red object. But here we do not find any grounds which would justify the preference of the concrete over the abstract. To that which characterizes the red object and the *abstractum* in relation to each other there correspond the concepts of the absolutely concrete and of the abstract, whereas to that which characterizes them outside of their reciprocal relationship there correspond the concepts of independence and dependence. But neither pair of concepts contains what is intended by intuitionality and non-intuitionality. Only the first pair seems to do so, and that, indeed, because the concept of the abstract representation, as will proceed from the reflections to follow in section 5 below, is equivocal. It is not merely representations of dependent partial contents that are called abstract representations. These contents are *abstracta* in the sense made precise above (I, section 2), which involves objective unity in the object and, with regard to the [106] act of representing, authenticity. But we also speak of abstract representations with regard to those unifications of partial contents, A, B, C ..., which are made possible merely by "relational thinking." These contents (if they are not themselves represented through mere words) are collected from separate concrete entities and formed together into utterly inauthentic representations. We express such representations in language as follows: "Something which possesses the properties

A, B, C ... ."; or "An A which simultaneously is B and C"; and so forth. Abstract representations in this second sense certainly are paradigms of the non-intuitional. And they obviously are regarded by us as non-intuitional solely in virtue of the fact that they merely intimate their objects, instead of genuinely placing them before us. But this is not true of abstract representations in the first sense, with which these latter certainly must not be indiscriminately grouped.

Also, in the case of abstract Moments it does sometimes happen that a *supposedly* intuited object is substituted for a genuinely intuited one. We find this, for example, with geometrical intuition. The geometrical figures and relations are not intuitable at all, if those are in the right who shy away from attributing ideal properties which the perception of spatial entities cannot detect to the corresponding phantasms. End points of idealizing, and hence conceptual, processes are *eo ipso* non-intuitional. The figures and relations actually before us "present us with" ("*stellen uns ... vor*") the geometrical figures and relations truly intended, and do so in virtue of certain analogies – infinitely coarse analogies, when measured by the Ideal of the geometrical definition. The intuited figures and relations in fact often act as surrogates for the geometrical ones in the process of living geometrical thought. Whoever gets clear on this fact will immediately reject talk of intuition of geometrical *abstracta*, on the basis that it is merely figurative language. Of the *drawn* figure, considered *per se*, there is, of course, intuition; but there is no *intuition* of the geometrical figure – with which the drawn figure is not identical, and which it merely Represents.

Let us now take, as our last example, representing in the sense of designating (*Bezeichnens*). Although not the rule, still one does occasionally hear it said that a sign represents what it designates. The mathematician sometimes says: "The sign '*a*' represents an arbitrary number, the root of this equation," instead of saying that the sign designates it. One will hardly be inclined to call *this* representing an intuiting, and obviously for grounds analogous to those discussed in the cases above. According to *Kant*, though, the symbolism of the [107]
arithmetician also is a construction upon intuition, like the diagram of the Geometer. He calls the former "symbolical" (*charakteristische*) construction, and the latter "ostensible." In truth, that which is designated is not constructed in either case. In the latter case there at first glance does at least exist the illusion of a construction, in virtue

of a certain analogy between sensible and geometrical figures. The construction would be actual if the intuited figures were geometrical in the strict sense – and thus were the genuine Representants of the geometrical concepts, in the sense of being their definitions. For this surely is the intent of all constructing: to take what is grasped only inauthentically, and indeed is grasped only *via* conceptual determinations, and to present it as an intuition; thus, to "intuitionalize" the concept. But to regard the inscribing of arithmetical symbolism as an act of construction is wholly inadmissible. The sign and what it designates are here totally different kinds of contents, and are only united by association. The sign thus does not render intuitive that which is thought of, but rather only *refers* to it. Moreover, in the case at hand, that of arithmetic, what is designated is almost always something which *cannot* be made intuitable at all.

Section 2: Provisional Delimitation of the Concepts

This series of reflections yields a division of representations into those which are intuitions and those which are not. Certain psychical experiences, in general called "representations"[13] [*Vorstellungen*], have the peculiar character of not including their objects in themselves as immanent contents (and thus, as present *within* consciousness). Rather, in a certain manner which still must be more precisely characterized, they *merely intend* their objects. Perhaps for the present it will suffice to state the following definition, which obviously covers the matter and deliberately states more than would be required: The phrase "merely intended" here signifies precisely that a content is a content not given *in* consciousness, but one aimed at, minded, or referred to with understanding, by means of some contents which are given in consciousness. These latter contents are used, with understanding, as Representants of the former; and, indeed, they are so used without the intervention of conceptual knowledge of the relationship which obtains between the representation and the intended [108] object. Such representations (*Vorstellungen*) we will call "Representations" (*Repräsentationen*). In contrast to them stand other psychic processes (*Erlebnisse*), likewise called "representations" in the language of many psychologists. But these processes do not merely

[13] Cf. Appendix IV below, on the classification of Representations. [D.W.]

intend their "objects." Rather, they *really include* those objects *within themselves* as their immanent contents. Representations in this sense we call "*intuitions.*"

But we still must remove another ambiguity here – one which infects our explanation as well as the common usage of the term "intuition." On the one hand, by an intuition can be meant the psychical process in which the ultimate objective of a Representation is attained. In such a process, an immanent object of the act simultaneously appears to us as that which is intended by the act. But on the other hand "intuition" can also be used to mean simply that we are "representingly" ["*vorstellend*"] turned to an object – not, as so often is the case, by a mere intention, but rather primarily as an immanent object, and without any regard to whether or not we are so in conjunction with a back-reference to a corresponding Representation, that back-reference being consciously known or otherwise indicated in (or simultaneously with) the intuitive act. A correlation of the concepts *intuition* and *Representation* obviously obtains in the first of these two interpretations of "intuition," and also with the second. Given the second, each Representation, whether direct or indirect, points to an intuition that corresponds to it but is not actual. By contrast, it is not true that every intuition refers back to a determinate Representation appertaining to it. Nonetheless, it is the clear intention of the term "intuition" to set a representation of the group covered by it into relation with some determinately or indeterminately conceived Representations, as when we say, for example, that *X* is intuited, and not merely represented, or that a given Representation passes over into its corresponding intuition, and so on. In order to stay in agreement with what, if I am not mistaken, is the prevailing usage, we will have to define the word "intuition" in conformity with the *second* concept. In fact, hardly anyone will object to the statement that the child has intuition of the very first sense contents to which it attends. In this case Representations, since they are not [109] even formed, also cannot come into play. Moreover, it may serve us as a welcome confirmation, that our definitions are in essential agreement with certain ones offered by *Kant* – thus, when he contrasts intuitions, as non-mediated representations, with conceptual and imagical representations, which are mediated.

As to the first of the above two concepts of intuition, with which we shall be much concerned in the following investigation, we can easily delimit it by means of a paraphrase. If a Representation goes over into its correlative phenomenon, e.g., into an intuition immediately intended by it, then the immediate psychical experience of the fact that the intuited *is also* the intended shall be designated as consciousness of the *fulfilled intention*. Of the intuition involved in such a case we say that it is borne upon a consciousness of *fulfilled* intention. Of the Representation we say, more simply, that it has found its *fulfilment*.[14] This latter term will be used by us in general to designate the non-mediated or mediated correlate of a Representation. It is commonly the case that Representations are only mediately directed upon intuitions, but are immediately directed upon other Representations. Thus, for example, only the concept of a product of n factors of a is directly Represented by the arithmetical concept a^n. If the former concept itself comes in turn into Representative representation, then the consciousness of fulfilled intention likewise sets in with regard to a^n. But by means of the new phenomenon the intention is extended onward; and so already with this simple case we have a whole series of Representations mediating between the original one and the full intuition (denied us in this case). With reference to such cases we will call the correlative phenomenon most nearly adjoined to the Representation its *proximate fulfilment*. The *ultimate* fulfilment of any Representation is *the* intuition proper to it. It is *pure* intuition – a term which expresses the fact that a content bears no Representative function whatsoever. In contrast to this, we speak of [110] an intuition which is *impure*, or is *deputizing*, where a Representing content, in virtue of the identity or similarity of its content with what is Represented, temporarily serves us as an equivalent, provisional replacement for the latter. In such a case, as is still to be discussed in section 5, we are then turned to the Representing content itself in precisely the manner characteristic of pure intuition. An impure intuition is said to be *incomplete* when the immanent content of the Representation consists of only a part of that which it Represents. The terms "immediate" and "mediate," as applied to the Representation, are clear without further ado in the light of the above example of a^n.

---

[14] Cf. Appendices V and VI below. [D.W.]

Section 3: The Immanent Content of the Intuition at One Moment, and as Continuous through Time

The provisional definition of the concepts *intuition* and *Representation* which we have attained to in the above still leaves much to be desired. In the first place, to speak of "immanent content" in relation to intuition may give offense: So the content of the intuition is supposed to be immanent in it? That surely can mean nothing other than that it is content of the intuiting consciousness. Do we not thereby come into an unbearable contradiction with the general, ordinary use of the term "intuition"? Have we not, with the goal of following its natural meaning as closely as possible, stretched it to the point of non-usability? We do with good sense say of a melody which we hear that we have an intuition of it. Yet the melody is not immanent content. As the melody is playing, the immanent contents of the sense of hearing which turn up differ from moment to moment; but in no one of these moments do we grasp the melody itself. Similarly, we commonly say that we have an intuition of visible *things*. We have conceded the impropriety in such talk in cases where we must be satisfied with a one-sided view of the object. But are we then also barred from speaking of intuition with full validity when, given normal conditions of perception, we observe the thing from all sides, viewing all of its visible parts and properties? But the thing also does not become an immanent object in this case. We intuit from moment to moment different aspects of the thing, but in none do we intuit the thing itself. And also – and here is a further misgiving – how are [111] we supposed to intuit the thing itself, this objective reality, in which according to the view of the common man all of the aspects, all of the parts and properties to be intuited, coexist?

Referring for the moment to the last example, the difficulties raised can be resolved in the following manner. First, we must distinguish between the *phenomenal* thing and the *transcendent* thing. To the concept of an objective unity of parts and properties, articulated in such and such a manner and coexisting independently of our consciousness, there of course corresponds no intuition, for that would be a contradiction. But *this* concept also has no currency in the domain of common thought. It is the product of reflection concerning the thing representations of the common life. Such reflection is something which remains foreign to most men, as to all animals, throughout their lives. To natural thought, that very sequence of reciprocally cohering contents (accompanied and encompassed by certain psychical acts), which we live through under normal conditions of perception in the "observation of the thing from all viewpoints," just *is* the thing itself. And it alone is what is meant whenever a thing – a house, a tree, or the like – is spoken of. If the sequence of contents occurs, then the thing-Representations oriented toward it find their *ultimate* fulfilment. There remains *nothing* more to be intended. We can also express our view in this way: Talk of an intuition always involves a reference to some sort of Representation. The question of whether intuition of *things* is possible therefore leads us back to the question about the *meaning* of the correlative Representations. If we take "thing" in the sense of a transcendental unity, then *eo ipso* we rule out talk of intuition. But if we take the word in the sense of the Representations of natural consciousness – as, for example, that sense adheres to the words "house," "tree," and the like – then the ultimate fulfilment of those Representations lies in a continuous course of intuited contents which is encompassed by one undivided act, enduring continuously the same throughout the successional variety of the content, and which thus is immanent in that act.

However, we still have to settle the associated doubt as to whether one can say of such a course of contents that it is intuited, since [112
course of contents and *immanent object* indeed appear to be incompatible concepts. Here, I think, it all depends upon how we agree to use our terms. Why should we hold ourselves to understanding by "immanent content" of an act merely the immanent content of that act in any *one* moment of its existence? We stipulate that this phrase also is to be understood to cover the immanent *course* of content, or the immanent continuum of content through the totality of its continuous existence. Then – depending upon the direction of interest and without any inconvenience – now the one thing, and now the other can be meant. And one will, accordingly, speak now of continuous, now of momentary, intuitions. In any case, the limitation of the concept of intuition to momentary acts would too much impair its common and useful employment. In a similar manner we also have, of course, an intuition of a melody, and in general of any phenomenal temporal unity (in which temporal relations themselves belong to the immanent content).

Section 4: The Immanent Content of the Intuition as the Content of an Encompassing Act

A number of more serious difficulties lie in another direction. The intuited is the immanent object of a certain "representing" act, while the Represented is the merely intended object of such an act. But what *kinds* of acts are involved here? Is it the intent of our definition that in both cases essentially the same psychic function, called "representing," is carried out, so that the distinction here in question – since it cannot be grounded in intentional content, that being non-existent – would lie in the respective immanent contents of the two acts, or else in conjoined thoughts and the like?

Let us temporarily leave the Representation to one side and ask, first, whether or not intuition is a psychical function that enters into relationship with *all* contents of consciousness, as is customarily maintained of "representing." Does there, thus, correspond to each content of consciousness an act which intuits it? If, for example, we have an intuition of a figure, is the color inseparably bound up with the figure then a content of the *same* intuition? It is certain that we do simultaneously have *some* consciousness of the color. Likewise for the background, so far as it is present, to howsoever indeterminate [113] a degree; and again for especially obtrusive objects that force us to notice them simultaneously. Should we perhaps say that all of these are also intuited, but not in one and the same act which intuits the figure? Is there perhaps *one* intuitional act in relation to the *totality* of the contents of consciousness at any given time, within which, then, a special act of intuition is directed upon the partial content?

Disregarding all other difficulties, it seems to me that these interpretations far remove one from the true meaning of the term "intuition." Only what is separately and specifically noticed (*für sich bemerkt*) can be designated as intuited. If I look at the knife lying here, then only the knife is intuited, and the unfocused background, which for its part does not advance to any sort of objectivity, is not simultaneously intuited. Only with what is picked out with precision am I especially engaged; only to it do I have regard; only it do I "look upon" ("*schaue Ich an*"). But this noticing of something separately and specifically, this peculiar disengaging[15] of a part from the

[15] It does not matter here whether it is a question of a simple disengagement, in the literal sense of the word, or rather one of parallel changes of content, in the part

encompassing whole, obviously still does not suffice. Often several partial contents are isolated off to themselves, and stand out from a common background, and yet we have regard only to one of them, "looking at" only one. This knife along with the inkwell, penwiper, and pencil standing by it form a cohering group of objects which are separately and specifically noticed.

Now it can happen that I, turned to this group as a whole, intuit it as a whole; in which case attention may be parceled out over the members of the whole more or less equally, with now this one, and now that one, variably favored. But it can also be that an exclusive attention to a definite member occurs, giving to it a peculiarly exclusive emphasis, thus marking it as *the* intuited one, while the whole indeed continues to be noticed. In attending to a melody many sorts of noises – unrelated human voices, the sounds of wagons rolling outside, and the like – may incidentally claim our notice, and yet without getting any sort of special attention, perhaps not even a moment's worth. We have heard the noise, but not intuited it.

Intuition is, thus, a peculiar type of engagement with, or a characteristic manner of being turned toward, a content that is separately and specifically noticed. It is a delimited and delimiting act, which cannot be said of the consciousness of the background or of other contents which, although separately and specifically noticed, are excluded from the content delimited by the intuitive act. This cannot be said, at least, if we are to be supported solely by the declarations of inner experience (*Erfahrung*), and not by psychological constructions. Contents of the latter kind just mentioned occasionally come to consciousness as incidental to encompassing acts, in that such acts have, when more carefully examined, besides their encompassed content – or *main content*, as we may say – yet an excluded or *incidental* content.[16] Only the former is intended when one speaks simply of "the content" or of the "immanent content."

or in the whole.

[16] William James would at this point probably speak of a "fringe," a "psychical overtone," or a "halo." Cf. his *Principles of Psychology*, Vol. I, New York, 1880, chap. IX, p. 258 *et passim*.

Section 5: With Regard to Its Immanent Content, the Repräsentation Is Not an Intuition, but Rather Is a New Mode of Consciousness

Now we will draw the correlative function, the Representation, into our reflections, and consider the question raised above: Whether perhaps (regardless of what the meaning of the word "representation" may be in our definition) it can be concluded that in both of the functions in question the same mode of consciousness is present. It is obvious that, with respect to its content (or perhaps its main content), not every psychic act is to be designated an intuition. No one will call an act of aesthetic disapprobation of a painting an intuition of that painting. The disapprobation is based upon an intuition, but is not itself an intuition. Likewise, it will not do at all to call a Representation an intuition, in view of its immanent content. This would in fact be to deny any essential distinction between intuitions and Representations. If the term "Representation" meant nothing more or nothing other than that, in addition to the contents which a certain act intuits, there are some others which it does not intuit, then certainly this would be a distinction drawn from quite incidental [115] points of view. In truth the situation is very different from that. A sharp descriptive distinction concerning the mode of consciousness (of "having in mind," of psychical involvement) separates Representations from intuitions. This is best seen when one contrasts the purest cases possible: namely, those where – so far as this is possible in a developed consciousness – the same content is at one time the basis of a pure intuition, but at another time the basis of a Representation. Consider, for example, the case where certain figures or arabesques first effect us purely aesthetically, and then there suddenly flashes the understanding that they could be symbols or word signs. Let us fix upon the latter, new state of affairs before it comes to explicit logical consciousness through thoughts and words – which often can happen only after some span of time. Thus, not even a single-termed judgment such as "A sign!" or "A symbol!" has intervened. Or take the case where one attentively listens to a word, one wholly foreign to him, as a mere sound-complex, without even so much as suspecting that it is a word; and again consider the case where one, in the middle of a conversation, consciously hears a word with whose meaning he is familiar, but hears it as utterly bereft of its significational content. Yet again, what constitutes the distinction between simply taking a *concretum* X just as it itself is, and grasping it as Representant for

any arbitrary X? Moreover, we also can very well use less pure cases (and even the previous cases are not wholly pure), as when we compare attending to a melody which we are presently singing with the consciousness of an understood word or sentence (where the accompaniment of significational contents is, again, excluded). Or when we compare the consideration of an arabesque with the very different type of consideration of an arithmetical sign in its living function. In these examples, which could be increased arbitrarily, what we have is obviously no mere distinction of content. It is also clear that an *increase* of content could justify only talk of more inclusive contents, but not of intuited contents in contrast to Representing ones. Perhaps someone will think that the difference lies in the accompanying "thoughts." But without stressing the fact that these thoughts would have to be assigned to the unnoticed "background of consciousness," which in this respect is no less longsuffering and efficacious than the "unconscious," I simply point out that there is no thought with- [116]
out a Representation. One must not say, referring to the example of the sign understood in spite of the absence of its significational contents, that to the sign belongs the significational representation: that a *sign* requires it, and that in virtue of accompanying thoughts we have an awareness, be it ever so obscure or fleeting, of that significational representation; for since we cannot actually represent the lacking contents as such, this means that they *must* be somehow Represented in those "accompanying thoughts." But then the difficulty is only postponed. What constitutes, and wherein do we notice, the Representing function of the "thoughts" in question? Mediation through new thoughts, new Representations, would be required; and so on *in infinitum*. Many insist that the significational contents in question are not unconscious, but are only unnoticed; and they are perhaps inclined also in other ways to find the distinction between the intuition and the Representation in some sort of unnoticed contents. But should it be true that unnoticed contents, of whatever sort, are present in the one case and absent in the other, *they* still do not constitute the distinctions between the contents *noticed*. At most they condition those distinctions. Obviously the same type of objection applies to those who focus exclusively upon the distinction between innate and acquired psychic dispositions, or upon the distinctions between the conditions which evoke such dispositions, or even upon the distinctions between the corresponding physiological processes.

If anywhere at all, it is here that the witness of inner experience is clear, and yields the conviction that there exists in the two cases in question a different mode of receiving the content into consciousness, a different manner of psychical engagement with or in the content.

It may be that in the examples considered above the Representation rests upon an immediately preceding intuition. But the Representation itself is not therefore an intuition, just as little as an aesthetic feeling is an intuition because it rests upon an intuition. In that instant in which the arabesque becomes a sign, and thus obtains the character of a Representing content, the psychic state of affairs has wholly changed. Indeed we do see the sign. But our focus is not upon it, and we do not intuit it. Similarly, heard words are also only intuited when heard without the thoughts – as when, perhaps, we are diverted from understanding by a fascination with a peculiar tone of the voice or an oddness in pronunciation. But if words have their natural effect, [117] then they are *not* intuited, though they *are* heard. At most one might ask whether or not in the instant of attending to the emerging word-content an intuition takes place – which then immediately, however, passes over into a Representation.

To this we add one comment. If in *these* examples it remains dubious whether or not intuitions and Representations show up in the most intimate of interconnections – the one introducing or taking the place of the other, where they are not completely interwoven – this certainly is not in any doubt with other examples, such as those discussed at the outset of our deliberations. When we 'intuit' a *thing* there rather is commonly present, when the matter is more closely considered, an intermingling of the two sorts of activities in question. What is *genuinely* seen serves as an instrument of the understanding in relation to the whole thing, which is merely Represented. But on the other hand we are, in such cases, also intuitively focused upon that aspect genuinely seen, so that here the two functions are combined. Again, phantasy representations have a Representative function in relation to the perceptual intuitions corresponding to them. But this does not prevent our also being able to focus intuitively upon such representations themselves, which often enough we in fact do. And when the geometer operates with intuition, whether with reference to a perceived or a phantasized figure, then once again both functions come into play. And likewise with all conceptual thinking which takes its start from intuitions or accompanies intuitions or reverts to them – as, for example, in "A flower like the (intuited) one before me." There quite certainly are many ways in which a Representation can attach itself to, or ground itself upon, an intuition. And this does not in the least contradict the above thesis, that the immanent content of a Representation is, as such, not intuited.

In this state of affairs there is also grounded (as already touched upon in section 1 above) the alternating application of the different terms on the basis of one and the same phenomenon. Of course the interweaving of the two functions does not mean that *every* designation of a complex phenomenon as an intuition or a Representation is vague or outright incorrect. If the whole has a decidedly Representative character – and, indeed, in such a way that the interwoven intuitions are *mere* presuppositions or bases of the Representation (mere means of Representation) – one can only designate it as a Representation [118]. Likewise, in the opposed case one can only designate the whole as an intuition; as when, in the process of psychological analysis, we say that we have an intuition of a Representation. But the term does become equivocal in application to cases where the two functions each preserve an equal right to autonomy, the one thus not being absorbed as a foundation for the other. In such cases it can even happen that the founding function determines the character of the whole in such a way that, as we have indicated earlier, an undoubted Representation forces us to designate it as an intuition. We spoke, in this regard, of impure or deputizing intuitions. In such cases the intuitional character prevails, in virtue of the fact that the Representing contents, because of their identity or similarity with what is Represented, are turned into foundations of a primarily intuitive focus (and thus are not mere means for a Representation) – in the course of which the Representative character totally fades into the background, and may even momentarily be shifted to the status of a mere "stimulated" disposition.[17]

Section 6: Some Supplementary Remarks

The foregoing considerations, however much they may require filling out, have surely evoked or strengthened the conviction that in the

[17] Cf. concerning this concept the instructive discussions of B. Erdmann, "Zur Theorie der Apperception," *Vierteljahrsschrift für wissenschaftliche Philosophie*, Vol. X, 1886, p. 343.

intuition and the Representation we have to do with characteristically different types of conscious states. But the solution to the problem of *what kinds* of conscious states they are is not yet attained. We still do not even know whether in these two states simple or composite phenomena are given – and thus also, how they are related to additional known phenomena, provided, indeed, that they are not wholly reducible to such phenomena. That the designation of them both as "representations" signifies nothing for the goals of descriptive analysis has become clear. However, our definitions surely will not be misunderstood as claiming to make, by using the term "representation," even the smallest of contributions to the analysis of the phenomena themselves. That was obviated by the entire manner [119]
in which we have used "representation." All along the term was unavoidable and, in spite of its equivocations, useful for the purpose of giving a provisional and intelligible delimitation of the groups of phenomena to be investigated – and thus of the concepts involved, from the viewpoint of their extension. But it could not evoke, and was not intended to evoke, a clearly determined concept, descriptively well-founded. For that purpose the term is all too equivocal, and the concepts drawn together in it have been all too little investigated. Moreover, it should be pointed out that when intuitions and Representations are both called "representations," there is as a rule also committed an equivocation which surely is for the most part overlooked. If these two are specifically different modes of being conscious, then at least one of them is not a representation – whichever specifically determinate phenomenon we do call by that name. But if by "representation" is understood a generic character, in which both participate as species – and I do not argue against this – then the term is indeed not equivocally used by the one who so understands it; but it nonetheless leads to an equivocation, and to one which seems to me to be very dangerous. It easily misleads one into understanding the genus term as a species term, and then on to holding the two ways of being conscious to be specifically, and not only generically, the same. And so it leads one to seek the distinction between the two phenomena in the mere content. But errors of disastrous consequence have been caused in epistemology and psychology by overlooking and falsely interpreting the distinction between intuitional and Representing consciousness, and therefore not sufficiently doing justice to that distinction. For this reason I have here avoided recourse to terms

composed from the term "representation" (*Vorstellung*). In fact, the distinction itself, although never sufficiently investigated, has long been known. Already with the Scholastics it had not entirely escaped notice. It is to Representations that terms such as "inadequate," "inauthentic," or "indirect," refer; and to intuitions the corresponding opposite terms. However, with this there was in view, in part, merely the distinction between intuitive and non-intuitive *concepts* (or that between concepts which could, and those which could not, be rendered intuitive). But also a part of what was intended was that [120] distinction (taken much too broadly) which can be designated, when carried over to contents, as the opposition between the authentic and the merely intentional contents of consciousness.

Section 7: An Excursus Concerning the Psychological and the Logical Significance of the Two Functions, and the Importance of Research Concerning Them

For the whole of psychology, and in particular for the psychology of knowledge and logic, the investigation of the two psychical functions under discussion here appears to me to be of fundamental significance. Especially is this true of the Representation, which is of extreme interest. Such investigation is so important for psychology because there is a series of the very most important psychical acts (I mention here only desire and will) which essentially require representations, in the sense of Representations, as bases or presuppositions, while other acts make no such requirement. It is so important for logic because concepts, judgments, and all other sorts of logical activities, themselves belong within the group of functions just singled out as requiring a Representative base. I believe that I can safely say, in particular, that no theory of judgment can do justice to the facts unless it is supported by a deeper study of the descriptive and genetic relationships between intuitions and Representations.

I just now described the Representation as a function of extreme interest. In various respects it actually is an occasion for astonishment. In and for itself it is certainly a fact most worthy of consideration that a psychical act can reach out beyond its own immanent content to another content which is not really met with [*bewusst*] at all. And yet it seems we do have consciousness of these latter contents in a certain manner. For – and this again is a fact most wor-

thy of our attention – while we are engaged with the Representing contents, we take ourselves to be employed about the Represented objects themselves. In the flow of conceptual thinking, optical and acoustical sequences of words, in most cases, do the Representing alone or almost alone.[18] The significational contents enter into consciousness either not at all or only in a quite rudimentary fashion. Occasionally wholly different contents, which stand in a distant rela- [121]
tion to them, will act as surrogates for them: as when at the mention of London merely the shape of Great Britain indistinctly comes to mind. But also, there where conceptual thinking retires, where in the interactions of everyday life linguistic Representations are directed upon the intuitive aspects of the external environment, matters are not much improved. What a sparse show of deputizing intuitions turns up for the unprejudiced observer in these cases! Here and there such intuitions are wholly lacking; and where they occur they are as a rule faded, mutilated, often ungraspably fleeting, and inadequate even in the typical aspects of the intuitions intended. But all of this does not trouble us. It seems as if the meant objects themselves underlay the sequence of words. The state of affairs in question here is so far removed from the customary direction of attention that it has been left out of consideration even by the best of psychologists and logicians. And yet here lie great, unsolved puzzles. We stand close to the most obscure parts of the theory of knowledge. By that I do not have in mind the psychological elucidation of this illusion and of the entire situation in general, although that has to be regarded as fundamental for all subsequent work. Rather, I refer to the possibility of knowledge in general. Scientific knowledge – the kind of knowledge which will first come to mind here – is totally based upon the possibility of our being able to abandon ourselves completely to thought that is merely symbolic or is otherwise most removed from intuition, or of our being able purposively to prefer such thinking, with certain precautions, over thought more fully adequated to intuition. But how, then, is rational insight possible in science? And how with such a style of thought does one even come to mere empirically correct results? Mathematics stands as the model of an exact science. But the controversies of its practitioners – dragging out through the centuries, and not even settled today – over the meaning of its ele-

[18] Cf. James, *loc. cit.*, chap. ix, especially p. 270 and elsewhere; and B. Erdmann, *Logik*, Vol. I, "Logische Elementarlehre," Halle a. S., 1892, pp. 299f.

mentary concepts and the grounds of the validity of its methods stand in a curious contradiction with the supposed universal Evidence of its procedures. The great mathematician *Cauchy* advocated totally different theories of imaginary numbers in various periods of his life.[19] Which of these theories really guided the Evident procedures that he developed? It cannot be objected that it is one thing to have [122] insight, and another insightfully to reflect upon that insight. With the theories in question it is no matter of mere "reflections," but of certain (indeed, very complicated) chains of inferences that must of necessity mediate scientific thought if it really is to be insightful. The theories change, but the practice remains the same. Its Evidence – and of this there can be no doubt at all – is mere delusion. But how is it, then, that it at least comes to results which, to an amazing degree, stand up to subsequent "tests," or which agree with experience, and thereby condition a corresponding measure of confidence in practice? In the light of the fact that analogous doubts concern all of science (and ordinary thought as well), should we revert to *Hume*'s skepticism as our basis, and then extend it out beyond its great author to take in even mathematics and all apriori science? Vainly we turn, for the resolution of such doubts, to the old logic or the new. They totally leave us in the lurch. Logic, the "theory of science" ("*Wissenschaftslehre*"), must concede, if it will be honest, that all science is a mystery to it. This is where we stand today, in spite of the efforts applied to logic in recent decades by a number of truly outstanding minds. Now I certainly will not deny that one can considerably advance logical understanding of the soundness of symbolic thought (and above all, of course, mathematical thought) without a more penetrating insight into the essence of those elementary processes of intuition and the Representation which everywhere make that thought possible.[20] But without such insight one surely

[19] Cf. Cauchy, *Cours d'analyse*, Part I, "Analyse algébrique," 1821, as well as his *Exercises d'analyse et des physiques mathématiques*, in *Oeuvres complètes*, First Series, Part IV, Paris 1884.

[20] Investigations which I have pursued for some years concerning algorithmic methods are of this tendency. However, only brief selections from these investigations, and that with reference to more specific problems, have been published. Cf. my *Philosophie der Arithmetik*, Vol. I (1891), chaps. xii and xiii. For example, see especially pp. 268–286, and 299ff. The cases there analyzed can, however, serve as typical only for certain classes of symbolic processes. For a *general* theory of the algorithm, quite different moves must be worked out.

cannot obtain a full and truly satisfactory understanding of symbolic thought or of any logical process.

That this insight into our two psychic functions is also of fundamental significance for the broader field of psychology has already been indicated above. I still would like to lay stress here upon one important problem toward the solution of which no serious step can be made without the insight in question. I have in mind the much [123]
discussed problem of the origin of the representation of space. If I am right, the masterly studies[21] that have been devoted to it have merely made it clear that certain theories, which are at first very promising, are definitely false. But to a proof of the correctness of a theory we have not come, and we *could* not come; and this because none of the presumed theories go beyond generalities, because none focus sharply enough upon the problem which really must be solved – the problem, namely, of explaining the actual psychological and logical state of affairs in all of its *determinate peculiarity*. The concept of space is the concept of a determinately articulated manifold. But what kind of manifold is it? By means of what logical properties is it definable? Space as it exists psychologically – from which the concept of space originates (In what manner?) – is a complex of phenomena and dispositions characterized by definite properties. (Which ones?) But where are the descriptions? Who has undertaken to fix upon these determinate psychological and logical states of affairs in order subsequently to explain them in their special character as *these* determinate states of affairs?[22] Now just as soon as one makes a serious effort in this direction, he runs upon the intuitional and Representative functions that work together throughout this area, and without the understanding of which one gropes helplessly in the

[21] From the psychological approach one of course thinks here of the admirable works by *Lotze* and *Stumpf*, as well as of the ingenious and stimulating writings of Th. *Lipps*.

[22] It is truly astonishing what receives the name of a 'theory' in this area. If some three series of graduated feelings or sensations (whose participation in the perception of space is made out to be probable) are brought together, and one is inclined to accept their fusion, then the problem is solved! But what is then explained? As good as nothing at all. Three series of sensations – three coordinate dimensions – this is the nexus. As if the concept of a threefold, graduated, homogeneous manifold were the same as the concept of a threefold Euclidean manifold (which is provably false)! And as if, even if that identity held, the question about the origination of the representation of space were already settled by that identity, instead of just being touched upon.

dark.

These indications will, in any case, suffice to make clear the importance of the problems here raised. I do not delude myself that I am capable of overcoming those most weighty difficulties barring us from their solution. Perhaps, however, my subsequent reflections may serve, at least on some few points, to clarify, or even merely to stimulate others.

REPORT ON GERMAN WRITINGS IN LOGIC FROM THE YEAR 1894[1]

The logical literature for this year is uncommonly rich in valuable publications. The much-contested domain of the theory of the elements of logic is the one most abundantly represented; and within this domain, as is to be expected, there is preferred that class of questions which either fall within the area of the "theory of judgment" or stand in close relation thereto. We follow the natural order by beginning with presentations of logic as a whole. Then there follow works on the theory of the elements of logic, in systematic order determined by their themes. We conclude with works in the theory of methodology. Two of the more important publications, namely, W. *Schuppe*'s *Grundriss der Erkenntnistheorie und Logik*, and K. *Twardowski*'s *Zur Lehre vom Inhalt und Gegenstand der Vorstellungen*, have received detailed consideration in P. *Natorp*'s yearly report on the theory of knowledge.

W. *Wundt*, *Logik: Eine Untersuchung der Principien der Erkenntniss und der Methoden wissenschaftlicher Forschung*, 2nd edition, revised, Stuttgart 1893, Vol. I, *Erkenntnisslehre*, (xiv and 651 pp.); Volume II, *Methodenlehre*, First Part, *Allgemeine Methodenlehre. Logik der Mathematik und der Naturwissenschaften*, 2nd edition, revised, Stuttgart 1894, (xii and 590 pp.)

In discussing the new edition of this work I perhaps can limit myself to an indication of the substantial modifications found in comparison with the first edition. In innumerable reviews, critiques, essays and texts, the characteristic features of this logic have been discussed from all quarters, and its importance suitably appreciated.

[1] First appearing in *Archiv für systematische Philosophie*, 3, 1897, pp. 216–244. Republished in *Husserliana XXII*, 1979, pp. 124–151.

A new review could therefore count on favorable interest only if in the new edition the total character of the work, its foundations or its structural plan, appeared to be essentially modified. But this is not so. The renowned author presents us with what is, on the whole, the old work carefully revised. Even the considerable growth in exterior dimensions must not mislead us on this point. At least in the two volumes before us, that growth is mainly a matter of the new layout of the print. To be sure, we do find in the details a new development inserted here and there. But much more often the solicitude of the author is shown through considerable abbreviations, increasing clarity, and synopsis. In order to spare those acquainted with the first edition from having to make a point by point comparison, let us go through the chief modifications in sequence: [125]

In Section I, the first two chapters – treating logical thought with reference to the *association* of representations as well as to their logically *apperceptive* combinations – exhibit those changes that one would have to expect in the light of the repeated reworkings of the same materials in the new editions of the author's chief works in psychology. The introduction to the first chapter, in which the author discusses (pp. 11–13) the distinguishing marks of the logical act of thought in contrast to all other inner experiences, is totally new. Further, the new opening of chapter 2 is worth noticing. There the apperceptive combinations of representations in formations of free phantasy are distinguished from those found in logical thought, and the closer consideration of the former is referred to aesthetics (pp. 31–32).

Much more profound modifications are found in Section II, "Of Concepts" (pp. 94 ff.). Right at the beginning we find the first subsection, dealing with the essential marks (*Merkmale*) of the *concept*, completely reworked, and the main content of the old subsection 2 inserted into it. The distinction between logical and scientific concepts – which p. 88 of the first edition of volume I had denied any deep justification – is now brought to the fore. A *logical* concept is defined as any thought content which can be derived from a logical act of thought (a judgment) through analysis of it (p. 95, etc.). On the other hand, the concept in the sense of a result of a *series* of judgments, and not in the sense of a mere element of a judgment, [126] is called a *scientific* concept. These two types designate the polar extremities in the development of thought. Considering logical concepts in this sense, we are then led to their fundamental marks. These are no longer, as in the first edition of volume I, determinateness and general validity, but rather determinateness and *logical connection* with other concepts (p. 96). These are the only marks belonging to the concept in all of its developmental levels, and thus to the concept as such (p. 98). Confusion of the logical concept with the scientific explains the insertion of other properties, and especially general validity and universality, into the content of the former. It is proven in detail that we must not consider general validity as a generally valid property of concepts. With that, therefore, we have a polemic against the presentation of the first edition, whereas the corresponding polemic against the mark of generality is taken over from the older presentation (subsection 2).[2]

The changes in the remainder of this chapter are yet more radical. The view expressed on p. 101 of the first edition volume I, that divisions of concepts into abstract and concrete, and into general and individual, are "of little theoretical value," is abandoned. The first edition subsection 3 (of chapter 1 in Section II), a part of which is inserted word-for-word into subsection 2 of the second edition, obtains a new function in the new context. The present subsection 2 treats the distinctions mentioned as "forms in the development of concepts." The development of logical concepts stands under the general law that our thought receives its first stimuli from the facts of experience, and that, therefore, it gradually progresses from the thought acts in which particular perceptions are conceptually articulated, first to collocation of a manifold of experiences which present certain interrelations, and then to a conceptual identification of the very relations in which various thought contents stand to one another. [127] Accordingly, the processes of progressive formation of concepts divide in turn into two stages: The first consists in the development of general concepts out of individual ones, the second in the development of abstract out of concrete concepts (p. 105). Corresponding to this the author divides this subsection 2 of chapter 1, Section II, into two parts: a). *Individual concepts and general concepts.* The individual concept forms the first rung in this scale of concepts. Issu-

[2] Probably in consequence of editorial oversight we find, at the opening of the last subsection on p. 99 of volume I, two short sentences ("It is otherwise ....") which have no sense in the total context, and also do not harmonize with what comes later. If they are simply stricken, then all is clear.

ing from the decomposition of the content of a particular perception, it is a constituent of a representation of an individual, but it only contains the motives toward its separation. The separation itself is realized in an act of judgment; and the individual concept is therefore, in contrast to the representation, already an offspring of logic (p.105). The individual concept becomes a general concept as soon as the judgment out of whose articulation the concept arises brings a number of interrelated experiences to expression (p.106). General concepts are not to be identified with species concepts. Rather, there exists besides species concepts yet a second form of general concept, concepts of relations (p. 108). The distinction between these two forms does not arise from divergent operations of thought, but rather from divergent motives tending toward the operations in question. If the operations are incited by experience contents which agree, then species concepts arise. But if by opposed contents within a stratum of experience united by relations of agreement, then relational concepts arise (p. 109). b). *Concrete and abstract concepts.* The author here sets out from remarks on pp. 97–98 of the first edition. But the distinguishing mark of the abstract concept given there – namely, that it can be represented merely through a word or other arbitrarily chosen symbol – does not now satisfy him. The author states that that can only be the expression of intrinsic characteristics which, in their turn, constitute the logical distinction of the abstract from the concrete concept. What are these *intrinsic* characteristics? (p. 112) The answer is provided by a comparison with concepts of relations. Whereas, with these latter, one determinate thought content is grasped in relation to another to which it stands in some relationship, in the case of *abstract* concepts the very relations (never carried over in [128] their details, and only retained in their overall effects through the mediation of deputizing words) which occur between the different thought contents form the content of the concept (p. 114). Examples of this: *substance*, *causality*, *just*, *virtuous*, and so on.

In the theory of judgment (section III), consideration of *Brentano*'s view of impersonal judgments is the occasion of a lengthy addition (pp. 178–179). Further, consideration of *Bergmann*'s and *Windelband*'s investigations concerning the negative judgment occasions some new discussions (pp. 213–217) – in place of the first edition presentations (187–191) with their frequent references to *Sigwart*'s *Logik*. I can only emphasize a few points here: The division of judgments into affirming and negating is, the author thinks, in truth no division of the judgments themselves, but rather just a distinction between certain annexed thoughts alone, which can combine with judgments. In the confusion of a reflection on the object from outside of it with the object itself, this view completely resembles that view – mentioned with reference to indeterminate judgments – which sees in the positing of a concept, whose existence is asserted or negated, the primitive form of the judgment in general (p. 213). Also, the thought function of *distinguishing*, which provides the negative judgment with its outward character, does not exhaust that judgment's signification, since it also is never lacking in the positive judgment (pp. 214–215). In consequence of its affiliation with the positive judgment, the function of denial extends to all judgment forms. More specific logical determinations are always linked with it in the suppression of the corresponding positive judgment. These derive in part from the context in which the negating judgment stands, but are above all directed by the logical function which comes to expression in the general form of the judgment (216). To this the author then adds a more detailed discussion of such logical determinations found in the enumerative, descriptive, and explanatory judgments (pp. 216–217).

In the exposition of the negative predicative judgment, the negative alternative judgment is counted (220) as one sub-form, although it was earlier (Vol. I, p. 196 of the first edition) treated in conjunction with the problematic judgment. [129]

Less important additions follow on p. 297 (the logical calculus) and p. 319n (the general principle of relation – *contra Sigwart*).

Totally new is the insertion, in section V, of a comprehensive discussion on the theory of probability (pp. 437–446). Two main forms are distinguished: qualitative and quantitative probability. The first is characterized by the fact that it never admits of a quantitative or, indeed, numerical evaluation of its degree (p. 437). It corresponds solely to a transitory phase of the investigation, or, where it is enduring, rests only upon special obstacles impeding the path of the investigation's completion. Quantitative probability, by contrast, has a character of finality to it (p. 438). Qualitative probability is either the result of an analogical inference or of an incomplete induction. A quantitative probability is present in all cases where, from the factual frequency of an event in given cases, the frequency of the same event to be expected in other cases is inferred. This definition leads immediately to the two forms which show up as apriori and empirical probability in the mathematical theory of probability (p. 440). The discussion of these two forms, in their reciprocal relationships and combinations, constitutes the content of the following pages (pp. 440–446).

There is an important change regarding the theory of the intuition of time. The psychological theory of the origin of time intuition, which is developed on p. 435 of the first edition, is abandoned. We can refer to the discussion on page 484 of the second edition for its total tendency. According to that discussion, psychological analysis agrees with the results of logical investigation in showing that a derivation of time intuition from other types of psychical elements is impossible. The discussion of the logical motive which determines us to separate (pp. 485–486) time intuition from the remaining perceptual content is also new; and also, on the whole, the discussion (up to p. 487) relating to the concrete content of time.

Also in the theory of spatial intuition we find (p. 514) noteworthy modifications. It sounds quite like a concession to moderate nativism when the author says: "Space can, just as little as time, be psycho- [130]
logically constructed . . . from some sort of non-spatial elements."

The theory of number comes before us in greatly modified form (pp. 521–532). Much of what I was obliged to criticize in my *Philosophie der Arithmetik* (Vol. I, Halle a. S., 1891, pp. 93–96) has been deleted or modified. Of the older presentation, on pages 468–474 in the first edition, there really remains only a fragment of the first subsection (from p. 468, first edition). All the rest is essentially new. Number has no special affinity with time (in contrast with the first edition, p. 468 of Vol. I). In its general signification it derives from the conceptual comparison of the two forms, space and time, and from the more inclusive universal concept of extensive formal manifold which is developed on the basis of that comparison (p. 522).

In Section VI the investigation of the logical axioms, in revised form, is assigned a chapter of its own. The introductory subsection 1 of the first chapter in this section is new, dealing with the task and the signification of the logical axioms (pp. 558–562), as well as providing a discussion of their relationship to the axioms of physics and mathematics. Then subsection 2 treats of the law of identity and of the principle of contradiction – and, indeed, of the former in a completely new fashion. A distinction is drawn between two meanings of the law of identity: In the first it expresses the requirement that, in any given context of thought, every concept must retain the characteristics attributed to it in thought. In the second, and more important, signification it expresses a function: namely, the recognition of what coincides *as* coinciding. Inasmuch as in the positive judgment emphasis is always mainly upon this positive aspect of comparison in thought (although the negative aspect, the differentiation, is in constant collaboration), the law of identity also can be designated as the axiom of the positive judgment (pp. 562–563).

In Chapter 2 the kinematic axioms are added into the discussion as something new, under the heading of "The Principle of the Degree of Velocity" (p. 583).

In addition to these larger modifications, many smaller ones are present in Sections V and VI – above all, innumerable and often not insignificant abbreviations.

The Methodology appears in two volumes in the new edition. The first comprises General Methodology (1st Section), the Logic of [131] Mathematics (2nd Section), and the Logic of the Natural Sciences (3rd Section). On the whole this volume is merely a re-edited imprint of the corresponding sections in the first edition. Mostly it is a mere matter of isolated additions or changes introduced to take account of progress, in the sciences considered, that has accrued since the time of the first edition. I cite in this connection p. 142 (the definition of continuity in *Cantor* and *Dedekind*), pp. 434–435 (atomic theory and the theory of continuity), pp. 454–457 (Principle of the Conservation of Energy), p. 490 (chemical energy), p. 512 (the spatial structure of atoms), pp. 542–545 (theories of heredity), p. 551 (principle of the summation of small effects over long periods of time), p. 557 (limits of the application in Biology of the principle of [the conservation of] energy), pp. 586f. (infectious diseases, etc.).

G. *Glogau, Die Hauptlehren der Logik und Wissenschaftslehre. Für den Selbstunterricht dargestellt*, Kiel and Leipzig, 1894.

"This book, intended for the self-taught, aspires to a forceful clarification of the motive toward the scientific mode of thought, and then further shows how that motive proves to be active in the specific forms and methods of thinking." Thus the author, early deceased, characterizes his goal in the Preface. But in the actual execution a different goal stands much more to the fore. It is one which the author expresses as follows, in the concluding words of the book: "In its mode of statement and inner economy, the work also aims, not, indeed, to offer a so-called 'brief summary of my system,' but to provide the student with an exact and concise characterization of my view of *all* the chief questions of science, and to do so in a form that is – so far as it may be possible and permissable – both didactic and intuitive. In this way the book also should smooth the way for entry into my main writings."

Part I (pp. 1–103) deals with the first principles of logic in four chapters which, in order, develop the logical and ontological presuppositions, the theory of the judgment, the theory of inference, and the forms of systematic thinking. The "Theory of Science" forms Part II, and is an abstract of metaphysics – attached to logic with the same sense and same intent as was customary in that traditional lecture on "Logic and Metaphysics," which in earlier times was a fixture of the [132] academic program.

This work shows how a broadly educated and strong-minded philosopher – but one whose strength does not exactly lie in the sober rigor and strictness of logical thought – conceptualizes the problems of logic and their solutions. We can extol the elegant presentation, and many fine individual observations. For the problems which dominate the logical movements of our times, the author obviously had no interest. He passes over them without even touching upon them.

F. H. Th. *Allihn*, *Abriss der Logik und die Lehre von den Trugschlüssen*, 3rd edition, edited by O. *Flügel*, Langensalza, 1894.

As we gather from the Preface, this is a newly revised edition of the *Antibarbarus logicus*, which *Allihn* published in the year 1850 under the pseudonym "Cajus." Its main purpose was to combat the Hegelianism still dominant in the higher schools of Prussia at that time. The present editor has brought the *Antibarbarus* up to date by choosing as examples for the sections on fallacies – which fill up half the work (pp. 56–110) – predominantly those that "define the thinking of the present." I hardly need say that we have in all of this a vindication of the *Herbart*ian philosophy.

E. G. *Husserl*, "Psychologische Studien zur elementaren Logik," (I. "On the Distinction between Abstract and Concrete," II. "Intuitions and *Repräsentationen*"), *Philosophische Monatshefte*, Vol. 30, 1894, pp. 159–191.

The first study seeks to trace the distinction between abstract and concrete contents back to the distinction between independent and dependent contents, which was previously noticed by *Stumpf*. Subsection 1 deals with this latter distinction, and leads to the definitional statement that a content is dependent if we have Evidence that it exists only in connection with other contents, and, thus, is conceivable only as a part of a more encompassing whole, whereas this insight is lacking in the case of independent contents. (162) This statement is then [133] supplemented by a distinction of relative from absolute dependence or independence.[3]

Subsection 2. The division of content wholes into disjunctive parts leads either to "fragments" ("*Stücke*") or to "abstract parts," depending on whether the disjunctive parts are independent or not in relation to each other. The divided whole can itself be either a fragment or an abstract part of another whole, for abstract parts can be divided into fragments, while fragments can be abstractively divided up. A content is designated as "concrete" in relation to its abstract parts. An "absolute concretum" is one which is no longer abstract. Subsection 3 of this first study contains some critical observations which contest the view that one really can find a characteristic psychical activity of "abstracting" which might supply the positive criterion of abstract representations.[4]

The *second* study is a piece of purely descriptive psychology. In it I take up, among other things, the fundamental distinction between representing (*Vorstellen*) in the sense of intuiting something, and rep-

[3] This definition of course does not intend that the contingent act of living through (*Erlebnis*) Evidence, which indeed only occurs upon subsequent reflection and on condition of favorable dispositions, constitutes the dependent character of a content. Hence, the definition must be given an objective turn, in a very obvious manner: – There obtains a law of objects in general according to which a content of the relevant type can exist only as a part of a whole, and thus only in connection with other contents. In the case of several contents which are dependent *relative to* one another, the law states that *such* contents, i.e., contents of the appropriate content *types*, can exist only in connection with each other. One sees immediately that the crucial distinction here is not one restricted to 'contents', but rather is one which applies to objects in general. That makes it metaphysically significant. But the same then is also true of the remaining distinctions connected with it which are dealt with in this study.

[4] Since the appearance of my essay, I have become aware of the essential distinction between abstract *contents* (as parts of intuitions) and abstract *concepts* – a distinction which, unfortunately, I totally overlooked. The occasional confusion of these two does not interfere with the investigation, provided only that in it one thinks only of abstract *contents*.

resenting in the sense of the Representation (*Repräsentation*). I try to work this distinction out clearly and establish its validity. The introductory analysis of examples in subsection 1 leads to a provisional delimitation of the concepts: "Certain psychical experiences (*Erlebnisse*), in general called 'representations' (*Vorstellungen*), have the peculiar character of not including their 'objects' in themselves as immanent contents (and thus as present *in* consciousness). Rather, these experiences 'merely intend' them, i.e., consciously refer to objects by means of certain contents which *are* given in consciousness. They *mean* them – and, indeed without any conceptual knowledge of the relationship which obtains between the Representant and the intended object. [134]

Representations (*Vorstellungen*) in this sense I call 'Representations' (*Repräsentationen*)." (subsection 2, p. 174) By contrast, representations in the sense of intuitions are oriented toward an immanent content, without that content serving as a Representant. (175) To this is connected the definition of other concepts from this area: "consciousness with fulfilled intention," "pure and impure, complete and incomplete intuition," "immediate and mediate Representation." Subsections 3 and 4 mainly deal with the delimitation of that which is to be assigned to the content of an intuition, while subsection 5 demonstrates the major thesis that the distinction between intuition and the Representation can be grasped, not as a distinction between contents, but rather only as a distinction in the "mode of consciousness" (of "having in mind," or of psychical involvement). After the interweavings of the two functions are also discussed, subsection 6 then stresses that recourse to the concept of representing (*Vorstellen*) can contribute nothing to the clarification of the distinction in question. Subsection 7 points out the psychological and logical significance of this distinction, as well as the importance of research concerning it.[5] [135]

[5] In the theory of apperception, dominant stress is commonly laid upon the fact that the sense content is not always the same given identical stimuli, in virtue of dispositions left over from earlier experiences. It is held that the contents truly conditioned upon the stimuli are overrun by content Moments which stem from the actualization of those dispositions – indifferently of whether all or some of them. But this is not the whole of the matter. Not only is a different content Present (*präsent*); rather it also is *grasped objectifyingly* – sometimes as the same thing, and sometimes as different. But in this 'grasp' we find a peculiar "relation of consciousness" to what is intuitively Present; we live through the consciousness of

W. *Jerusalem*, "Glaube und Urteil," *Vierteljahrsschrift für wissenschaftliche Philosophie*, Vol. 18, 1894, pp. 162–195.

The author here publishes some of the basic ideas from his theory of the judgment, holding forth the prospect of a full treatment of them in a larger work – which has appeared in the meantime. This he does as a contribution "toward clarification of the rather confused relationship between belief and judgment." In the judgment, the complex of representations, given as a whole, is articulated and formed in virtue of being apprehended by our consciousness as the activity of a *thing*. The representation of a tree in bloom, which formed an undifferentiated totality prior to the judgment, then appears as *this* determinate activity of *this* determinate entity. Simultaneously with this formation and articulation – through which nothing is added to the represented content, it being instead only isolated off to itself from other processes and, as it were, put into finished condition for consciousness – *belief* is brought forth in the judgment. In the judgment,

---

"fulfilled intention," which includes an act of Representation. The descriptive and genetic investigation of this psychical phenomenon – in the course of which one must above all take into consideration the fact that it is not resolvable into mere complexes of contents – is the main task for any theory of *apperception*. This latter in turn is the essential foundation of any theory of judgment. It is easy to give a critical proof that whenever in the theory of judgment reference is made to "representations" which "merely represent" the same thing that the relevant judgment accepts or rejects, nothing other is thereby meant than Representations, and never intuitions as acts of direct perception. Thus it certainly is true that every judgment has a "representation" for its "basis," and that in virtue of this there arises a descriptively new relation to the "represented" object. But one must also truly understand by "judgment" the *act* of passing judgment, and not, with *Erdmann*, the predicative relation – which, in the sense here in question, is "mere representation." And with this there also is resolved the disputed question about whether simple perception is a judgment or not. The answer reads: Yes and no, depending on whether or not by "perception" is understood the "grasp" mentioned above or, rather, direct intuition (perception). The distinction under discussion also is of essential significance for the broader field of psychology. If one has distinguished between primary contents and psychical acts – in that the latter is defined by the positive criterion of having a relation to an *immanent* (not intentional) object – then the *acts* divide into two groups: (i) Those which are Representations or which have Representations as a basis and, in virtue of that basis, obtain an intentional relation to objects. Here belong, e.g., affirmation, denial, conjecture, doubt, interrogation, love, hope, courage, desire, will, and so on. (ii) Those with which this is not the case; e.g., sensuous pleasure and revulsion ("feeling tone"). These latter are, so to speak, lower modes of consciousness, and also are the genetically earlier and more primitive. The objection that in the life of the *developed* mind contents are never found without Representations attached, would not do away with the value of the proposed distinction between primary contents and psychical acts.

"The tree is in bloom," the tree is viewed as something independently existing and causally efficacious. It is considered as a center of force. Thus it is, so to speak, 'released' from my representation and objectified (pp. 165–166). The most important element in the act of judgment is, precisely, the introduction of willing into the sensation complex given through the mediation of the senses (p. 170). Belief is nothing other than a feeling which accompanies the judgment's holding of something to be so.

One can hardly expect advancement of the theory of judgment from such fictions as these.

A. *Marty*, "Über subjectlose Sätze und das Verhältnis der Grammatik zu Logik und Psychologie," 4th and 5th articles, *Vierteljahrsschrift für wissenschaftliche Philosophie*, Vol. 18, 1894, pp. 320–356 and 427–471. [136]

After a long pause the author again takes up, with these two essays, his critical investigations into the theory of judgment. In these essays he first considers the more worthwhile of the attempts published since his first series of articles toward a clarification of impersonal and existential propositions. For those acquainted with his literary style there will be no need to say that the author's expositions are distinguished by their trenchancy, clarity and thoroughness. That this may earn him little love and gratitude is to be expected, given the predominant disinclination toward critical interchanges. It is certainly hard to see how philosophy is to draw near to the ideal of unanimity and come into the pathway of solid progress, when the individual investigators do not engage with each other, and, as it were, bypass philosophizing *with* each other. Instead of seeking a critical adjustment, they anxiously seek to avoid it. In any case, those with an objective interest in the psychology of the judgment will find rich stimulation in *Marty*'s works, especially if they study those works in conjunction with the authors criticized, and as thoroughly as they deserve.

H. *Cornelius*, *Versuch einer Theorie der Existentialurteile*, München, 1894.

The author has a higher goal than the title leads one to suspect. He aspires to a theory of the judgment in general, rather than a mere theory of the existential judgment (p. 5). But it seems preferable to him to set out from the more specific problem because in that way the demonstration (of his view) takes shape with considerably more ease.

The work falls into six chapters. The first is to consider the activity in virtue of which we directly cognize the existence of the present contents of our consciousness, as well as the forms in which we bring such cognitions to expression. The question about how the existence of further contents can in some sense be known and expressed is dealt with in chapters II–V, while chapter VI draws the consequences for the general theory of the judgment.

Perception, the author says, is obviously nothing other than a finding or *observing* (*Bemerken*) of a phenomenon factually occurring in our consciousness (p. 10). It is, thus, identical with the *analysis* of [137] the current content of our consciousness. But when one considers that the analysis of simultaneous partial contents leads back to the analysis of successive ones, then one immediately sees that the perception of a content consists in the *discrimination* of it from the previously perceived total content (p.13).

In this relationship of the perceived content to the foregoing total content, and indirectly to earlier contents in general, there obviously is also contained everything that can serve toward characterization of the present content. Any naming – and, in general, any cognition ("knowledge about," in *James*' sense) concerning the present content – can only be based upon its relation to earlier contents, can only consist in a cognition of that content's interrelations with data preserved in our memory (p. 15–17).

Perception is the simplest act in which we become aware of the existence of a content, and with it a knowing (*Wissen*) – even an infallible knowing – is given. It therefore is to be characterized as a judgment, and, more precisely, as an existential judgment (p. 18). Being represented is all the same with being perceived. The distinction of a representing act from a perceptual judgment seems to the author a pleonasm (p. 20).

Since along with any naming there occurs a comparison of the present content with memory images of similar but earlier contents lived through (the result of which consists in a predication), a sentence cannot be regarded as an adequate formulation of the elementary judgment of perception. Along with the naming, those relations referred to would already be co-asserted. Only the designation of the content as one which is present (*gegenwärtigen*), as by the mere "*this*," can be allowed. But when we give an *explicit* rendering of that which the demonstrative expresses, we must state: "*This* is something other than what went before." The elementary perceptual judgment, therefore, also is not a single term. In it too we find, as object, a *relation* between two representation contents (p. 23); and it, like the other types of two-term judgments, is a relational judgment.

The more fully determinate perceptual judgments, in which the named content appears, are, accordingly, complex judgments. E.g., the judgment "This is red" comes together in the following stages: 1) The elemental perceptual judgment (*This*). 2) The elemental judgment of comparison. (See above. The author means the comparison [138] alleged to occur when naming occurs.) This judgment finds expression in the copula, or possibly in the ending of the verb. 3) The emergence of the denomination "red." The true subject matter of the assertion is obviously this total procedure (p. 24). The analysis of the explanatory judgment yields a similar result (p. 27).

We accordingly must not concede that in a question or in a negative judgment the same *combination* of representations is present as in the positive judgment. The question has only the sense of referring the hearer to the representations named, and inducing him to actualize the elemental judgment of comparison between them. He judges only when he perceives and compares the representations which up to that point where merely referred to. The judgment can maintain the sense of affirmation or denial only in relation to a question posed (p. 28). If the relation perceived is not one of similarity, but rather one of significant difference, then there occurs – depending upon the question or the subject raised – either the positive judgment, "*A* is something other than *B*," or the negative, "*A* is not the same as *B*." The latter says that the relation of the two contents is other than the identity relation, which was only feigned in the question (p. 29).

After the author discusses the relationship of sensation and phantasy representations in chapter II – the two are held by him to be "different *toto genere*," their relationship to be "*relatio sui generis*" – he turns in chapter III to the analysis of the symbolical existential judgment. We can make a judgment concerning objects not actually present to us only if they are Represented by means of contents that are so present. If the object is a sense content, then the corresponding phantasm serves as the Representant thereof. The principle in virtue of which *this* phantasm appears as the symbol of *that* content is the specific relationship obtaining between them, which, like any relation of comparison, permits the indirect determination of one of its terms by means of the other. In this case the judgment asserts that a sense content corresponding to the presently given phantasm is to be met with under certain conditions. We then have a discussion of how still other symbols also can function in place of the phantasm. Yet, in the case of "association symbols" as well (e.g., words), phantasma intervene as "comparison symbols." The upshot of this discussion is "that the assertion of the existence of an object not presently perceived has no other *signification* than this: that certain perceptions *could* be [139] realized, the more simple and summary expression of which is given precisely in the assertion of the existence of the object."(p. 55)

The IVth chapter contains polemical discussions with *Hume*, *Brentano*, and *James*, while the Vth deals with the analysis of the memory judgment. The author adopts *James*' theory of "*fringes*," holding that the phantasm which shows up as the after-effect of a past lived experience takes on a specific "coloration" ("*Farbung*") that is lacking in the product of pure phantasy. (p. 89) Yet we also are told that the *Gestalt* quality is different when arising from past and from present perceptual contents; that in the case of the memory judgment it is a question of the cognition of the relation of a present content to another which is not present. And so on. – Pronouncements which, like so much else in the author's expositions, really make no sense to me.

The memory judgment is assigned that class of Evidence which *Meinong* accords to judgments of differentiation, since a memory judgment is essentially based upon the *differentiation* of the memory image from a corresponding product of phantasy. The respective judgment is passed with complete certainty, if the *fringes* are noticed at all (p. 99). "The question of whether that which we judge to be the after-effect of past lived experience, on the basis of this characteristic mark [the "fringe"], really corresponds to past lived experiences is likely to degenerate into a mere dispute about words" (p. 100).

From the concluding observations I emphasize only the consequence that, in general, we are to reject any truly fundamental opposition between the activity of representation and that of judgment (p. 101). If I correctly understand it, the essential idea of this book is to be sought precisely in the claim that all "representing" and "judging" (denials just as well as affirmations) – is mere perception, a simple finding of a content; whereas all distinctions which we allude to by choosing among those various terms must lie in the contents found before us. It is noteworthy that the author, who at the outset of the treatise emphasized that perception is an *infallible knowing* (*Wissen*), comes here in the conclusion to the claim: "If one understands by the activity of judgment that psychic function with reference to which there arises the question of truth, then obviously perceptions are not to be considered as judgments. For it makes no sense to [140] inquire about the truth of a perception. Perception is simply given to us. We can just as little doubt it as we can do away with it." "The concepts *true* and *false* first find their place when one content is used as a symbol of another."

Admirers of the author's earlier works are in for some disappointment with this new one. It testifies to great mental agility; but unfortunately it far too often operates with dubious hypotheses, which are usually put forward as "obvious," with no serious attempt at proof. The details also leave more to be desired in conceptual sharpness and clarity than we would have expected from this gifted author. Here we can go into only a few points. The chief foundation of the present theory is the identification of perception and differentiation. This is not supposed to be a mere terminological convention, but to constitute an important bit of knowledge: *viz.*, that the perception of a content consists in the distinguishing of it from the foregoing total content. That it does so consist is why the "elemental perceptual judgment," identical with the perception, finds its adequate expression in the sentence, "This is something other than what went before." I cannot but see in this the confusion of two fundamentally different significations of the word "distinguish." There is the distinguishing that goes into *relating* things, and that which goes into *analyzing* something. The first certainly presupposes, as does any act of relating, that the terms of the relation are separately and specifically noticed (*für sich bemerkt*), and thus "perceived" in some sense *other* than "being distinguished." Wherever a content is separately and specifically perceived, then it is of course correct to judge that it is distinct from the total content perceived immediately before it. But it by no means follows that that perceiving *consists in* this relational judgment, which is introduced only by reflection after the fact. If one takes the detaching of a content from the background to be an irreducibly peculiar type of relationship (one which even admits of degrees), then I find nothing against that. But the perception of the content is not the perception of that relationship. Rather, the *objective* existence of the relationship (i.e., *not* the realization of a relational and perceptual *activity* in which we may now and then grasp it) is only the pre-condition, cognizable in psychological reflection and induction, for the perceptability of that content. How ready the author is to transfer his reflections into the primitive fact [141] is also shown by his repeatedly representing even the cognition of the temporal relation to what is past as a constituent of the simple perception. He further fails to see the infinite regress in which the "distinction" theory involves him: If a content has not already been noticed, then we also cannot distinguish another from it in relational cognition. Much worse still is a second regress: The author clearly substitutes for the simple noticing of a content the perception of its relationship to what is past. To be consistent he must then for this latter itself substitute, in turn, the perception of the relationship of *that* relationship to one in the past. And so forth.

This same error in the distinction theory also plays a role, of course, in the absurd classification of the memory judgment within the same Evidence class as judgments of distinctions (see p. 232 above).

Objections of no less serious a nature arise for pretty well all of the positions taken by the author. How, for example, could we accept his theory of the function of linguistic symbols, since he maintains that each word is accompanied in living discourse by phantasms (as symbols of comparison), and since he moreover identifies these phantasms with the signification of the words? The former contradicts all unprejudiced experience. But as to the latter, I ask whether the phantasm of a temple in ruins, which for me comes into play precisely with the word "culture," constitutes the signification of this word. And further, what is one to say of a theory according to which with every "naming" – that is, with every use of linguistic expressions – "a predication is carried out," and the relation between the name and the named is also asserted (See p. 230 above); a theory, moreover, according to which affirmation and negation are essentially distinguished only in that in one case similarity, and in the other "significant difference," is asserted. (See p. 231 above.) This latter point would have the remarkable consequence that affirmation and denial are not cleanly separated, but lead over into one another through a continuum of nuances. How astounding that the evaluation of the memory judgment as admissable or erroneous should depend upon whether or not the *fringes* have really been noticed. One would suppose that with the *fringes* unnoticed the memory judgment simply could not be made. Are not these *fringes* that which descriptively characterizes the memory judgment as such? Between the correct and [142 the false memory judgment does there exist the faintest descriptive distinction? – Thus we find one dubious point after another.

Th. *Lipps*, "Subjektive Kategorien in objektiven Urteilen," *Philosophische Monatshefte*, Vol. 30, 1894, pp. 97–128.

The author provides us with a serious bit of systematic, logical work in this treatise, originally drafted as one chapter of his *Grundzüge*. My brief report can have no thought of exhausting its full conceptual content, which would provide the starting points for many stimulating discussions. The task here can only be to characterize the main ideas and their systematic interconnection.

We begin with Section II, for the first section only draws together a few theorems from his *Grundzüge*.

Paragraph #7. The categories, i.e., the predicates possible in a judgment, divide into those which are objective and those which are subjective. The latter are determinations of our manner of representing objects. They tell us nothing about objects, but rather about the manners in which we represent or must represent objects. To them belong: (A) unity, plurality, wholeness, singularity, multiplicity, totality; (B) identity, numerical distinctness, equality and inequality.

Paragraph #8. These "manners of representing" can mean nothing other than modes of "positing." And here there are, right off, only two possibilities: either *one* or *a number* of positings occur. Unity and plurality therefore must be the fundamental subjective categories. In fact, in the others they are already presupposed. E.g., the identity of two persons signifies that the two are "one." And so forth.

Paragraph #10. Two types of subjective categories are now differentiated: (A) Categories of positing, (B) Categories of comparison. The purport of the former is that a simple or multiple positing has occurred, and, of the second, that it must have occurred if we are to measure objects of consciousness against one another in some determinate manner. With regard to the allocation of these types, cf. the enumeration under (A) and (B) of the paragraph above.

Paragraphs #11 and #12. The application of the categories of positing (unity, multiplicity, etc.) is chiefly an arbitrary matter, also depending upon the contingent course of our representations. (I can arbitrarily select these or those stars or groups of stars out of the night [143] sky.) These categories are simultaneously conditions *and* factors of our aesthetic and practical behavior (e.g., in the unity of a melody).

Paragraph #13. By contrast, the categories of comparison are on all points bound up with the consciousness of objective necessity. In identifying or distinguishing I know myself to be conditioned merely by the objects compared. This consciousness is, as consciousness of objective necessity, a judgment – and, to be sure, a subjective judgment, because a subjective category constitutes its predicate.

Paragraph #14. The application of the subjective categories is grounded with complete objectivity – that is, is required independently of all caprice or subjective involvement with the respective objects – there and only there where those categories are *conditions* of the predication in objective judgment. But thereby they themselves participate, in a certain manner, in objectivity; and it is in this sense that the expressions "objective unity," "objective plurality," and the like, are to be understood.

The more exact consideration of, chiefly, the categories of positing is the task of Section III.

Paragraph #15. The author distinguishes and discusses: qualitative unity or uniformity, and quantitative unity or diversity (a man – *one* man); likewise (in paragraph #16) qualitative and quantitative plurality, as pluralness and as set (*Menge*) respectively.

Paragraph #18. In the same manner, qualitative and quantitative totality (*Totalität*) are dealt with as wholeness (*Ganzheit*) and as completeness (*Allheit*). The latter is defined as the closed sequence of positings of a determinate object.

Paragraph #19. The totality as *concretum* is equivalent with number in the sense of a closed *set*. E.g., the representation of three and *only* three men is a number. Three men are all the men which the concept comprehends and requires to be posited. "All men in general" is not a totality that can be consciously realized, and therefore is also no number realizable in consciousness.

In the IVth Section we first encounter the systematic main idea of the whole investigation: To each subjective category definite types of judgments are intrinsically conjoined. The category displays the condition or necessary presupposition of the relevant types of judgments, and thus achieves objective signification for itself. For each [144] class of category we have two possibilities: that the category forms the predicate, or that it does not. To these possibilities there correspond two correlative groups of judgment types. In the latter case, once again, the category can simultaneously appertain to the logical and to the psychological subject of the judgment. In the correlative groups, therefore, the judgment types always show up in pairs.

In Section IV this idea is first carried through with respect to *qualitative* unity, plurality and totality. If these categories are expressed in the predicate, then there arise judgments of unity, plurality and wholeness (paragraph #27).

In the other case, one finds for each category two types of judgments: To unity there corresponds the "logically simple judgment" and the "psychologically simple judgment" (paragraphs #20 and #21). For example, the judgment, "This is a tree," is logically simple, because the unity of the subject, as condition of the predication, belongs at the same time to the genuine act of judgment. On the other hand, the judgment, "This sphere is white," is only psychologically simple, because the parts of the sphere to which the predicate belongs are united not by that predicate, but rather by the concept of the sphere. – Likewise and for similar reasons there corresponds to qualitative plurality the "logically disjunctive – or, for short, disjoining – plurality judgment." (E.g., This surface is white and red.); and, on the other hand, the "psychologically disjunctive – or the disjoined – plural judgment." (E.g., That surface, luminous in certain areas, is the surface of the sea.) To the category of wholeness there belongs the pair: "logically disjunctive and psychologically disjunctive judgment," or "dividing and divided plurality judgment."

The grouping is completely analogous, then, in the case of the categorical classes of "*quantitative* unity, plurality and totality" (paragraphs 29–36). We have the logical and psychological singular judgment, the disjoining and disjoined, the dividing and divided judgment about sets. And in the other group we have the judgments which contain the categories under consideration as predicates (paragraph #35).

In the Vth Section the systematic concept for the second main class of categories, that of comparison, is worked out, and these categories themselves are examined in detail. I omit further statement of the forms which develop, since without the corresponding definitions [145]
and discussions – indeed, without the fullness of detail given in the treatise itself (which is, moreover, quite concise) – those forms

certainly would not be intelligible.

H. *Rickert*, "Zur Theorie der naturwissenschaftlichen Begriffsbildung," *Vierteljahrsschrift für wissenschaftliche Philosophie*, Vol. 18, 1894, pp. 277–319.

The author takes as his problem the task of exhibiting the essence of the concept; that is, of showing its specific accomplishment in the attainment of the goals pursued in the work of the natural sciences. He develops (pp. 279–288) the basic conception for his solution through the following line of reflection: If we set for knowledge the task of imaging the world as it is by means of our representations, and of then expressing in judgments what is represented, we encounter the extensive and intensive inexhaustibility of *things*. (By the latter the author understands the infinite fullness of what there is to be known *in each* individual intuition.) The problem would therefore be insolvable. – Hence, if there really is to be knowledge of the corporeal world for the finite human mind, it can only come about in such a way that the extensive and the intensive complexity of things is somehow set aside or overcome. It is overcome by means of general representations. *Via* the significations of words we after a fashion, and at one stroke, absorb a mass of intuitive forms, and yet we represent only a small portion thereof, and perhaps nothing at all of the infinite intuitive content in it. The surmounting of the extensive and intensive complexity of the thing for the purposes of scientific knowledge of the corporeal world, *that* is the logical essence of the concept of natural science. Extensive complexity is overcome by its extension, and intensive complexity is overcome by its content.

It would seem to me that this line of thought is not wholly correct. Consider the point of discontinuity marked by the dash above. The possibility of knowledge of the world is dependent upon the "overcoming" of that inexhaustibility of the *thing* emphasized, only if, now as in other times, the goal of that knowledge lies in a completed knowledge of singular facts, and could, rationally, only lie therein. But if that is not its goal, or if it has other possible goals which are not effected by the inexhaustibility of singular facts, then [146]
the conclusion fails. And in fact it does fail, since *that* goal of world knowledge is in nowise one that can be maintained. Indeed, the author himself bears witness to that fact by later saying that the ideal of an all-inclusive knowledge of singular facts can have no place at all in logic (p. 288). It is, moreover, clear that that goal – even under

the assumption of the exhaustibility of the world by a human mind adequate to it – would just not be interesting. And it is, further, clear that, on that assumption, the knowledge of universals and of the laws grounded in them could retain its full sense and could constitute the finest goals of our theoretical endeavors quite as they do with the world as it now is. All of this immediately makes it clear that the task of conceptual knowledge has no essential connection with the "extensive and intensive infinitude" of the world. One also sees, with only a little attentiveness, that the images of "overcoming," of "setting aside" and of "simplification" of the complexity in question, immediately lose their proper sense – one prescribed for them in the original line of thought and determinative of their orientation – all the while remaining rooted in that line of thought. And so, unfortunately, a streak of unclarity runs though the entire treatise, which is all the more harmful as it step-by-step accompanies the *proof* that the essential accomplishment of concept formation in the natural sciences lies precisely in the "overcoming" of the infinite complexity of the corporeal world. If this basic thesis is lacking in a content that can be grasped with precision, then we can no longer speak of a genuine proof of it. Here I must omit to go into the details. I will only mention that the author, in elegant expositions, seeks to exhibit three levels in which concepts respond to the task proposed, to an ever higher degree, and thereby of necessity also must assume those attributes through which the essence of the logical concept has usually been characterized by others. The first level is formed by word significations naturally developed, to which (empirical) generality already belongs. The second is attained in defined concepts, which are equivalent to judgments and thereby acquire the character of determinateness. The last level is formed by concepts that enter into *laws*. These, however, will permit the attainment of the ideal of that world simplification only if there is *one* ultimate law concept [147]
which encompasses all possible natural laws as species – the functioning of this supreme concept being therefore a postulate of logic (p. 310).

Any fruitful theory of concept formation in the natural sciences must, in my opinion, be a theory "from below" ("*von unten*"), arising out of the workings of natural science itself. The author's theory moves so much in the domain of general constructions, is so much a theory "from above" ("*von oben*"), that in the entire treatise not a single example is to be found, and the absence is not even felt.

B. *Erdmann*, "Theorie der Typen-Eintheilungen," *Philosophische Monatshefte*, Vol. 30, 1894, pp. 15–49 and 129–158.

This stimulating essay introduces us to an interesting and little-known area of methodology. Traditional logic imposes on any correct division the requirement that its members be sharply delimited over against each other. But, as *Whewell* already showed, a glance at the disciplines of the natural sciences reveals that this requirement is not compatible with the factual way in which scientific method is practiced. On a broad scale – not merely in the natural sciences, but in all domains of our theoretical and practical knowledge – well justified divisions turn up whose members flow over into each other through manifold intervening stages, and accordingly stand in a fluid interrelationship. The logical examination of these divisions is the author's goal (Section II). The fluid interrelationships in question need not necessarily form a continuum. In them we are, quite generally, dealing with interrelationships of objects which present no fixed boundaries for division. That is because these objects, from each possible way of viewing them, blend with each other across numerous intervening stages that are unspecifiable under the circumstances, it not being possible to formulate distinctions that are exclusively linked to a single one among the differentiations within these stages (Section II). The whole to be divided, in all of this, can also be a single object – if, namely, we consider the development of a singular totality (the articulation of it into developmental periods, and the like, Section IX, p. 147). The author introduces the word "*Typus*" as a logically [148]
suitable term for the genera which stand in this fluid relationship. It is a word that has gradually come into vogue within scientific usage to designate such genera, while in practical knowledge it essentially possesses the signification of a Representative member.

In painstaking analyses that burrow deeply into the substance of the particular sciences, the following are distinguished: A). Schematic types, where the specific differences are arbitrarily stipulated, i.e., for reasons of expediency. E.g., the temperature scale or the degrees of solidity, and so forth (Section III, pp. 18ff.). B). Representant types (Section IV). These are models, but not in the normative sense. Rather, in the sense of Representants. For example, heat and cold, poverty and wealth, segments of the color spectrum in red, orange, and so on. C). Developmental types among organisms (Section VI).

D). The types of languages (Section VIII). E). Epoch types. E.g., epochs or periods of development in science, art, religion, and so forth (Section IX).

Not all talk of types in the sciences refers to types in the logical sense formulated here. Thus, what were originally called types in Zoology and Botany were simply Representative genera having a fixed delimitation (Section IV). For analogous reasons one must also exclude the "types" of Chemistry and Crystalography (Section X).

E. *Mach*, "Über das Prinzip der Vergleichung in der Physik," *Verhandlungen der Gesellschaft Deutscher Naturforscher und Ärzte*, Vol. 66. From the Vienna Congress, September 24–28, 1894. Published by order of the Directorate and the executive officers, V.A. *Wangerin* and O. *Taschenberg*. Part I, *Plenary Sessions*. Leipzig, 1894, pp. 44–56.

The theme of this brilliant lecture is much more general than the title might lead us to think. *Kirchhoff* dealt with mechanics in a wholly new spirit, designating its task to be "the most complete and simple description of movements occurring in nature." Thus he gave expression to a philosophical concept that at first found acceptance with only a few investigators, and even now makes only slow headway. "In *Kirchhoff*'s elegant manner of expression there [149
is embedded much which he never intended; and, on the other hand, it leaves out much that until now had been regarded as an essential attribute of scientific knowledge. What good is a mere *description*? What becomes of *explanation*, of insight into causal connections?" To resolve such doubts, to make clear the deep meaning of *Kirchhoff*'s idea and present the justification for it, is what *Mach* here aims to do.

The chief ideas are the following: Communication is what lifts us above immediate sense experience. It makes the experience of others useable by us and renders science possible. But comparison is what makes communication possible.

From comparison there arises the abstract concepts by means of which direct description of facts is carried out. By "direct description" the author understands "a linguistic communication concerning a fact which utilizes only purely conceptual means" (p. 46). In the case of rather more extensive facts, such a description is a laborious task. What a relief if one can simply say that this fact A is in many or all respects the same as a fact B, already known? (E.g., the moon behaves like a heavy body in opposition to the earth.) The author calls a description of this latter kind "indirect." Now under this category of indirect descriptions there also falls what is usually characterized as theories or theoretical ideas. For in their case also it is simply memory and comparison that comes into play. But in them our memory seizes upon entire systems of similarities instead of upon a single point. Confronted with a well-known configuration of features, the new fact immediately becomes one with which we are familiar. In fact, the general idea can yield more than we can *see* in the fact at the moment, pointing us on to new aspects, which we then look for and often actually find. This rapidity of the extension of knowledge is what gives theory its *quantitative* superiority over simple observation, whereas the former is essentially distinct in *quality* from the latter neither in the manner of its origination nor in the end result (p. 47).

In place of a fact *A*, theory sets before us in thought another fact *B*, simpler or more familiar to us, which can stand for the former in certain respects, but of course not in all. In that this limitation is overlooked, as may easily happen, the most fruitful of theories can [150] occasionally turn into an obstacle of research. This is demonstrated in detail by the author through many examples. It therefore seems imperative that, while not rejecting the powerful assistance of theoretical ideas in our research, still – in the measure in which one becomes familiar with new facts – direct descriptions of them should progressively replace the indirect ones, where the direct descriptions contain nothing extraneous and are restricted simply to the conceptual comprehension of the facts (p. 50).

We are in no position forthwith to describe every fact. The abundance of facts is too great. We advance by degrees only, again and again comparing the new with the already known, and thus adding it to the known. The domain of facts compared becomes evermore inclusive, and the concepts ever more abstract. In this manner there at some point will develop "a general phenomenology of the physical, encompassing all physical domains" (p. 41).

If this ideal is attained for a domain of facts, then description accomplishes all that the investigator can demand. Description is a construction of facts in thought. For the physicist in particular, units of measure are the building blocks, concepts the plan of construction, and facts the result of building. Our conceptual formations serve us as almost a complete substitute for facts, in which we can ascertain all of their characteristics (p. 53).

People say that description leaves the requirements of causal analysis unsatisfied. The author hopes that the science of the future will lay aside the concepts of cause and effect, because of their formal obscurity. He finds in them a strong element of fetishism. He recommends, rather, that we regard the elements conceptually determinative of a fact as interdependent simply in the purely logical sense, as is done by the mathematician: by the Geometer, for example.

Finally, the author discusses the relationship of physics to the remaining natural sciences. The special superiority of physics consists in the fact that it deals with large domains of qualitatively homogeneous facts, which are distinguished only quantitatively – that is, only through the number of equivalent parts into which their distinctive features are analyzable. If physics seems to accomplish so much more than other sciences, we must take into consideration the fact [151]
that also, to a certain extent, it meets with far simpler tasks (p. 55).

For those who are familiar with *Mach*'s writings it will hardly need be said that this lecture is rich in examples, which are always instructive and carefully analyzed.

P. *Biedermann*, "Die wissenschaftliche Bedeutung der Hypothese," *Programm der Annenschule zu Dresden-Altstadt*, Dresden, 1894, pp. 1–40.

J. *Henrici*, "Einführung in die induktive Logik an Bacons Beispiel (der Wärme) nach Stuart Mills Regeln," *Festschrift zur Einweihung des neuen Gebäudes des Grossherzoglichen Gymnasiums Heidelberg*, Heidelberg, 1894, pp. 15–28.

These works lay no claim to scientific importance. The latter pursues pedagogical goals, while the former illustrates, in a large number of interesting cases, the "significance of hypotheses," and also contains all sorts of nice quotations from works in the natural sciences.

REVIEW OF MELCHIOR PALÁGYI'S *DER STREIT DER PSYCHOLOGISTEN UND FORMALISTEN IN DER MODERNEN LOGIK*[1]

One does not need to read very far in this work to see that its content deviates fundamentally from its title. Namely, it does not really give a general treatment of the quarrel between the psychologistic and the formalistic logicians. Rather, it deals only with *my* quarrel with Psychologism in my *Logical Investigations*, appearing since 1900. It is the author's intent to oppose the "formalistic tendency in modern logic" (p. 5), the "precipitous, retrograde movement which writes on its banner the warcry – 'Away from psychology!'" He will oppose the "formalistic danger" of an "unfruitful and sterile formalism" (p. 1). Thus he puts his task, quite generally, in his "Introduction." But then, as we proceed, we immediately hear also that *Bolzano* is to be regarded as "the true originator of modern formalism in logic"; and we are reminded that *Husserl* is, so far, the only modern logician who has linked his positions to *Bolzano*'s *Wissenschaftslehre* in essential points. Then we further notice that the author mentions no other formalists. Rather, his attacks – some in particular chapters, and some in the form of sorties interspersed throughout the text – relate solely and only to *Husserl*. So we must, so far as this work is concerned, form the equation: Modern Formalists = *Husserl*. If, now, the malice in the tone preferred by the author does not suffice to keep me from complying with the kind invitation of the editor of this journal to review this work, it is in order that I might fulfil the [153] obligation, which every serious worker has, not to let his work fall victim to attacks that would distort it.

[1] First appearing in 1903, *Zeitschrift für Psychologie und Physiologie der Sinnesorgane*, Vol. 31, pp. 287–294. Republished in *Husserliana XXII*, pp. 152–161

My logical investigations place themselves against the dominant current of logical and epistemological persuasion – at least as an irksome impediment. It is indeed conceivable how a work which, like the one before us, energetically guarantees to have done away with the impediment, or proven it to amount to nothing, can calculate on easy laurels of victory, and how it might easily transmute present inclination into agreement with itself, and thereby mislead many readers into orienting themselves about my point of view according to the author's statements, instead of according to my writings. But I must keep this from happening. In his very orientation toward my views Mr. *Palágyi* employs a peculiar, though certainly quite convenient, method. He is satisfied with a cursory reading of a few chapters or subsections of volume one of my *Logical Investigations.* All of the rest is, for him, non-existent. Of volume two, which appeared fully three-quarters of a year before his work, he says not a word, regardless of the relevant points it might bring to bear on the discussion. The astounding thoroughness of the author, together with an equally astounding incapacity to grasp the plain sense of any sort of conceptual distinction, brings it about that he – and this is to be taken quite literally – also is unable to report on even *one* of my or *Bolzano*'s theories without twisting it into something fantastic.

This is immediately revealed by his general characterization of my position. He repeatedly ascribes to me the tendency "to allow Logic to sink, where possible, into Mathematics" (p. 9), the "struggle" to "tear Logic loose" (p. 43) from Psychology and to "split the blanket between Logic and Psychology" (p. 37), and so on. Naturally he is silent about the distinction – one which is decisive for the simple understanding of the sense of my tenents, and which has been justified in detail – that I make between "Logic" in that most inclusive sense which it has as a practical discipline, and "pure Logic," as the theoretical system of purely formal (categorial) truths. He also does not mention that I abundantly approve of a logic that has the methodological tendency exhibited in the logics of *Mill*, *Sigwart*, and *Wundt*. Nor does he remark that I in nowise contest the founding of logic, in *this* common sense, upon empirical psychology, but even require that such a founding be extensively provided. He does not mention [154]
that "pure" or "formal Logic" is for me only a title introduced to aid in understanding historical traditions and tendencies. It is affixed to a certain class of propositions that are called 'logical' in the fullest

sense, of which I sought to show the following: – that they belong to a distinct discipline which is apriori and independent of all psychology, and that this discipline, through a natural extension, also takes in formal mathematics (its theories being apriori, and foreign to psychology in the same sense), and is ultimately identical with *mathesis universalis* in the generalized sense coined by *Leibniz*. Whoever takes his directions from the author must assume that I intend to reject Logic – in the common sense of "logic" – and aim to reduce it to mathematics – "mathematics" in the common sense; or that I intend, in any case, to reduce it to a 'class logic' ('*Umfangslogik*') in the style of the Boolean school. That I laid bare the follies of extensional logic over twelve years ago in a very detailed critique[2] might be unknown to Mr. *Palágyi*. But he cannot be ignorant of the distinctions just stressed, the neglect of which can only have the effect of transforming the sense of my views into nonsense.

Mr. *Palágyi* is also quiet about my distinction between the *mere technique* that is associated with theories of formal logic (and which can be consigned to the mathematicians), and, on the other hand, the sphere of genuinely philosophical tasks, *viz.*, epistemological illumination (*Aufklärung*) of the primitive concepts and propositions of pure logic through comprehensive, descriptive-psychological (or "phenomenological") analyses. He is silent on how – as is to be seen from volume two of my *Logical Investigations* – I have worked at these philosophical tasks, and especially at the descriptive phenomenology of the experience of thought, at lengths scarcely ever gone to before. My work shows that my struggle against Psychologism is in nowise a struggle against the psychological grounding of Logic as methodology, nor against the descriptive- psychological illumination of the origin (*Ursprung*) of the logical concepts. Rather, it is only a struggle against an epistemological position, though certainly one which has had a very harmful influence upon the way
in which *Logic* is done. – So *that* struggle is what Mr. *Palágyi*, [155]
manifesting an exemplary reliability, characterizes as "splitting the blanket" between Logic and Psychology.

If, now, we go in sequence through the four polemical sections of the work which follow the *Introduction*, we find that the first

[2] Review of: "E. Schröder, *Vorlesungen über die Algebra der Logik (Exakte Logik)*, Vol. I, Leipzig, 1890," in *Göttingische gelehrte Anzeigen*, 1891, pp. 243–278. [Translated above. D.W.]

bears the title, "Facts and Truths." All of my arguments against Psychologism rest – so the author says – on the proposition *that truth is no fact.* I.e., they rest upon the proposition that the act of judging correctly is indeed a temporal process, but truth is not. Now, out of this mere conceptual distinction between the act and the content of the judgment – thus, that distinction in virtue of which we speak, for example, of *the* proposition, $2 \times 2 = 4$, whoever may express *it*, one and the same thing – there arises, under the hands of the author, a real (*reelle*) separation between two allegedly connected Moments (*Momente*): "*Husserl* supposes that, after subtraction of the judgment content, he can retain as *remainder* a psychic act" (p. 14, cf. p. 47 among others). And now the critique is worthy of the interpretation: – Not a psychic act, but rather a "mechanical process," is what is left over after "subtraction" of the judgment content; and so my view wavers, "in an unceasing equivocation, between the physical and the psychical" (p. 15). After this specimen there probably is no need to go into the other twists through which *Palágyi* seeks to support his favorite objection, to the effect that I confuse psychology and physics (or science of facts). Having previously decided to read all sorts of absurdities out of my statements, he is no longer able to read them according to their sense and interconnection.

In the next two sections of his work *Palágyi* concerns himself with damning up the source of "modern Formalism." These sections are directed against *Bolzano*'s theories about concepts (*Vorstellungen*), propositions (*Sätze*) and truths (*Wahrheiten*) 'in themselves.' Here I must first mention the suggestive manner in which *Palágyi* conceives of my relations with *Bolzano*. In a series of hints which would be insignificant taken separately, but which are efficacious when taken in sequence, he gives the reader no less a notion than that I have exploited *Bolzano* in a dishonest manner and have kept silent about my dependence upon him. Suppressing judgment on the author's procedure, I note, for the benefit of the uninformed, that not only – as the author once mentions (p. 16) – have I "commemorated" [156]
Bolzano, and "named" him one of the greatest of logicians. Rather – in an 'Appendix' specifically devoted to this purpose[3] – I have pointed out the significance of the *Wissenschaftslehre* as one of the foundation works of logic, and have stressed the necessity of building

[3] Appendix to chapter 10 of *Logical Investigations*, Vol. I.

upon this work and studying it with the greatest care. This I have done in such a detailed manner, and with such emphasis, as has never before occurred, either in earlier times or contemporary. And, not satisfied with that, I expressly designated *Bolzano* as the one (along with *Lotze*) by whom I have been "decisively influenced." These are my own words.[4] – With specific reference to my concepts of the "Ideal" significations, and "Ideal" contents of representations and judgments, they originally derive, not from *Bolzano* at all, but rather – as the term "Ideal" alone indicates – from *Lotze*. In particular, *Lotze*'s reflections about the interpretation of *Plato*'s theory of forms had a profound effect on me. Only by thinking out these thoughts from *Lotze* – and in my opinion he failed to get completely clear on them – did I find the key to the curious conceptions of *Bolzano*, which in all their phenomenological naïveté were at first unintelligible, and to the treasures of his *Wissenschaftslehre*.

If, like all earlier readers of *Bolzano*, his "propositions in themselves" previously appeared to me as mythical entities, suspended between being and non-being, it then became clear to me, with one stroke, that here we basically have a quite obvious conception which traditional logic did not adequately appreciate. I saw that under "proposition in itself" is to be understood what is designated in ordinary discourse – which always objectifies the Ideal – as the "sense" ("*Sinn*") of a statement. It is that which is explained as one and the same where, for example, different persons are said to have asserted the same thing. Or, again, it is what, in science, is simply called a theorem, e.g., the theorem about the sum of the angles in a triangle, which no one would think of taking to be someone's lived experience of judging. And it further became clear to me that this identical sense could be nothing other than the universal, the species, which belongs [157] to a certain *Moment* present in all actual assertions with the same sense, and which makes possible the identification just mentioned, even where the descriptive content of the individual lived experiences (*Erlebnisse*) of asserting varies considerably in other respects. The proposition thus relates to those acts of judgment to which it belongs as their identical meaning (*Meinung*) in the same way, for example, as the species *redness* relates to individuals of "the same" red color. Now with this view of things as a basis, *Bolzano*'s theory, that propo-

[4] From *Logical Investigations* I, lst German edition, 1900, pp. 224–226. English edition pp. 222–224. [D.W.]

sitions are objects which nonetheless have no "existence," comes to have the following quite intelligible signification: – They have the "Ideal" being (*Sein*) or validity (*Gelten*) of objects which are universals ("*allgemeiner Gegenstände*") – and, thus, that being which is established, for example, in the "existence proofs" of mathematics. But they do not have the real being of things, or of dependent, thing-like *Moments* – of temporal particulars in general. *Bolzano* himself did not give the faintest intimation that these phenomenological relationships between signification, signification Moment, and full act of signifying had been noticed by him. And this notwithstanding the fact that he treats the psychology of knowledge in great detail in volume 3 of his *Wissenschaftslehre.* Indeed, everything indicates just the contrary: – that his conception wound up unclarified, in spite of all his efforts to avoid it.

Now as to Mr. *Palágyi*, he has adopted my view, or that from it which happened to be intelligible to him, without qualification. He interprets *Bolzano*'s concepts through my ideas and expressions, but does it as if he drew directly upon *Bolzano*, and also as if the corresponding statements by me were only just borrowings – underhanded borrowings, moreover – from *Bolzano*. He adopts my theory of the identical, Ideal sense (*Sinn*), without mentioning it in his ever so friendly statements. But this *sense* he makes out to be an identical, Ideal *Moment* of the lived experience, because he overlooks or does not understand the distinction, stressed by me, between the species and the particular, between the sense as the *Idea* (*Idee*) which becomes an object for us through species-abstraction and the sense-Moment of descriptive psychology. Then, since I find a conceptual distinction between the identical signification and the signifying act (in the sense in which, say, the quality-species *redness* is distinct from a red thing), he has the concrete, psychic experience of judging to consist of *two Moments*: – of the supratemporal (!) sense-Moment and the act. After this distortion he objects to *Bolzano* and to me that we would tear the sense-*Moment* away from the act. As if *his* sense-Moment (which, inconsistently with its being a *Moment*, he characterizes as Ideal and supratemporal) were identical with *Bolzano*'s "proposition in itself," or with the sense as species! Obviously the author could have spared himself the effort of emphasis – hauled out with ceaseless repetition – upon the indissolubility of the sense-*Moment* from the act, by which he thinks to refute us so decisively, if he had merely quoted from my *Logical Investigations.*

Of the same nature and value are the other objections which *Palágyi* advances in these sections. Thus, for example, the reference to propositions of a kind with "I am now thinking" (25ff.), which include in their objective content a reference to the one judging – in a word, the "occasional" propositions already treated in detail in volume II of my *Logical Investigations* – is of no use whatsoever in somehow calling into question that "independence" which is characteristic of the species *vis-à-vis* the singular case, and thus of the proposition *vis-à-vis* the fortuitous act of judging. It is of no use, that is, unless – deceived by the ambiguity of the phrase "content of judgment" – one follows *Palágyi* in confusing *proposition* (*Satz*) with fact (*Sachverhalt*). Again, the "intrinsic contradiction in *Bolzano*'s philosophy," which with no little pathos is brought forth in subsection 4, is in truth a contradiction between what *Bolzano* is supposed, according to *Palágyi*, to have maintained, and what he really did maintain. One is immediately convinced of this by comparison of *Palágyi*'s remarks with the passage concerned.[5] While, according to the clear sense of his repeated statements, *Bolzano* taught that to the "truth in itself" (i.e., the truth in that quite common sense in which we designate, not an act of judging, but rather a *proposition* as a truth) being thought or, more precisely, being judged or known is *non-essential,* the author has him teaching that being *un*thought of or *un*known is *essential* to it. *Bolzano*'s careless use of the ambiguous expression "proposition thought" ("*gedachte Satz*") offers the author a handhold for ascribing, here and further on, a series of crude inconsistencies to this exceptionally clear thinker. Of the duties of a fair interpretation, or of a careful reading which compares passage with passage, Mr. *Palágyi* is ignorant. And so, throughout two long sections, [159] one must patiently endure such utterly untenable statements: statements which, of course, are supposed to uproot right off the whole of *Bolzano*'s philosophy and my *Logical Investigations* – as if they did not each go their own peculiar ways.

In section IV *Palágyi* again turns himself directly and exclusively against my *Logical Investigations*, and especially against my distinction between laws of the real and laws of the Ideal, which I bring to

[5] B. Bolzano, *Wissenschaftslehre. Versuch einer ausführlichen und grösstentheils neuen Darstellung der Logik mit steter Rücksicht auf deren bisherige Bearbeiter*, published by several of his friends, Vol. I, Sulzbach, 1837, sec. 25.

bear upon logic. The basic error of Psychologism consists, according to my view, in its obliteration of this fundamental distinction between pure and empirical generality, and in its misinterpretation of the *pure* laws of logic as empirical laws of psychology. Of course our author finds obvious absurdity in this view: So our actual course of thinking is supposed to be governed by two sorts of laws, is it? And by laws which belong to two worlds separated by an "infinite abyss"? But it is simply impossible to see how extra-temporal laws of the Ideal could come to have some sort of causal efficacy. Such an estrangement of the real and the Ideal means the utter impossibility of any knowledge at all (pp. 41ff.). Or so says Mr. *Palágyi*.

Unfortunately, the author has read too selectively. Otherwise he would have stayed on guard against misinterpreting the *opposition* between Ideal and real as *lack of relation*. Since the Ideal significations instance themselves in acts of signifying, each pure law of logic expresses a universal connection which *eo ipso* can be referred to the Ideal extensions of the respective signification species, and thus to possible acts of thought in the realm of the real. Thus, as I have sufficiently explained, there can be derived from *any* Ideal law whatsoever (e.g., any arithmetical law) universal truths about Ideally possible or impossible connections of psychic fact. The character of these Ideal laws, as truths of reason (*vérité de raison*), is transmitted to such derivations and is not affected by them. All of the forceful expressions with which the author embellishes his criticisms cannot change the fact that he is passing judgment upon matters which he has not sufficiently studied.

It is hardly worthwhile to enter into his further objections, which often evince a striking lack of intelligence. Thus, a contradiction of my opposition between real and Ideal law is supposed to proceed from the fact that I myself allow law contents to function as thought motives in judging. Through this function, *Palágyi* supposes, the Ideal law in fact attains the significance of a real law of our thinking (p. 45). Thus, according to this way of reasoning, the law of gravitation, where it guides the thinking of the engineer, and likewise any law that guides [160] us in practice, would attain the significance of a law of thought.

And what is one supposed to say to the objection that, through my separation of the laws of pure logic, as laws of the Ideal, from laws of psychology, as laws of the real, these latter laws "appear to fall into the same category as the laws of mechanics, and that then one at

least no longer knows wherein the psychical could be distinguished from the physical" (p. 43)? Likewise for the objection that the truth is simply one, so that the separation of truths into two classes, separated by an "unbridgeable abyss," is impossible (p. 52)? Likewise to the misconstruction, related to these objections, that – as it of course would not be possible for me to do otherwise – I base that distinction of laws just referred to upon a distinction between modes of human cognition (namely, knowledge through "induction" and knowledge through "insight"); and that I thus perpetually confuse two sorts of laws with two ways of knowing one law (p. 53)? I need not say that I do hold the distinction between the two kinds of laws to be grounded *solely* in their significational essence, but that with that essence there is connected a phenomenological distinction in the mode of knowledge of the states of affairs (*Sachverhalte*), of the one and the other sort, corresponding to the laws. In the context of the critiques in my "Prolegomena," the phrase "law of the real" does not signify just any arbitrary universal proposition referring to the real, but rather a general fact, or at least a proposition which, in the manner of our assertions of natural laws, is weighted with factual content. So my point here essentially comes down to the distinction between factual truths and purely conceptual truths (i.e., Ideal laws, laws in the most narrow and rigorous sense of the word). Were the world so made that all of the spheres in it were red, then, arriving at an inductive belief that this is so, we would speak of a "natural law" to that effect. However, in itself it would be no (genuine) law, no proposition grounded in the conceptual essence of "sphere" and "red"; rather, it would be a general fact. It is precisely this objective distinction, fundamental for both logic and epistemology, between what *Hume* called "relations of ideas" and "matters of fact," that our author surely rejects as invalid. But the inadequacy of his polemic against this great thinker works an immediate embarrassment, since he does not once grasp the sense of the distinction in question. *Palágyi* objects, against *Hume*'s well-known statements, that even facts come under the principle of [161]
contradiction. For, he says, the opposite of a fact is *never* possible. The fact can never be undone. As if *Hume* doubted *that*! Is it really so difficult to understand that *Hume*'s referral of the principle of contradiction to his "relations of Ideas" is intended to mean no more than that truths about relations of Ideas are rooted precisely in the Ideas (*Ideen*) alone (i.e., merely in the relevant concepts), and therefore

cannot be denied without contradiction; whereas negations of factual truths are indeed false, but not self-contradictory?

After this fruitful search into the innumerable absurdities which he imputed to his adversaries, in the concluding section of his book the author explains his own views about the relationship between logic and psychology. On page 72 he distinguishes awareness (*Wissen*), as a general or "abstract" (!) psychic function – to the character of which it belongs to stand in the most intimate of relations with the remaining psychic functions – from sensing, feeling or willing, as "concrete" (!) psychic functions, or as "isolated psychic functions," that lack the capability of relating to other psychic functions or to themselves. The investigation of the former function falls to Logic. The investigation of the latter functions falls to Psychology. In other words, what we have here is a division of what everyone calls psychology into "Logic" and "Psychology" – namely, into the psychology of awareness and the psychology of the remaining psychic functions. The author is thus so naïve as to suppose he can settle, through a slight displacement of terminology, such a weighty epistemological question as that about the relationship between Logic and Psychology. A formula with the ring of profundity to it is supplied to the author, for the expression of his views, by the new terminological distinction between "unreflected" and "reflected" consciousness (e.g., the seeing of the red and the awareness of this seeing), and by the – not precisely unheard of – introjection of the notion of the furtherance of knowledge into the concept of Logic. Then it is said (p. 81) that Logic "reflects on the reflected consciousness, and endeavors to raise our knowing processes to a higher power by investigating the laws of our reflected consciousness. Psychology, on the other hand, will try to carry reflection raised to a higher power by Logic over into the investigation of unreflected consciousness."

A REPORT ON GERMAN WRITINGS IN LOGIC FROM THE YEARS 1895–1899[1]

First Article

J. Bergmann, *Die Grundprobleme der Logik*, second and completely revised edition, Berlin, 1895.

Among the writings from the years here covered, this profound work, rich in ideas, occupies an especially prominent position. In it the author sums up his logical views, which have been repeatedly presented in well-known writings. On all points he subjects those views to painstaking scrutiny and, where necessary, modifies them for the better. Nothing else is to be expected, given his great critical discretion. It also goes without saying that all of the merits of the older works return in the new: seriousness and rigor in scientific attitude, the most fastidious attention to bringing out the essential concepts, problems and theories – all accompanied by critical observations on the worthwhile tendencies in the older logic, showing great originality in the interpretation and reformulation thereof. As with the *Reine Logik* of 1879, which was a more detailed treatment, so also this briefer presentation in the *Grundprobleme* belongs – in contrast to the line of logical literature inaugurated by *Mill*, *Sigwart* and *Brentano* – to that "conservative tendency" which has undeniably constricted the range of influence exercised by *Bergmann*. Those empiricist inclinations so preponderant in contemporary logic – connected, as is well-known, with the favored position granted to empirical psychology in providing a foundation for logic – are alien to Bergmann. He firmly holds (and this we take to be no small merit)

[1] This is the first of five "Articles" which first appeared in *Archiv für systematische Philosophie*, Vol. 9, 1903, pp. 113–132, 237–259, 393–408, 503–543, and Vol. 10, 1904, pp. 101–125. Republished in *Husserliana XXII*, pp. 162–258. The other four articles are translated in sequence immediately below. [D.W.]

to the Idea of a pure logic as an apriori discipline. However, in his treatment that peculiar kind of psychological foundation which truly *is* indispensable for the illumination of the sense of the pure concepts [163]
and laws of logic is somewhat neglected. I refer to those descriptive analyses of the lived experience of thinking (*Denkerlebnisse*) which, as prior to all empirically explanatory or genetic psychology, we better characterize as "phenomenological analyses."

The author (in subsection 1 of the Introduction) specifies the subject of logic to be "thought, with regard to its natural aptitude for the goals operative in knowing and understanding." But logic has to do only "with the general mode of thought" (or, what is the same for the author, with *the judgment*), as well as with those *specific* modes of thought which can be derived from or understood in the light of this general mode, without reference to the characteristics through which their objects are distinguished from each other. Strictly speaking, if I understand correctly, this takes in only "general" or "elementary" logic, which the author distinguishes – in the sense of Kant's well-known comments – from "special" logics, oriented in terms of special knowledge goals or of the methods of the various particular sciences. The author also takes from *Kant* the distinction between pure and applied logic. Kant's definition of pure logic – according to which it isolates the understanding off from the other mental activities – is interpreted by Bergmann to mean "that it considers thinking only in its *products*, or in the thoughts thought"; or, yet more suitably, that it considers thought, not as an activity of the mind, but rather with respect to that which is thought through such activity. Against the many definitions of logic as a technology (*Kunstlehre*), frequently employed even by himself, he directs a polemic: As the technology of an activity, logic would be required to show what one must *do* in order to carry that activity through correctly. But then, to speak in the Kantian manner, logic would not be the science of the employment of the understanding itself, but rather of the subjective and empirical conditions of that employment.

It seems to me that the *Kantian* distinctions certainly are justified in themselves, but that they do not bring to complete conceptual clarity the various natural specifications to which the Idea of logic can be subjected. This even remains true given the assistance of the author's explanations. The contrast between pure and applied logic can be taken as synonymous with that between theoretical and [164] practical logic. If, on the other hand, "pure" is understood to mean "apriori," then we have a second contrast, one which must be properly aligned with the first: that between apriori and *a posteriori* logic. To begin with the first contrast, the long-favored concept of a practical logic, a logic as *Kunstlehre*, is fully justified. Taken in a quite natural sense, that concept delimits an art of thinking, which is founded upon empirical psychology, and is therefore both empirical and concerned with specifically *human* thought. But this does not presuppose psychology alone. Rather, it also is based upon "pure logic," a theoretical and apriori discipline, whose pure laws are normatively applied by the *Kunstlehre*, and made to bear practical results under the empirical conditions of human (and especially scientific) thinking. Now this pure logic has nothing whatsoever to do with thinking and knowing, much less then with the normative regulation and guidance thereof. Rather, it deals with the units of thought and knowledge, with the significations and propositions to be actualized in thinking and knowing, with the Ideas and laws of Ideas that constitute the Idea of "understanding." It therefore does not concern itself with representing, judging, inferring, comprehension, and the like, *as* psychical experiences, but rather with certain Ideas (with concepts, propositions, implications, and the like, as particular species) that can be extracted from such acts with insight by means of (authentic) abstraction. And it is to these particular species that those general laws, "logical laws" in the strongest sense of the term, have reference. To have clearly seen that the core content of any normative and practical logic consists in propositions and theories that do *not* deal with acts of thought, but rather with those Ideas instanced in certain of their Moments, seems to me to be one of the most momentous of logical insights. It lies hidden in *Bolzano*'s *Wissenschaftslehre* – which unfortunately has only restricted the appreciation fitting to his discovery through the manner in which it introduced his fundamental conceptions of the "representation in itself" and the "proposition in itself," omitting any examination into their phenomenological origin.

The *Wissenschaftslehre* has remained unused by *Bergmann*, as well as by almost all contemporary logicians; and hence he also is lacking in that essential Idea of formal logic which was at least foreshadowed [165]
in *Bolzano*'s unique treatise by his theory regarding representations and propositions "in themselves." Also *Kant*'s definition of pure logic as merely *formal* – profound, though open to misunderstanding

– is rejected by Bergmann. He fails to appreciate the valuable content of the Idea, which proves to be efficacious and fruitful precisely in his own treatment of the discipline. Certainly the manner in which the distinction between material and formal logic is *traditionally* drawn, and set into relationship with the distinction between formal and material truth, is not to be condoned. When (in subsection 2, devoted to the critique of formal logic) *Bergmann* rejects the restriction of the task of logic to mere freedom from contradiction, or to "formal truth," instead of to agreement with the object (or material and real truth) – and when he declares inadmissible the opinion that rules which subserve the goals of knowledge could intend only the formal correctness of knowledge – then we are in complete agreement with him. He rightly points (pp. 10ff.) to the fact that the demand for consistency finds its justification in the demand for truth, and that in all presentations of our discipline – even in those of the formalistic logicians – we see rules formulated which could never derive their justification from the goal of mere consistency. In spite of all this, there is at work in the assertions and logical presentations of the formalists a valuable Idea – only it is one muddled by the unclarities of tradition. And, with specific regard to what the author himself quotes from Kant's *Critique of Pure Reason*, that Idea – appropriately interpreted – touches upon what is correct. It is solely a matter of how we take the concept of logic; and, if it is to be taken as *pure* logic, we must be more attentive than is usual to its theoretical character. As soon as, in logic, we place the goal of knowledge to the fore, and require of logic rules and norms for the advancement of *that* goal, we stand on the terrain of the logical *Kunstlehre*. But then we must also be consistent, making this practical logic as practical as is possible, and as practical in every respect as the furthering of the goal of knowledge makes seem desirable in any way. But if we restrict ourselves to general and pure logic – understood as a theoretical discipline – then there is no talk here of goals and norms, or at least there should be none. It is here the same as in [166] pure and theoretical arithmetic, which formulates laws for number relationships, but no norms for calculation. It is obvious that such laws at any time admit of transformation into norms, by introducing the concept of the regulation of practice, originally foreign to them. It is the same way with pure logic and those rules immediately to be formed from its laws by putting them into relationship with the

goal of knowledge. What the formalists usually say of logic is, then, true of this pure and theoretical logic, but not of logic in general. Pure logic is "formal" logic, as the science of the "forms" of thought. The "matter of knowledge" does not fall in its domain. It precisely investigates those laws alone which are grounded in categorial forms, and thus provides truths which are indifferent to how those forms may be filled out with sensuous "matter." Those derived truths referred to above no longer belong, then, to the content of pure logic itself, but rather to the range of its possible "applications." The generality of form, and of the lawfulness relevant to form, has, given all this, precisely the sense of mathematical generality – as the most general concept of the purely mathematical coincides with the categorially valid or the purely logical. In the purely logical or mathematical sphere is to be found no "material" truth, i.e., none containing sensuous concepts, because it belongs to the sense of the purely logical to leave all matter indeterminate – in the manner of the undetermined *a*, *b*, *c*,... of algebra. The contrast between the "material" and the "formal" is, as one sees, identical here with that between the "sensuous" (perceivable in outer or inner sensibility) and the "categorial." Formal logic in the broadest sense (or, also, pure *mathesis*, taken in its widest extension) is the total range of the purely categorial, i.e., of all laws and theories that stand free of sensuous concepts. That the domain of these laws is one simply free of contradiction, and that it includes not a single existential proposition about the real world, no single assertion about "facts," but rather only propositions, and all propositions, which belong to the sense (*Sinne*) of thinking as such (i.e., precisely to the categorial forms): – All of this is now clear. And it is likewise clear that, for any thinking which follows these laws as norms, only "formal" truth is guaranteed; and not without further considerations is "material" truth secured. For only such material truth is guaranteed through [167]
those laws as is grounded, not in the determinate matters given in the particular case, but only in their forms, and which therefore remains valid under arbitrary variation of the matters. It is not just any error that is a "*logical* absurdity." It seems to me that this interpretation well describes the proper tendency of the doctrine of the formalists, but now freed from inadmissible constrictions or confusions – a tendency with which precisely our author, given his mode of treatment for logic, could not be wholly unsympathetic.

There is a very close connection between the question about the correct ways of defining the concept of logic and that about the relationship of logic to the other philosophical disciplines. The author gives a detailed discussion of its relationship to metaphysics in subsection 3. He disputes the view that metaphysics and logic are basically only *one* science (namely, *Wissenschaftslehre* in Trendelenburg's sense). Rather, we have in them two sciences that are to be kept distinct. However, they are so related that each must deal with the problems of the other. (p. 21) To metaphysics, as the science of being as such, he assigns the task of clarifying the *concept* of being and of resolving the problems inherent therein. (p. 20) But the concept of being is said to be inextricably involved with that of thinking: a relationship that in no wise entails the *identity*, or even the partial identity, of the two sciences. In the system of philosophy metaphysics is prior to logic. The latter must take the solution of certain problems over from the former – problems which, "along with being as such, simultaneously have as their concern thought in relation to knowledge as a goal." (p. 23)

Bergmann's position on these matters is one to which I cannot assent, or can accept only in a highly qualified manner. In any case, everything turns upon which Ideas of logic and metaphysics are presupposed. If one is restricted to that pure logic mentioned above, as a theoretical system of purely categorial truths, then *this* is certainly prior to all sciences precisely in virtue of the fact that it holds itself removed from all matter of knowledge, leaving it completely undetermined. Thus there is also (and especially) no need in pure arithmetic for premisses stemming from any other sciences, and in particular from metaphysics. But then pure mathematics in the usual sense, [168 as well as in the broadest of senses, requires a supplementary philosophical investigation that illumines the difficulties with respect to the sense and objective validity of its fundamental concepts and laws. What is in issue with that investigation are not inadequacies in the interrelations of theorems or in their establishment through axioms. To fill out such deficiencies would be a matter for mathematical work, not philosophical. Rather, that investigation concerns the satisfaction of epistemic needs which would be left unsatisfied even by the ideally completed mathematical theory. These needs do not bear primarily upon the sense and accomplishments of particular mathematical theories, but rather upon thinking in general and upon the elemental

forms and elemental laws belonging to the logical content of thought. In short, a critique of knowledge is required. If one takes the concept of pure logic so broadly that it includes a philosophically illuminating critique of knowledge, the answer to the question of how, then, logic and metaphysics are interrelated depends upon how we specify the concept of metaphysics in relationship to the theory of knowledge. One will scarcely wish or be able to separate the elucidations (*Aufklärungen*) relevant to the *concept* of being off from other epistemological investigations. Metaphysics, as *Bergmann* defines it, thus would be integrated into what we here (and, I hope, quite naturally) designate as critique of knowledge. It, along with pure logic, would be prior to practical logic and to all of the sciences. But the ordinary signification of the term "metaphysics" is, if I am correct, something quite different. According to it metaphysics has the task of extracting the ultimate and absolutely valid determinations of *real* being from the jumble of suggestions, merely provisional and relatively valid, to be found in the empirical sciences. While these sciences are more interested in law-like formulae for use in orienting us within the world of appearances, and while assumptions about existence within them have the status of mere instruments to that end, the theoretical interest in metaphysics is primarily directed upon being itself. It aims to seek out the ultimate sense of the world. It will not be satisfied with those formulae of mere economy, which allow us to construct, forwards and backwards, the course of phenomenal existence and process. It does not have an eye to the practical domination of nature. It merely wishes to know what is true and actual. Metaphysics in *this* sense is the science of what can be definitively asserted about that which [169]
really exists in the domain of actuality, given all of the particular sciences, and constantly utilizing the elucidations won through the critique of knowledge (concerning the *concept* of being, concerning the concepts of the *forms* of being and of thought intimately interwoven with that concept, and concerning the sense and possibility of objectively valid knowledge). *This* "metaphysics" obviously is something subsequent to all of the sciences. The same is clearly true of the logical *Kunstlehre* that is based upon pure logic and the critique of knowledge, upon empirical psychology and, as methodology of the individual sciences, upon all of the particular sciences. As to what we have said in attempting definitions of the relationships between those disciplines to be singled out under the titles "logic," "critique

of knowledge" and "metaphysics," I think we may assume that the author is basically not far removed from us, and that, from his point of view, he could well be quite satisfied with those definitions.

In subsection 4 the author deals with the "kind of knowledge" that we have in logic, and with the "division" of logic. Logic is, according to him, totally devoted to the consideration of the *concepts* of thought, of truth, and of knowledge. Norms which have to do with the "understanding in and for itself," with thought considered independently of all other types of mental processes, are one and the same thing as norms that have their grounds in the goal of thought and therefore can be understood only in terms of that goal. So the author is in agreement with *Kant's* view that pure and general logic is independent of all contingent experience, and that it is a demonstrable doctrine, completely apriori. But he wants to maintain this same position on the nature of applied logic as well. The rules of this applied logic are still to constitute knowledge apriori, even though the objects in the domain of application are given through experience. For those rules do not assert that certain objects – or, in general, objects with such and such characteristics – exist. Rather they only assert that, *whenever* the understanding engages with objects so characterized, it must do so in conformity with those rules.

In all of this the author proves himself an advocate of formal logic (rightly understood). Only in one thing is he incorrect, and that is in taking the laws of formal logic to be *essentially* normative; whereas on my view, indicated above, the thought of norming is originally foreign to them. But, in any case, the laws in their normative formulation [170], and the propositions obtained from them by a mere transference ("application") to determinate objects, remain apriori, just as in the case of arithmetic. But the natural expansion of normative and practical logic beyond the sphere of such *mere* transference to determinate subjects of knowledge – its expansion, for example, into a scientific methodology – of necessity abolishes its exclusively apriori character. Reference to the contingent particulars of human nature also will introduce into logic, directed in general to the advancement of knowledge and science, rules which do not draw their justification from pure logic alone. Already with the theory of how, partly helping and partly hindering, language influences thinking – for example, in the distinguishing of the most important classes of harmful equivocations – we clearly are dealing with matters that belong within a range of *a posteriori* logical doctrines. – We essentially agree when, finally (p. 28), the author objects against *Kant* and his followers that they merely lay down logical rules without deriving them from the concept of knowledge, or without showing how they arise out of the concepts of knowing and of truth; and when he demands that logic not rest satisfied with the immediate certainty to be found in logical principles, but instead arrive at "insight into thought." That is, we agree with the interpretation according to which logic must not be a mere formal (mathematical) theory of the purely categorial laws, but, as philosophical logic (and independently of its extensions bearing upon the practical aspects of knowing), requires phenomenological and epistemological elucidations in virtue of which we not merely are completely certain of the validity of its concepts and theories, but also truly understand them. The "empty formalism" so often scorned does not satisfy the philosophical interest – which proves nothing against its intrinsic value, for it does satisfy instead the "mathematical" interest. The critique of knowledge illumines the objective sense of the "empty" forms. It constitutes the specifically philosophical task.

The author closes his Introduction with a justification of his "division" of logic. He distinguishes two main parts, the first of which deals with thinking and knowing in general, while the second deals with *progress* in knowing. The first falls in turn into two sections. [171]
The earlier, under the title "Forms of Thought," is mainly to deal with the *forms* of thought common to both true and untrue thoughts, on which rests the natural aptness of thought to the goal of knowledge; whereas the second section aims to specify the concepts of knowledge and truth, along with the criteria of truth and falsity grounded in them. We shall first go point-by-point through the content of the earlier section.

In subsection 5 the author prefaces his investigation into the "general modes or forms" of the judgment with the observation that "in any simple judgment something is affirmed or denied of something, the object." And he declares that the more exact analysis of the sense of the impersonal, the hypothetical and the disjunctive judgments will make clear that they are merely *apparent* exceptions. The representation or concept is explained (somewhat too summarily, and not quite with clarity) as "the having of an object in consciousness or thought; or the positing of an object in such as way as it forms the foundation for any judgment." (p. 31) What is meant, as the following pages make clear, are all acts which can function as complete subject acts for predications. Thus we cannot speak, according to the author, of predicate representations; for predicates, he holds, "do not have the signification of objects for consciousness." The narrow delimitation here placed upon the concept of the representation (and it is a delimitation that certainly coincides with a concept essential to logic) requires as its supplement a broader concept or term, suited to all parts of the signification of an assertion that cannot themselves be regarded as full assertions. If we are not permitted to speak of *predicate* "representations" – and similarly for the representations that correspond to the words "beside," "right of," "and," "but," and so forth – then what else are we to say? Obviously there is lacking here distinctions between broader and narrower concepts of representation, along with their essential modifications. There is at the same time lacking the *phenomenological* investigations suited to clear up the immense difficulties inherent in the Ideas *signification* and *representation*.

There then follow two generic distinctions between kinds of "representations": the distinction between singular and general, as well as that between concrete and abstract representations. The singular [172] representation has for its object an individual thing in its individual peculiarity, while the general representation has a class or genus of things. Concrete representations are those whose objects have, by nature, the signification of independently existing beings: *things* or classes of things. By means of abstract representations, by contrast, there is signified or represented something existing only *in* or *on* things, or by means of such things, in that a thing-like being is projected for it. (pp. 32–33) – According to these explanations by the author, the first distinction would fall wholly within the framework of "concrete" representations. But then how are we to accommodate the distinction between the representation of the individual color *in specie* ("red," let us say) and the representation "color in general"? Further, where do representations like "one unit" and "the number two" belong? In fact, I then do not even know what to do with representations such as "a tree." Likewise with totality representations, such as "Socrates and Plato," or "one number and another number"; and plurality or universal representations, such as "several men" and "all numbers in the number series"; and numerical representations (two propositions, three horses); and so on. It seems to me that

the number of essential logical distinctions within the concept of the representation must be very considerably increased. It also seems doubtful whether abstract representations in which we 'substantialize' the dependent Moments of *things* are to have the character of "fictions," which impute a thinglike being to such Moments. Do we not here, as well as with forms of thought in general, have to do with a primitive and essential form of thought, which is therefore wholly indispensable? If this form had the sense of a mere fiction, then the related forms of assertions would have to have merely non-literal significations; and it would have to be possible, in every case, to provide equivalent reformulations through which the general forms would be completely dissipated. I cannot believe that such transformations are possible. I am unable to make any truly general proposition coincide in its signification with one that is not general. I think that one can never get by without conceding that propositions (universal or singular) concerning numbers, figures, and colors (all taken *in specie*), are completely univocal and correct expressions of their sense, and that in this sense there is to be found nothing of the *fiction* of thinglikeness [173]
as to their species objects. I cannot even imagine what such a fiction could ever accomplish in thought. I would further hesitate to identify the objects of *abstract* representations with the individual, dependent Moments in which they are only instanced.

In subsection 6 the author begins with a discussion of the concept of the determination (the mark or characterizing property (*Merkmal*)). According to him a determination (*Bestimmtheit*) belongs to a thing at a certain time. It is what may be predicated of the thing in relation to that time. This narrowly confined concept of *determination*, although important in itself, suffers from the fact that it cannot prevent the broader concept – the concept of predicate (or predicable) *simpliciter* – from inexorably intruding upon it. We indeed do speak, and scarcely can avoid speaking, of determinations or characterizing properties of algebraic numbers, of hypothetical propositions, of analytic propositions, and the like, wherein there is no talk of "things" and of temporal relations. How unsuitable it is so greatly to constrict the term "determination" is shown by the role which it, in fact, plays in all of the author's further discussions. All of the purely *logical* distinctions are drawn under it; whereas the greatest areas of such distinctions, e.g., the entire syllogistic, are totally indifferent to the distinction between the individual or real predicate and predicate in general.

There follow (pp. 36ff.) discussions of the concept *the content of a representation* and of the distinction between constitutive and supplemental determinations. Each object represented is *this* particular object here, for the one doing the representing, in virtue of the determinations under which it is represented; which determinations, as is usually said, form the content of his representation. To the content there belong, according to the author, at least so many determinations of the object as are requisite for the representation to be precisely of *this* object and no other. But representations (such, for example, as "Berlin, the capital city of Prussia") can also contain more determinations than are required for that purpose. If that is the case for a given representation, then it is possible that the one doing the representing should credit a part of those determinations with the role of making the represented object appear to him as this determinate object and no other, while the remaining determinations are taken as supple- [174] mentary to that part. "If a representation is developed in this fashion, then we shall call the former determinations constitutive . . . and the others supplemental." The constituting content of a representation is then defined as the totality of the constituting determinations of the object. (pp. 37–38)

Perhaps the reviewer will be permitted here, once again, to add a few reflections. To begin with, as concerns the expression "content of a representation," it can be interpreted by means of various important concepts. If by "representation" we understand the act, the lived experience (*Erlebnis*) of the representing, then we distinguish the content in the psychological sense (all that enters into the psychological make up of the lived experience) from content in any logical sense. But with respect to logic "content" can mean several different things: 1. The object of the representation. 2. The sense of the representation, e.g., the identical signification of the verbal expression. The same object can be meant by means of different significations. But the usual language used for this "sense," or this "signification," is not "content" of the representation. Normally we just speak of "the representation" itself. 3. Content in contrast to categorial form. This "content" is the totality of the partial representations (partial significations), whether they bring the object itself to *direct* expression, or do so for the objects, characterizing properties or forms of complication related to it. Thus we would have the same "content," in this sense, for the two representations, "Socrates, the teacher of Plato" and "the teacher of Plato, Socrates," which differ in their senses. The content in both cases is the *group* of representations (significations): Socrates, Plato, teacher. The distinction resides in the logical "form." 4. If one grounds talk of the "content" of the representation on the content of the object, then it can mean the total set of parts and Moments in the object which find their explicit or tacit expression in the representation (or signification).

This latter concept of content appears to be what the author has in mind, whereas the other concepts – certainly of no less importance – are not considered. As concerns, then, the associated distinction between constitutive and supplemental determinations, I must confess that I am unable to get completely clear about it. If I have got the matter right, the author has important distinctions in view, but in his [175] presentation they do not stand out with complete clarity. The same objectivity can be represented through distinct – that is, logically distinct, in terms of signification content (cf. #2 above) – representations. In these matters an essential distinction is to be drawn. On the one hand we have the "direct" representation (e.g., the signification of a proper name), which directly refers to a determinate object without favoring, *in* its sense, any among the determinations that are co-intended in its sense. And on the other hand we have the "indirect" or attributive representation, which, engaging in such favorings, refers to the object *as having* these or those determinations – and is otherwise indifferent as to whether it also simultaneously refers to the object directly, or else refers to it indirectly in the fullest sense, i.e., *only* as "something" thus and so determined. In the univocal usage of the proper name "Socrates" there may be psychologically favored, depending on the occasion, now this and now that property or relationship of Socrates. But none are favored *logically*, for the sense of the word remains unchanged. It is otherwise with representations such as "the teacher of Plato," "the son of Phanarete," and the like. The apprehension of this same object *as* an object having this or that property belongs to the content or sense of these representations, which are neither identical in signification with one another, nor with the representation in the proper name. The constituting content (or, better, the attributive) would then be the totality of determinations attributively related to their objects *within* the sense of the representation concerned. Whether the object is a *thing* or not, or whether the representation of it contains more determinations than are requisite to distinguish it from all other objects – and whatever else might yet be introduced into the discussion by the author – all of this, it seems to me, does not matter. This concept of content also in fact covers representations that refer to objects in an "indeterminate" fashion: that is, those which result from the concoction of attributions out of the representation *something*, together with representations derived by the addition of purely categorial forms. (For example, something *and* something.) With these distinctions, as all similar ones, we are moving within the sphere of pure forms of signification. The general task in this domain is to determine all distinctions grounded in the Idea of the "sense" (or of the signification content) of a representation, abstracting from all that is material – in other words, to determine the general modes, distinguished solely on grounds of pure logic, in which representations can have reference to objects. –

Subsections 7–11 lead us into metaphysics. To begin with, the author (in subsection 7) briefly but very instructively discusses the relationship between thing and determining property. He explains that no *thing* is more than the totality of its determinations, or is more than all of that which an all-knowing mind could predicate of it. Therefore, nothing which is in an absolute contrast with all that is perceptible can be attributed to a self-existent thing as its determination. The author further treats, in an admirable fashion, of how the relationship between whole and part (= fragment [*Stück*]) is distinguished from that between thing and its determination. Thus is obtained the fundamental definitions: A thing is what is capable of an isolated (real) existence. The decomposition of a thing into parts signifies a decomposition of it into things. By contrast, determinations cannot exist by themselves, but rather only in or with something else: namely, only in their union with other determinations to form a thing. The author manifests his connection with *Berkeley* by always giving definitions with double meanings. For in each case (and this we have suppressed in the previous sentences) he mentions that which can and cannot *be represented* by itself right along with that which can or cannot *exist* by itself: a complication this reviewer finds superfluous, and, in fact, misleading. The dependent Moment in an object also can be represented by itself. It can be taken as subject of a predication and thus become objective for thought. But it cannot be represented or posited in a true judgment as "something in itself" – i.e., as something existing in isolation. And this is because it is, precisely, something dependent, and in virtue of its very nature cannot exist by itself.

The author then proceeds to discuss the contrast between the dependency of determinations and that wholly different kind of dependency of *the thing* itself within the unity of the encompassing network of things. He then comes at the last to the question of how the determining properties of a thing interrelate with each other. He responds in a fashion that certainly is pertinent and significant (given a proper grasp of the concepts), by saying that this relationship is that of consequent to ground. [177]

The next section (#8) elicits one doubt after another. Here the author tells us that, in virtue of the fact that we represent it, we represent everything which we do represent *as* a thing with determining properties, or *as* a class of things which have determinations; and likewise, that we represent everything *as* contained in the world – as present, past or future constituent of the world – the idea being that consciousness of the world is contained in *all* representing as presupposition and foundation. But we have consciousness of the world in conjunction with consciousness of our self. We are unable to posit the ego otherwise than as a something-or-other in the world, or the world otherwise than as that wherein our ego exists, and as that wherein our ego posits itself as being. This being contained within the world, or standing in relation to our own ego, is, the author further claims, the same thing as *existing*. When we say that something exists we mean that it belongs to the world of our ego. Thus, all that we represent is represented in virtue of the fact that we represent it *as* existing (or as having existed, and so on). This is true even when the representing is a mere fictionalizing by phantasy. From this it further follows that existence or being does not belong among the determinations of things represented as existing. Thus, existence is not a constituent of the *content* of the representation of any thing; and, this being so, it also makes no sense to predicate it of a thing. When we assert that a thing exists, what we are making more determinate is not the representation of that thing, but rather that of the world which is the world of our ego, or the world which our ego represents itself as belonging to. That is, we posit the world as containing the object of which we assert existence. The world of our ego is the subject of all our existential judgments, and that it includes the thing is the predicate. The author resolutely draws (p. 46) the consequences of this view: In order to be able to represent anything as existing, I must already have represented my ego as existing. But that my ego exists means (since the concept of existence is the same in relation to the ego as in relation to other things) that it stands in relation to the ego already represented by me as existing. The representation of my ego accordingly has itself for a presupposition. In order to be able to acquire it I must already have it. Or perhaps I never at any time *begin* to have it, but rather just have it continuously. [178]

There is not a single point in this part of the author's expositions that I can accept for myself, and, indeed, I find nothing to recommend what he is saying. To the question of what we represent when, for example, we represent the number π, it seems to me that a prejudice-free absorption in the sense of the representation yields the simple answer: precisely the familiar transcendental number, which surely is no thing or class of things. And in vain do I endeavor to introduce the thoughts of world and ego without modifying the sense of the representation. And is not, I would further inquire, the thought of the world the same as the representation of the world? Thus there would be true of it what is true of every other representation: we would not be able to represent the world without representing it *as in* the world. Further, if I say "Socrates exists," then this means, according to the author: Socrates is contained in the world of my ego. If someone else says that Socrates exists, he therefore means that Socrates is contained in the world of *his* ego. The two assertions, though verbally identical, would differ in their signification, even if the world of the one and the other ego is the same. The existential sentence would of necessity have as many significations as there are persons which assert it. And even with this I do not altogether exhaust the difficulties.

If we forgo closer consideration of subsections 9 through 11 – in which the author deals with the relationship between temporal statement and "determination" (in the narrower sense that he adopts), with the possibility of changeable things or of singular representations, with the (possibly objective or real) significance of the singular and general forms of representation for delimiting individuals and genera, as well as with such a possible significance for the distinction between the constituting and the supplementing determinations of an object – then we are led back once again, by subsection 12, more into the central questions of logic. This section is concerned with the "extension" of a representation, and with the relationships of representations with respect to content and extension. The concept of extension is based upon that of the constituting content. The fundamental definition resides in the following proposition, here given a somewhat shorter formulation: The object of the representation *B* belongs to the "extension" of representation *A* – this latter being assumed free of contradictions – provided that the constituting content of the former is contained in that of the latter. Now it seems to me that the dubious concept of *constituting content* does essential [179] damage to these conceptual definitions and to all others constructed upon them, and, above all, constricts them in a logically inappropriate fashion. Are we not to be permitted to say, for example, that the object of the representation "Socrates," or perhaps "this," belongs to the extension of the representation *man*? It would be most natural, by the way, to link the concept of extension primarily to the concept of *predicate*, and not to that of the "representation" in the sense defined by *Bergmann*. This should be done in such a way that by the (ideal or real) extension of a predicate assumed to be in itself "possible" there would be understood nothing other than the totality of the (possible or actual) subjects of that predicate.

In his discussion of the subordination relation, the author seeks to revive the old theory of the reciprocity of extension and content, by understanding the content to be constitutive and the extension to be ideal. I cannot see that in this way one escapes the objections, which likewise are already old ones. The enrichment of the constituting content will certainly not narrow the extension if the enrichment is by a predicate necessarily connected with one of the original predicates and nevertheless not contained in it. At the conclusion of subsection 12, the sense of negative predicates is discussed in connection with the theory of contradictory opposition. Such predicates express no negative determination, we are told; for there *are* no negative determinations, just as there are no positive ones. Affirmation and denial are not properties of things, nor are they constituents of the contents of representations. Rather, they are modes of judgment. The sense of the negative predicate lies in a relationship to a negative judgment; namely, in this, that a certain negative judgment holds true of the object. This is the author's view. The contention that a negative predicate incorporates reference to a negative judgment must, in my opinion, be set precisely on the same level with the parallel claim that a simple positive predicate (such as "red") incorporates reference to a positive judgment ("*G* is red"). But this already involves the author's general theory of judgment, to which the next subsections (#13–17) are devoted.

A REPORT ON GERMAN WRITINGS IN LOGIC FROM THE YEARS 1895–1899

Second Article[1]

J. Bergmann, *Die Grundprobleme der Logik*, second and completely revised edition, Berlin, 1895.

The essential content of *Bergmann*'s theory of judgment is as follows:

In every judgment there is contained, as its presupposition or foundation, a simple representation of an object. The question, then, concerns what is added in the judgment to the simple representation of its object. The author first criticizes, in a quite decisive fashion (subsection 13), the older conception according to which in any judgment a certain *relation* is supposed to be thought. On that view the judgment is to be understood as a specific type of mere representation – one, namely, which refers to the relationship mentioned. That "many, if not most, logicians of the present day" stand under the influence of this viewpoint, as the author says, is something one will surely find hard to affirm now. To the contrary, no view could be more strongly discredited currently, given the influence of *Brentano*'s penetrating critique in his *Psychologie vom empirischen Standpunkt*, of 1874. In any case we are happy to agree with the author on the substance of the matter. He rightly emphasizes that it is impossible to explain the concept of the judgment simply by means of the concept of representation, or, in general, to derive it from other concepts. The concept of judgment can be cleared up in only one way: by presupposing the peculiar character of the judgment to be something

[1] This continues the review begun in the previous article. It first appeared in *Archiv für systematische Philosophie*, Vol. 9, 1903, pp. 237– 259. Republished in *Husserliana XXII*, pp. 180–200. [D.W.]

given.[2] [181]

This clarification is what is now at issue. The author begins with the proposition that every general affirmative judgment (including the singular judgment) refers a determination to its object, and, in that it does so, represents the object under that determining property. Thus the determination predicated always belongs to the content of the representation which the one judging has of the object of his judgment. But it never belongs to the "constituting" content of that representation. In other words, there are no "tautological" or "enantiological" judgments. Expressions of the forms, "*S* is (or is not) *S*" and "The *S*'s that are *P*'s are (or are not) *P*'s," are not expressions of judgments – or else they are not accurate expressions of what the speaker meant. All *judgments*, rather, are heterological (p. 73). In the interpretation of tautological assertions the author follows *Beneke*'s well-known theory.

On this point I cannot agree. No doubt in tautological assertions, such as "Napoleon is Napoleon" and "Blame is Blame," there usually is a modification in the sense of the term repeated. But it does not follow that that modification *must* occur. I deny that such expressions – and especially the partially tautological among them – *cannot* be understood in the literal sense dictated by the form of the expression. Since I, for my part, can carry through with genuinely tautological assertions as often as I wish, I have to contest the explanation – at first given for the general affirmative judgment alone – that the judgment is "a reference of a determination to an object, through which the determination is given the signification ... of a supplemental determination." (p. 75)

Subsection 15 advances the author's view on the essence of the negative (categorical) judgment. In such a judgment, first of all, we do not have a negative determination, non-*P*, predicated of *S*; as if its contrast with the judgment *S* is *P* lay in the predicate. For there are no negative determinations. If one then puts forward the view that negation modifies the copula – that to the positive mode of referring predicate to subject there is a co-ordinate negative one – it turns

[2] Decisive for the delimitation of the true sense of such theories is, certainly, the establishment, from among the many concepts of "representation," of the one that comes into play here. The greatest of difficulties reside in the relevant phenomenological analyses. With regard to these tasks, theories advanced up to now have, in my opinion, omitted much.

out that such a co-ordination does not exist; and negation is found, instead, to be an action that incorporates the positive referral which results in the affirmative judgment and that simultaneously rejects the judgment as invalid. I cannot deny *P* of *S* without referring it to [182] *S* in the manner of the thought that *S* is *P*. Negation consists, then, in the fact that I make that thought itself the object of my judgment and hold that very thought to be incorrect. The obvious objection of an infinite regress does not escape the author. If the assertion that *S* is not *P* does in truth say that the judgment *S* is *P* is *incorrect*, then it would seem that this interpretative (and, once again, negative) judgment requires the same interpretation. And so on *in infinitum*. The author settles this objection by considering that the incorrectness of a judgment – its property of being in contradiction with its object – is a positive determination of the judgment, so that the statement of that incorrectness is an affirmative judgment.

Since, like an affirmation of incorrectness, an affirmation of the correctness of a judgment that *S* is *P* is obviously possible, and since in this latter we have an affirmative judgment concerning *S* (that *S* is *P*, is actually and in fact *P*), then there are two types of affirmative judgments: the simple *S* is *P*, and the emphatic *S is P* – i.e., that *SP* is, is correct. However the author wishes to call only the latter "affirmative," in which the affirming – just like the denying in the negating judgment – is a peculiar thought action. And at the same time he wants to subsume this affirming judgment *and* the negating judgment under a narrower signification of the word "judgment." The wider signification for this word also is to cover the relation, one neither of affirming nor of denying, of a determining property to an object – the relation of "mere predication."

Thus, any judgment in the narrower sense contains 1. a representation, i.e., a positing of an object, 2. a predication, and 3. a critical attitude toward that predication – an acknowledgement or a holding of it to be valid, or else a rejection or a holding of it to be invalid. (p. 78) Finally, the author extends to *all* judgments (p. 79) the results of his analysis, arrived at in the first instance by considering the general judgment alone. For, he says, every judgment, as a product of thought, and subject to appraisal as true or false, is *eo ipso* either (in the broadest sense of the word) affirming or negating, and therefore leads back to one of the cases distinguished. This, in short, is *Bergmann*'s theory of judgment.

If, as a co-laborer in this area, I may be permitted to give a somewhat more detailed critique, I would object as follows: The inter- [183] pretation of the general denial, *S* is not *P*, by means of the assertion, "That *S* is *P* is incorrect," falls prey to an objection already raised (if I remember rightly) by *Bolzano*. Namely, that one of these assertions can be true and the other false. They are therefore not even equivalent as to their senses, much less identical. If I reject the judgment, "The prime numbers are even numbers," I do not therefore *assert* that the prime numbers are not even numbers – which would be false with reference to the number two.

The author's pertinent idea, that in the negative predicate "incorrect" there comes to expression a peculiar type of consciousness of contradiction, can serve as a starting point for the analysis of the sense of *any* arbitrary negative predicate. According to *Bergmann*, in judgments of the form, "That *S* is *P* is incorrect," we are dealing with a contradiction between the object and the judgment; or, as we would prefer to state it, between the true state of affairs (*Sachverhalte*) – i.e., one that could possibly be given with Evidence – and the judgment. But if we inquire about what lies *within* the consciousness of contradiction and what it presupposes, we will be forced to look at the cases in which that consciousness is brought to genuine and full realization, and in which the contradiction therefore becomes Evident. The judgment (as the judged *of* [*beurteilte*], and hence not as something that is of necessity always judged *by* [*geurteilte*] the one who is making the judgment) presumes *S* to be *P*, when in truth that is not so. The perceived, demonstrated, firmly believed state of affairs disagrees with the merely presumed one. But this disagreement presupposes that the meaning of that judgment has an inherent relationship with the state of affairs that is given or definitive. This is a relationship that as it were lays upon the judgment the requirement to do justice to the state of affairs, instead of disagreeing with it as it in fact does. Obviously the judgment meaning must in some part impinge upon the given or firmly established state of affairs. A part of its content must correctly "express" or "agree with" (or presume to agree with) a part or aspect of the state of affairs. But it does so in such a way that yet others of its parts, instead of likewise agreeing with correlated parts in the state of affairs, rather stand in contradiction to them.

Thus we find that a presupposition for the possibility of a contradiction between a judgment and a state of affairs definitive for it is contradiction between certain parts of the judgment which are no longer themselves judgments and parts of the state of affairs which [184] are no longer themselves states of affairs. These contradictions are, in their own right, expressed by *simple* negative predications in which some determination is denied of the given object as subject, posited as existing. For example, if someone says, "This surface is green, whereas I see it as red," then the "is green" is imbued with a consciousness of contradiction, which I express as "It is not green." This form, *S* is not *P*, is utterly primitive, and is also to be considered as the correlative of the affirmative form, *S* is *P*.

One should not object that the former conception, in the sense laid out, is a special case of the latter, and is so because it *asserts of* the *S* – thus *affirms* of it – that being *P* contradicts it. One must beware of taking "circumscriptive" expressions (as *Marty* on one occasion has nicely put it), which refer to an idea in an indirect and paraphrastic fashion, as identical in signification with that idea itself. Certainly there is present in a negative proposition an expression of the contradiction of *P* with *S*, but it does not affirm the determining property "contradictory to *P*" of *S*. Rather, it expresses contradiction in the same sense in which "*S* is *P*" expresses agreement: namely, expresses the fact that *P* belongs to *S*, fits, agrees with *S*. Here too it is *P*, and not the belonging of *P* (or, rather, the "having" of *P*) that is the determination asserted. In reality, the judgment in its primitive and simple form first of all posits *S*; and upon the basis of that posit, intended to assign to the object the status of something existing, the supplementary and non-independent act "is *P*" is carried out. This dependent act is to be designated as the act of "predication" in the pregnant sense of the term, while in the wider sense the complete, whole act of holding *S* to be *P* bears that name. In the Evident judgment that which the judgment intends, the state of affairs, comes to "perception" (a much broadened concept of perception, not to be confused with the ordinary concept of *sense* perception). That *S* is *P* is viewed (*erschaut*), and, in this unity viewed, there then lies the basis for the abstraction of the relational concept of "belonging to" and of the other concepts joined with it. In reflection upon it we grasp and denominate the "subject" *S* as "having" the "predicate" *P*, and *P* as "belonging to" the subject *S*. Such is the origin of the [185] paraphrastic concepts and assertions. It is exactly the same in the case of the negative judgment with its negative state of affairs. If this latter is Evidently given (or, what also suffices, is imagined with a more graphic fullness), the circumscriptive concept of contradicting can arise, and then can function as a predicate.

One can moreover quite well hold that, in a certain fashion, negative predication does refer back to the affirmative, and thus presupposes the concept of it, without therefore deviating from the position that negative predication stands on an equal footing logically with the positive. In that the *P* stands in opposition to the *S*, it nevertheless raises the claim, as it were, of belonging to *S*. But however one may interpret this "claim" to belong, the signification forms in the two cases are essentially different, and are different in such a way that in no possible literal sense is one contained in the other. It would indeed be quite absurd to say that the negative *judgment* includes the affirmative; for whoever makes the former cannot, in the same act, make the latter. Thus it is only to the signification content of the judgment, abstracted from the *belief* character, that talk of containing or being contained could have any reference – and even then it must not be taken literally.

This view aims to do justice to the sound nucleus of the two ever recurrent theories: that negation modifies the copula, and that it appertains to the predicate. It would reconcile them with each other. The "not" certainly does belong with the "is." But the "is" is no copula, no connective bond between *S* and *P*. Upon the positing of the subject "*S*" there is grounded, in an immediately transparent manner, the "is *P*," which is non-independent – as has already been mentioned – but is nevertheless one unitary act. It is not that to the "*S*" there is linked by the "is" a "*P*" of equal standing, just as autonomous as is "*S*" itself. Rather, if one just *will* speak of a "linking," there is linked to "*S*" the unitary "is *P*." If within the categorial assertion we distinguish, following the natural articulations of the signification, between that whereof the assertion is made (the subject) and that which is asserted of it (the predicate), then not "*P*," but rather "is *P*" or "is not *P*," is the predicate. Obviously this is what is had in view by the theory that locates the "not" in the predicate. But, misled by the ambiguity of the word "predicate," it confuses predicate *signification* [186
(in the sense of the "is *P*" just specified) and predicate *concept* ("*P*"). It interprets – fundamental error! – "*S* is not *P*" as "*S* is non-*P*," and hence as an affirmative judgment with a modified predicate concept.

In reality the "not" belongs, just as the "is" always does, within the predicate in the sense of the predicate *act* earlier referred to, or within the signification content of that predicate.

It is already clear, given these statements, that my conception of the sense of the categorial assertion deviates quite substantially from that of the author. I also do not understand why that reflexive judgment which asserts correctness or incorrectness of a judgment that *S* is *P* is to be regarded as "a judgment in the narrower sense," whereas the simple *S* is *P* is to be assigned the name "judgment" only in the broader sense. It would almost seem that the author has been misled by a confusion of two essentially different relationships: 1. the relationship between the mere representation underlying the judgment and the *belief*-Moment consummating the judgment, and 2. the relationship between a plain and simple judgment and the judgment which may be directed upon it in acknowledgement or rejection. If in the theory of the judgment we contrast mere representation with judgment, and start off with the question of what is to be "added" to the "mere" representation to turn it into a judgment, then, in my opinion, we have to include under "representation" the *total* signification content of the judgment, the whole of the meaning of the assertion. Thus we cannot, as in *Brentano* and *Bergmann*, restrict "representation" to the (nominal) representation of the object taken as subject, even though it were to include in one content the representations of the determining properties predicated of the object. Only *one* Moment must remain excluded: that of the "belief," of the characteristic of certainty or conviction. Whoever merely hears and understands the message has grasped its entire sense. However, he lacks "belief." He merely represents, but does not judge. He represents. That does *not* here mean that he has the subject of the judgment (laden with the determinations predicated) before his mind in a unitary intuition or in a nominal representation. Rather, he actualizes the total signification of the assertion, with all of its articulations and forms: thus, for example, the whole of "*S* is *P*," "*S* is not *P*," and the like, exactly in the manner in which the assertion means it. Only he does not *believe* that which he hears. He judges nothing whatsoever concerning it – neither simply assuming it, nor accepting it, nor rejecting it. To each judgment, *S* is *P*, there corresponds the mere representation of [187]
S being *P*. But this latter is in nowise intended as a representation of the *judgment*, but rather as a representation of the (judged or judge-

able) *state of affairs*. If we call the sense of this representation – which encompasses within itself the totality of *what* is believed in the judgment, just as it therein is believed, only abstracted from the *belief* – the "matter" of the judgment, then, from the formal point of view, we come to the same division between matter and quality in every judgment which *Brentano*, and (in a different terminology) also *Bergmann*, has taught. But for us the matter is not the representation, possibly as it existed prior to predicative articulation, and it is no representation expressible by a *name*. And again, the quality is, in our view, no acknowledgement or rejection directed upon such a representation. The thought of a reduction of every judgment to the (alleged) normal form of the existential proposition is far removed from our view. For not only the matter in the traditional sense (*S*, *P*), but also all of the categorial forms – such as the "a" and "some," the "if ... then," and especially the "is" and the "not" – belong to the "matter." The "is" therefore is nothing less than it is an expression of "belief," and much less then is the "is not" an expression of a co-ordinate "unbelief." Rather, the positing and 'certainty' characteristic belongs to the matter as a whole, regardless of however it may further articulate itself into parts. The usual expressions for this characteristic – "holding to be true," "believing," "consciousness of validity," and the like – all suggest the erroneous view that we have here a predication of truth, validity or correctness upon the matter, and moreover, that we must here distinguish *two* co-ordinate qualities: a holding-to-be-true and a holding-to-be-false. Even this latter point does not seem to be absolutely beyond all doubt. Every (normal) assertion expresses a judgment, but every judgment also finds its expression in a possible assertion. Now there certainly are assertions which express disbelief, a rejection, a declaration of non-validity. But this occurs solely in modes of predication which either make an assertion concerning the judging subject and his act ("That such and such is not believed, or is rejected, by me or some other person"), or else, objectively directed, make an assertion concerning the relevant proposition (that it is untrue, incorrect). In each case the expression of the rejection, of the non-belief or the untruth, pertains to the matter of the assertion ; and what makes it an assertion is not [186]
the non-belief predicated, but rather the character of conviction or of "believing" which as it were animates the matter. Every asserting is a "believing." Whether this view, to which (after long vacillation) I

am inclined, is definitive or not, it in any case is grounded in facts that demand attention and clarification.

In concluding, I still must point out that I cannot accept as proven the author's reduction of all judgments to predication. From the fact that all judgments as such fall under the contrast of the true and the false it does not follow that they are either affirming or negating, or that they therefore assert something of something. The question about the fundamental forms of the judgment (or, rather, about the fundamental forms of *the matters* in the judgment) seems to me still not adequately answered after all. That all propositions can be put into equivalent categorical form does not prove, as seems commonly assumed, that the categorical form is essential to every judgment and is only, upon occasion, hidden by grammatical form.

In subsection 16 the author takes a position on the disputed question of whether in every categorial judgment (or, what for him comes to the same thing: in every judgment whatsoever) the existence of the subject is co-asserted. He holds that it is. "Every judgment presupposes the existence of its object." Those common assertions in which we seem to carry out with full consciousness predications upon imaginary or impossible objects do not genuinely state anything about those objects, but rather about their representations or their names, unless they really are to be interpreted as hypotheticals (If *S* exists, then it is *P*). – I strongly suspect that this viewpoint does not satisfactorily explain all of the types of cases. Suppose that, absorbed in contemplation of a painting, we arrive at an explicit judgmental consciousness of the subject. Do not we judge concerning the object painted exactly as we do concerning real objects, while we nonetheless do not for a moment take it to be real? And how do matters stand with objects which appear to us, but we are *undecided* as to whether they are real or hallucinated?

In subsection 17 the author treats of that classification of judgments which wholly "disregards the distinctions between the things judged about" and "concerns simply the form." According to him there are [189]
only three such classifications: those in terms of quality, quantity and modality. (cf. p. 88) By "judgment" *Bergmann* understands – as is obvious from the entire content of his presentations – his judgment "in the narrow sense," and hence decisions concerning the validity of a predication. Such a decision can either be affirmative or negative (declaring for validity or for invalidity: distinction of quality). The

decision can further, in the subjective or in the objective respect, be completed or uncompleted (distinctions of modality and of quantity). It is completed in the subjective respect if it is certain – if it is as it were a definitive decision. It is uncompleted when it is not so. In the objective respect it is completed if it extends to the whole, and uncompleted if it extends to only a part of the whole, of whatever is in question. And as concerns this latter contrast, in the authentic sense it has to do, according to the author, only with predications whose subject representation is a *general* one. Such a predication can, namely, be acknowledged or rejected for the whole of the real extension of the subject representation or for a part thereof; and judgments accordingly divide up into the universal (to which, given an extension of its sense, the singular judgment can be assigned) and the particular – these being commonly expressed by the formulae "all *S* are *P*" and "Some *S* are *P*."

According to this account, therefore, the sentence "All *S*'s are *P*" says the same thing as "The predication '*S* is *P*' is valid for all *S*'s." In the same fashion the forms, "Some *S*'s are *P*" and "The predication '*S* is *P*' is valid for only some of the *S*'s," are identical in signification. I am unable to corroborate this identity of signification. It seems undeniable to me that, for example, in the first identity claimed the complex internal structure of the interpretative proposition contains within itself the concept of the simple proposition equated with it, or at least that of a simple proposition closely related to it. The complicated formula does indeed contain within itself the thought, "The *S*'s are P," to which there corresponds (with the same wording) a possible simple judgment. But "the *S*'s" in that thought are, of course, "all *S*'s"; and so the judgment is either identical with All *S*'s are *P*, or else something to be made identical with it by means of an equivalence transformation applied to the subject representation. In any case it is these primitive forms of judgment which have first claim on being analyzed and accommodated within a classification of judgments, whereas with *Bergmann* they totally disappear in favor of [190]
the secondary forms of judgments about judgments. Obviously this contra-natural treatment of the theory of the forms of the judgment arises out of the erroneous or misleading account of "judgments in the narrower sense," as decisions concerning the validity of predications.

It also is not exactly a happy solution to derive the distinctions of modality from the contrast between the subjectively completed and the subjectively incompleted resolution of judgment. Even worse since their equivalence with the distinctions of quantity is admitted. According to the author the contrast in question, taken in the original and authentic sense, concerns apodictic and problematic judgments. But then he – in an obviously broadened sense – classifies the assertoric judgment on the side of completely resolved judgments. The most concise formulation of the meaning of *Bergmann*'s interpretation lies in the following identities of signification: *S* is actually *P* = That *S* is *P* agrees with reality. *S* is necessarily *P* (or is possibly) *P* = That *S* is *P* is a necessary consequence of what we know already (or is compatible with our present knowledge). The negation of the apodictic proposition therefore expresses non-confirmation, while the negation of the problematic expresses impossibility, i.e., incompatibility with our present knowledge.

One sees that in this classification we have something essentially different from the previous one. If the previous one focussed upon distinctions between judgments as propositions, quite independently of the judging subject and his dispositions, in the latter the opposite is true. It is a viewpoint foreign to pure logic to consider judgments as to whether they follow from someone's knowledge, or are compatible with it or not. The author has, unfortunately, overlooked the objective and purely logical concepts of possibility and necessity which are in addition to the subjective concepts. "It is a necessity that *S* be *P*" certainly can also have the objective sense: It conforms to law that *S* be so, or its being so is a law or a consequence of a law. But we also find that with the corresponding formulations in pure logic of judgments concerning necessity, possibility and so forth, there arise no forms which could be set equal with those treated under the heading of quantity. For they do not, like the latter, express simple forms of judgment. In the nature of the case, the very first distinction within the pure theory of the logical forms of the judgment (or better, [191] of the logical proposition) must be that between simple forms of propositions – i.e., those forms with no parts which are themselves in turn forms of whole independent propositions – and complex forms of propositions, which do contain such forms as parts. On the side of the former certain primitive and intersecting distinctions come into consideration, above all, the distinction between singular and plural judgments (*S* is *P*, S_1 and S_2 are *P*, and the like) and the distinction between judgments with or without terms for indefiniteness (e.g.,

Something is P, An S is P, An S stands in a certain relationship to an S_1, and the like). The further distinction between particular and general predication concerns, then, propositions with indefinite terms: namely, every indefinite term receives or possesses, in the proposition considered, either the value of a certain singular thing not more closely characterized (or else a plurality of such singulars) or the value of a generality (of an A in general, an arbitrarily selected A, any A; or, in the plural, of A's in general, perhaps with the complex supplementary concept of *without exception*, or *all included*). One and the same proposition can possess several such indefinite terms – and, indeed, in each main member (in the predicate also) – and can simultaneously be general with respect to one and particular with respect to the other. (It is advantageous here to speak of "particular terms" and "universal terms.") The systematic treatment of simple proposition forms should be followed by that of complex proposition forms. In this latter would fall the treatment of forms of propositions about propositions and of compound propositions.

Some important classes of judgments from this latter class are dealt with by the author in his subsection 18, under the title, "judgments of logical content." In the cases in question he teaches that not the mere form of the judgment is to be considered, but rather particular elements of the content – which are, however, particularities that are themselves of a logical sort. In the discussion of the hypothetical and the disjunctive judgment, the author does not remain with mere forms of expression, such as "if ... then" or "either ... or." Rather, he finds significant distinctions among various judgments commonly expressed by means of these particles. If we leave aside for the sake of brevity all negation, then there is within the hypothetical assertion first of all a group of judgments defined as "determinative" by the [192]
formula: An S is P if (or in case that) it is Q. In this formula is implicit, according to the accompanying explanations, the particular judgment, "Some S are P" (namely, all of those which are Q). The author designates as "conditioned" the judgments of the form: S is P, if A is B. To them he gives the sense: S is possibly P, and certainly is so provided that A is B. Finally, the author comes to include under the heading of "hypothetical judgment" judgments which have as object certain relationships between judgments; and, namely, the "hypothetical determinative" judgment as specified by the formula: In all cases where S is Q, it is P, which contains the particular judgment

that in some cases *S* is *Q*. Further, the "hypothetical conditional" judgment is specified by the formula: Necessarily (or possibly) if *A* is *B*, *S* is *P*. Included within the latter are the problematic judgments: *A* is possibly *B*, and *S* is possibly *P*. One is easily convinced that the 'suppression' of the particular or problematic judgments which are implicit in all these forms yet leaves them remaining as full and independently realizable judgments, and that we therefore here have to do with *complex* forms which, as such, are secondary to the primitive forms, as to theoretical interest, within the context of a systematic theory of forms. To study the primitive forms of the judgment in their natural sequence and their natural interrelationships, and to exhibit synoptically the possible types of forms in their step-by-step complication: that is the task of such a theory of forms, and one not yet adequately taken in hand. This general observation also bears upon *Bergmann*'s treatment of the disjunctive judgment, which I shall not go into. With regard to the reduction of all these judgments to the allegedly basic form, that of the categorical proposition, I am in the end unable to repress the suspicion that the quest for simplicity has misled us into transformations which are no mere transformations of linguistic expressions, but are at the same time also transformations of the original conceptual content. The temptation to pass off directly and evidently *equivalent* forms for *identical* ones is, as already observed above, one of the main hazards in the formal theory of significations.

In the Second Section of the First Part, the investigations relevant to the highly controversial distinction between *analytic* and *synthetic* judgments are especially valuable. (Subsections 20–23) The ultimate [193] foundation of certainty can only be immediate, deriving from comparison of a judgment with its object. With such immediate (that is to say, *insightful*) comparison, two cases are possible (pp. 100ff.): either, in order to know a judgment to be true or false, one needs to draw the object into comparison only insofar as one already has it before one's eyes through the constituting content of the subject concept (example: $2 + 3 = 4 + 1$); or else that does not suffice, and one must obtain yet further determinations of the object from intuition of it (example: This apple is sweet.). The author defines judgments of the first type as analytic, those of the latter type as synthetic. Among analytic judgments he then counts as well all judgments that can be known to be true or false through simple comparison with analytic ones. Analytic judgments divide into "identities" and "contradictions." The identities are those affirmative judgments where the predicate determination can be derived with insight from the constituting content of the subject concept, and, on the other hand, those denials in which the predicate determination can be seen to be irreconcilable with that constituting content. If we interchange in this definition the affirmative judgment and the denial with each other, then there results the definition of the contradictory judgment. There then results these propositions: Every identity judgment whose object exists is true, and every contradictory judgment is false. The author interprets the traditional principle of identity in the sense of the first of these propositions, and the principle of contradiction in the sense of the second proposition.

In connection with these penetrating expositions there arises the important problem of how analytic judgments can be "heterological" (augmentive of knowledge), since they are mere tautologies (enantiologies, if they are false). The problem retains an obvious sense and importance even if one does not share the author's view that this heterological characteristic belongs to the essence of every judgment. For certainly the innumerable judgments of which the analytical sciences are constructed are no empty tautologies. If we pass over the discussion of the *Kantian* theory – which, in my opinion, the author misinterprets – the following is the sense of his solution. The possibility of heterologically valid analytical judgments presupposes that the predicative determination is indeed "contained" in the constitut- [194] ing content of the subject concept, but not in the literal sense, not as a constituent. These judgments, to the contrary, are possible in that the predicate indeed does differ from the content of the subject concept "subjectively, or as it is apprehended," but is identical with a part or aspect of that content as it actually is.

These exceedingly important and fruitful ideas, which were already stated in the earlier writings of the author, have unfortunately not received the attention which they deserve. They require, to be sure, further investigation, and especially certain restrictions. I hardly need to say that I would replace the "constituting" content – in the sense of the critique attempted in my first article – with the attributive content. But the following supplement (formulated here in a rough and succinct manner, for the sake of brevity) would be even more important to make. There are, in my opinion, two essentially different manners in which a judgment can be found true "by direct comparison with its object." *Either* the object must be given in perception; and only thus can not merely its existence be known (in case individual existence is asserted along with the rest in the judgment), but rather, as well, the adherence of the predicate to it as a subject thought of under such and such a concept. *Or* the intuition which brings matters to Evidence need not have the character of a perception in order to make the adherence of the predicate knowable. In this latter case there are, once again, two possibilities. In one case what is predicated of the object (or the objects) is "contained" among the determinations of the subject: that is, the predicate introduces nothing "objectively" new. Its determinations refer us to no new objects or objective Moments, and at most introduce new modes of apprehension or new categorial forms. In this case, as described by the author, those partial significations in the proposition which express objective Moments are totally variable without restriction. The proposition does not hold true of just *these* objects alone, as bearers of just *those* determinations. Rather, it holds true of any objects whatever, indifferently of what determinations they may have, and strictly upon the basis of categorial form. (Obviously it holds true without qualification and as a categorical proposition provided that objects with such determinations exist, and otherwise it is valid as an hypothetical proposition.) The second possibility is where the determinations predicated are not "contained," in the [195] sense just discussed, among the determinations referred to by the subject term, but rather are *necessarily linked* to them and known with Evidence to be dependent in relation to them. The judgments in these cases are grounded in the generic features of the determinations concerned, which therefore are not variable without restriction, but only within their genera. Now these distinctions (appropriately modified) hold not only for judgments of the simple categorical form "*S* is *P*," but rather also for judgments built up of any arbitrary form and upon arbitrarily many positings of conceptually represented objects. Suppose into sentences of the first class – the analytic in the fullest sense, but determinate as to actual content – we introduce unrestricted variables for the actual elements, for the expressions for objects and determinations of objects (e.g., into the proposition: These 7 apples and those 5 apples are, taken together, 12 apples). There then result universal (or also "general") and therefore "pure" propositions, free from all individual or existential content and all sensibility (e.g., 7 + 5 = 12). These are propositions built up exclusively from categorial concepts and forms (such as unity, number, multiplicity, relation, object, determining property, consequence, etc.) and containing not a single sensuous concept (one to be derived through abstraction from sensation) – and thus no concept such as color, sensation, house, etc. Finally, they also do not contain any positing of an individual as existing, as when we speak of "this," of such and such a thing, and so forth. These purely analytical propositions are totally apriori. The propositions of the second class are also apriori (if we exempt those possibly asserting existence), but they are *synthetic* propositions apriori. If we purify them of all that is empirical, and of all positing of the individual, however indeterminately, then there result synthetic laws which are apriori. They are laws of real synthesis, governing the necessary cohering of dependent Moments within the unity of the sensible object, and being grounded in genera of those Moments. Here, for example, we have propositions such as these: No color without extension, no extension without some qualitative overlay, no tone quality without a tone intensity, and the like.

The author has here omitted this latter class of apriori propositions or laws completely, no doubt in part because of the lack of a rigorously unequivocal terminology for use in the explication of analytical judgments. The intensity of a tone does not reside analytically in the tone quality, and yet it suffices to think in a determinate fashion of an [196]
object merely as a tone quality in order to think of it also as possessor of a tone intensity. We see that these determining properties are, precisely, necessarily connected; and this necessity is an apriori one, as belonging to the generic essences of quality and intensity. We consequently cannot at all hold it for obvious, as *Bergmann* seems to do, that every apriori judgment must *eo ipso* be an analytic judgment. And when he attacks *Kant*'s theory of the synthetic apriori judgment (in subsection 25), we may concede his claim, with respect to the *Kantian* examples, that the judgments in question are either analytic or not apriori. (Certainly I would object to several points in the content of Kant's *Critique*.) On the other hand, it is quite beyond doubt that implicit in the general essences of all Moments of the real there are laws without which no unity in the real would be conceivable; and these laws, so far as they belong to the genera given to our sensibility, provide utterly unquestionable cases of synthetic judgments apriori.

In connection with these questions the author deals in subsection 23 with the basic forms of logical relationships between judgments: the relation of *consequence* and that of *contradiction*. While earlier, with the analytic judgment, one spoke of the predicate being contained in the subject, now, in the explication of "consequence," it is a matter of a judgment content *B* being contained in a judgment content *A*. The judgment *B* "follows" from the judgment *A* if the content of the former judgment is identical with that of the latter, whether wholly so or in one part. With this the judgment *B* may or may not also be identical with *A* in its "mode of apprehension" ("*Auffassung*"), and not only in its "subject matter" ("*Sache*"). The former is substantiated by reference to the (hardly admissible!) example of the relationship between "All *S*'s are *P*" and "Some *S*'s are *P*," and the latter by the relation between $a>b$ and $b<a$. – I regret to find here no definition of the crucial auxiliary concept, *judgment content* or *state of affairs*, which the author obviously intends not to identify with the concept of the object of the subject representation, since he ascribes the same state of affairs, for example, to the judgments $a>b$ and $b<a$. – The author formulates, in conjunction with these analyses, two pairs of "general rules for the ascertainment of truth and falsity of a judgment through mere comparison of judgments." The [197] one pair he lists under the Principle of Identity, the other under the Principle of Contradiction. They are the following theorems: A consequence of what is true is true. Whatever has a false consequence is not true. Whatever contradicts the true is not true. Whatever the true contradicts is not true. There then follow in subsection 24 "specific rules of comparison" derived through application of these principles to judgment combinations of particular types. Of course this "derivation," which concerns such propositions as the *dictum* [*de omni et nullo*], the Law of Excluded Middle, and the like, is no deduction, and probably should not be designated as a derivation at all, since it concerns immediate axioms – of which some, at least, must function as irreducible principles in theories of pure logic.

The next two sections, #26 and #27, break away from the natural progression of the formal expositions, being devoted to the metaphysical Principle of Sufficient Reason. For the author, in some dependence upon *Leibniz*, it has the sense that all determinations of a thing are contained in its "peculiar nature" or "substance," the past determinations as past, the present as present, and so forth. The author holds this to be an "identity" proposition, which includes as special cases the Principle of Identity, in restriction to singular predications, as well as the Causal Principle. – No doubt many valuable ideas of the older metaphysics are brought back to life here – especially in the author's more detailed discussions – but certainly also some which provoke grave doubts. *Bergmann*'s concept of the *thing* contains within it more of the Leibnizian monad than I can admit to be correct. But we prefer to follow the movement of the purely logical investigations resumed by the author at the beginning of Part II. (pp. 137ff.)

The first main section of this Part deals with conclusions (*Folgerungen*) and inferences (*Schlüssen*), under the title, "Extension of Knowledge by Means of Thinking." *Bergmann*'s stimulating expositions link the concept of inference to that of consequence (*Folgens*), by taking it in a very narrow sense, as already done with the latter concept, discussed above. Only conclusions that "broaden our knowledge" are called inferences by him. Thus the contents of the derived judgment, in comparison with the contents of the judgments taken as premisses, must not only, in general, be "identical" as to "subject matter", but, with that, "different as to the mode of their apprehension." According to the author, the deduction of "some" [198 from "all" yields no *inference*. In general he wishes to exclude "empty tautologies." (It would seem to me more convenient here to hold to the broader concept of inference and then, perhaps, to distinguish between tautological and heterological inferences.) As predications of a "conclusion" ("*Folge*"), inferences are *identity* judgments of the hypothetico-conditional type. But not all identity judgments of this type are inferences. There are, namely, cases where the truth of such a judgment becomes clear without a relation of "concluding from" being present. For example, if the judgment "*S* is *P*" is valid, then of necessity there is also valid the judgment "It is true that *S* is *P*." The content of the latter is not contained identically in that of the former, and thus is not "concluded" ("*folgt*") from it – in the sense of *Bergmann*'s definition. However, an error must be present here. If this identity is lacking, then the connecting hypothetical judgment is, precisely, no identity. But if it is an identity, then it *eo ipso* expresses a conclusion. The error here probably is bound up with the incomplete development of the concept of the analytic, for the full development of which the author has cleared the way in such a significant manner. According to what was explained above, we can grasp this concept so broadly that it takes in the total sphere of what is valid in purely categorial terms. By "categories" can be meant the concepts of the non-sensible forms of objects and states of affairs (such as, for example, object, totality, unity, determining property, relation, and the like) just as well as the concepts of the forms of significations (such as, for example, proposition, concept, validity, and the like). Both types of concepts can, evidently, be implicated in any state of affairs, as when we say, in place of "*S*," "the object *S*," or "the object of the concept *S*," and the like. This does not modify the real content of the proposition or the state of affairs. Nothing is changed in the sum total of sense matter, nor, when considering something in a relation, in the totality of what, so considered, is given to us directly as objectively obtaining. I refer here to a relativization, very often misleading, to which the distinction between the analytic and the synthetic is also subjected. If we make a judgment, for example, concerning pure numbers, as found in pure arithmetic, then they are the "facts" ("*Sachen*"). In relation to *these* facts the judgment can be analytic as well as synthetic; whereas with respect to the possible real units of applied arithmetic the judgment is in all cases analytic. [199] *Bergmann* notices, but does not clearly grasp, the distinction between the case where the deduced state of affairs is "contained" in the state of affairs presupposed, and the case where the former is indeed analytically derivable from the latter, and yet is not "contained" in it (a distinction which one also finds in the relationship between predicate and subject). Now this distinction comes down to the fact that in the derivation of the second, but not of the first, type, circumscriptive *signification* categories come into play. We have here a mere modalization of the facts by relating them to significations, in which the factual content essential to the governing interest remains unchanged. But both types of derivations are analytic. Or, both are derivations of "pure logic" – for, after all, the basic concept of the purely logical is rigorously specified by that way of putting it.

After these reflections there follow some excellent discussions in which *Bergmann*, like *Beneke* (and *Bolzano*), opposes the view that all inferences "consist of logical relationships," or, as we now can more precisely formulate it, that all inferences are analytic judgments in the above sense. As examples he takes inferences concerning magnitudes; and he shows that their reduction to purely logical inferences comes about only by the introduction of the principles of inference peculiar to them as new premisses. (pp. 141ff.) Accordingly, there results an essential division of inferences, which we – in consequence of our distinction (attempted in the critique above) between analytic and synthetic judgments – had to designate as the division between the "analytic" (categorial) and the "synthetic" (real), terms which certainly have a wholly different signification in our author. (pp. 179ff.) In fact, those inferences banished from pure logic by the author – such as inferences concerning magnitude, position and the like – are grounded in the real contents of the states of affairs concerned, and, more precisely, in the respective conceptual relations obtaining in reality (greater, to the right, more intensive, and the like). In the purely logical inferences, by contrast, all of the real is concentrated within the "terms," and is unrestrictedly variable with them. By the way, I fail to see in *Bergmann*'s logic, as well as in the whole of contemporary logic, the formulation of that fundamental truth belonging to the Idea of inference, and with which any theory of inference must begin: namely, that each inference has its "law of inference" or its "inference form." In other words, that it has its "terms," which, in [200] the hypothetical formula of the inference, are unrestrictedly variable, and thus can be replaced by general, indeterminate signs.

Here I shall not go into the further content of the presentations of the author concerning conclusions and inferences. They contain – and we could expect nothing else – many valuable ideas characteristic of him, but ideas which can neither be clearly presented nor weighed critically without lengthy discussions of detail. The author, like most German logicians, has not appropriated the modern achievements of mathematical technique for the elaboration of the theory of inference. The philosophical confusions of mathematical logicians – although they are not of a different or more harmful type than the philosophical errors of arithmeticians and geometers (in their occasional reflections concerning the origin and significance of fundamental mathematical concepts, and the like) – have restrained him also from doing justice to the intrinsic worth of mathematical techniques. Remembrance of *Berkeley*'s attack upon the infinitesimal calculus should serve as a warning to us to trust the mathematical instinct, which is at first uncritical, and yet in essentials is a sure guide. But of course the technical exposition of theories of pure logic is one thing, and their philosophical clarification is another. The following from among the further investigations by the author seem to me especially worthy of notice – and, indeed, in this *philosophical* respect: the analysis of the categorical inference as a process of substitution (subsection 31); and further, the presentation of inferences in the forms of incomplete induction, incomplete analysis or analogy, as categorical *probability* inferences.

The short final Section concerning the "Logical Development of Knowledge," of less importance than the preceding Sections, treats of definition, division and proof.

I turn from the present work with the wish that it may be studied with all the enthusiasm demanded by its rich and uniformly important content.

A REPORT ON GERMAN WRITINGS IN LOGIC FROM THE YEARS 1895–1899

Third Article[1]

Julius *Bergmann* has followed up his overall presentation on the fundamental principles of logic with several more detailed logical and logico-metaphysical treatises. Since in the foregoing article we have entered into point-by-point discussion of the main ideas of his logic, let us here only make mention of these treatises, which obviously merit the most careful consideration.

"Ueber Glaube und Gewissheit," *Zeitschrift für Philosophie und philosophische Kritik*, Vol. 107, 1896, 176–202.

"Ueber den Satz des zureichenden Grundes," *Zeitschrift für immanente Philosophie*, Vol. 2, 1898, 261–345.

"Zur Lehre Kants von den logischen Grundsätzen," *Kantstudien*, Vol. 2, 1898, 323–348.

These are carefully reworked and incorporated into the author's *Untersuchungen über Hauptpunkte der Philosophie*, Marburg, 1900.

L. Rabus, *Logik und System der Wissenschaften*, Erlangen and Leipzig, 1895.

Professor *Rabus*, author of many works in logic, offers this system to us "as a well-ripened harvest and his life's work." "I have made it my business," so he states in the Forward, "to reconcile the apparent conflict between empirical and speculative, and between material and formal, logic, to establish the value of logic as the immanent organon common to the sciences, and to develop logic in itself and out of itself as a system – and, indeed, as a model of what a system is in general." (p. vii) I must decline to go any further into the content of this work,

[1] Continuation from previous article. First appearing in *Archiv für systematische Philosophie*, Vol. 9, 1903, pp. 393–408. Republished in *Husserliana XXII*, pp. 201–215. [D.W.]

elaborated with such loving care. It so much belongs to a range [202] of ideas and historical involvements of a past period in philosophy, that a common ground of understanding between it and the present younger generation is lacking. But we point out the work's richness in historical expositions and bibliographical references.

R. Heilner, *System der Logik im Sinne eines allgemeinen Organon der menschlichen Erkenntnis*, Leipzig, 1897.

The author indicates that this work is the fruit of the logic taught in Tübingen. But in its content it reveals nothing of *Sigwart*'s style and direction of thought. The critique of the organon transmitted from the middle ages – and *Sigwart* is held by Heilner to be the most serious champion of this critique – has done its work. At present, the task is to see whether or not logic can be brought into the secure pathway of a science. (p. vi) Now what the author presents to us, intending a *new* organon, are rather cursory and somewhat vague sketches of ideas that do not lead back to any foundation for serious scientific investigation. He seeks to define the task of logic in the following way (pp. 15ff.): The human mind requires among its ideas a particular sort of interconnection, that of functionality. It also endeavors to extend this type of interconnection over the domain of ideas generally, by setting up a certain correspondence between its ideas, on the one hand, and that which is in any way accessible to them, on the other. Correspondingly, the theory of the logical elements has the task of bringing before us the means with which our mind realizes this functionality within its realm of ideas, and methodology has the task of finding general procedures through which we secure the relationship of our ideas to objects.

The manner in which the author, furthermore, seeks to interrelate the concepts of *judgment* and *concept*, on the one hand, and the concept of *function*, on the other, recalls the work – probably unknown to him, and for that matter incomparably more valuable – of the ingenious mathematician G. Frege, *Funktion und Begriff* (Jena, 1891), which unhappily has not found the attention it deserves from professional logicians.

The First Section deals with the theory of logical elements (pp. 13–29), the Second (pp. 30–47) with methodology; and as a conclusion [203] there follows a collection of illustrations. (pp. 51–58)

The following over-all presentations of logic are adapted to the instructional requirements of Catholic theological schools:

C. Braig, *Vom Denken: Abriss der Logik*, Freiburg i. Br., 1896.

C. Frick, *Logica in usum scholarum*, Freiburg i. Br., 1896.

E. Commer, *Logik. Als Lehrbuch dargestellt*, Paderborn, 1897.

Elemente der Logik. Nach Dr. Stöckls Lehrbuch der Philosophie (Course for students of Catholic theology, Vol. 1), Mainz, 1899.

Th. Elsenhans, "Das Verhältnis der Logik zur Psychologie," *Zeitschrift für Philosophie und philosophische Kritik*, Vol. 109, 1897, pp. 195–212.

The final words of this essay indicate the viewpoint of the author: "There is no other route toward the resolution of the problems of logic than that of a comprehensive 'psychologism'." The main ideas serving to found this claim are as follows: Since logic deals with certain mental processes, its subject is (the author supposes) obviously a part of the subject of psychology. But this certainly does not mean that logic itself forms a *part* of psychology, inasmuch as essential distinctions could indeed be grounded in differing methods of treatment of the common subject matter. That this does not prove to be the case, however, the author seeks to demonstrate. (pp. 197ff.) To that end he contests the conception of logic as a *Kunstlehre*. As a *Kunstlehre* it would have to be an effective guide into a thinking adequate to thought's goals. But in truth it is far removed from being sufficient to this task. Yet today there is dispute concerning the principles of logic, while thinking has itself brought infinitely rich results to light. (p. 198) Such a *Kunstlehre* presupposes, further, a criterion of its results. How are true and false laws of thought to be distinguished?

As ultimate criterion we must accept the direct consciousness of Evidence which accompanies necessary thought. (p. 199) This consciousness is to be characterized as a qualitatively peculiar feeling, which is to be thought of as analogous to the so-called "feeling for language" (*Sprachgefühls*). As the correctness of linguistic expression is not, in the usual case, governed by a conscious knowledge of [204]
grammatical laws, but rather by an immediate feeling of correctness that arises with increasing practical exercise in the language, so thinking is governed by, so to speak, a "feeling for thought" (*Denkgefühl*). Certainly this consciousness of Evidence also can turn up in the individual case without the judgment realized being objectively valid. But ultimately the norm of assessment also, and our knowledge of deviation from it, rests upon the factual psychological necessity of thinking one way and not another: a necessity which is presupposed

in any mentally sound human being and which can be tested. (p. 202) In the light of this no other task can fall to logic than that of determining – through a scientific analysis of the mental processes in which this feeling shows up – the conditions under which it occurs. It does not instruct us in the art of thinking correctly (the technical perfection of thinking being only a *by-product* of logical science, pp. 202ff.), but rather it takes correct thinking as the subject matter of its purely theoretical psychological investigations.

A critique of the popular attempt to ground the distinction between logic and psychology upon the concept of *ought*, and to characterize its laws as norms (pp. 199ff.), leads to the same result. For science as such there are no norms, no oughts, no technical guidance, but only facts, laws and hypotheses. But the group of facts which logic treats lies within psychology, of which logic therefore forms one part. (p. 203)

Also, how little those logicians who want radically to cut genuinely logical reflections away from psychological ones will succeed in removing themselves from the criterion of psychological analysis, is something the author attempts to show through discussion of the problem of the judgment. For example: Consider the question of whether the relation of subject to predicate in the judgment is to be regarded as "unification" ("*Ineinssetzung*") or as "equalization" ("*Gleichsetzung*"), as "coordination" or "subordination." The logician will vainly struggle to make this process of judgment intelligible so long as he does not proceed, in the manner of the psychologist, from the observation that every psychical process is conditioned by the immediately preceding state of the soul. (p. 207) In conclusion the author distinguishes (p. 210) genuine logic – which, considered [205] fundamentally, is a part of psychology, and is treated as a science in its own right only with respect to its use – from methodology, on the one hand, as one "application" of logic (as "a technique derivative from logic as a science"), and from the logic which describes itself as "epistemology," on the other hand, "as represented by Schuppe," where the subject matter is not thinking as a determinate set of mental processes, but rather *knowing* as interaction with a real or apparent external world.

In order to settle the issue of Psychologism one must, as I have elsewhere tried to show, separate into different spheres the material that stands under the title of "Logic." If we consider the theoretical

foundations of logic in its common acceptation as a normative and practical discipline (which, with this author, also comes at last into its own under the name of "methodology"), then we certainly find among those foundations bits and pieces of psychology also. But we find there as well – and, indeed, find it as the most essential foundation – "pure logic," i.e., the totality of those thought-laws which are grounded in pure categories, together with the "formal" theories built thereupon. Here belongs, for example, the traditional syllogistic, but also the pure theory of cardinal numbers, the pure theory of ordinal numbers, of *Cantor*ian sets, and so on. Moreover, in spite of its rather isolated status, here also belongs the pure, mathematical theory of probability. No one wants to be responsible for the absurdity of classifying this realm of pure logic (Pure *Mathesis*) under psychology. It is precisely to this realm that the task of "critique of knowledge" has reference: the task of rendering "intelligible" the possibility of a knowledge which is delimited by concepts and laws of pure logic, by tracing these back to their "origin"; the task of resolving, in this way, the profound difficulties which are tied up with the opposition between the subjectivity of the act of knowledge and the objectivity of the content and object of knowledge (or of truth and being). This task does not fall to pure logic itself (in the restricted sense of pure *Mathesis*). The critical illumination of pure arithmetic as knowledge is no arithmetical task, that of pure syllogistic is no syllogistical task, and so on. On the other hand, it also is no task for psychology – at least not so long as we use this term in its usual sense. As physics, [206]
or natural science in the ordinary sense, is the empirical science of physical facts, so psychology is the empirical science (the natural science) of mental facts. Both sciences proceed from the "world" in the common, pre-critical sense of the word, with its division of facts into the physical and the mental. Both remain uncritical, howsoever much they may modify the content of the original world-idea. As *explanatory* sciences they presuppose a prior objectification, whose sense, whose *illumination* in terms of what makes it possible, they can dispense with – as in fact is shown by the advanced state of these sciences without any help from the critique of knowledge. Nevertheless, it is precisely from these pre-critical objectifications (with their distinctions between ego and non-ego, between "one's own" ego and "other" egos, with their interpretive introjections of the direct givens of consciousness, as "psychical activities and states," into one's own

ego, and their interpretive expulsions of physical things and states, of "other" persons and lived experiences (*Erlebnissen*) and the like, by placing them outside of the ego) that the difficulties of the metaphysical problem of the possibility of knowledge arise – which, for its part, requires an illumination of knowledge in general, independently of any metaphysical intentions. This illumination requires a *phenomenology* of knowledge; for the lived experiences of knowing, wherein the origin of the logical Ideas lies, have to be fixed upon and analyzed in the illumination, but in removal from all interpretation that goes beyond the real (*reellen*) content of those lived experiences. And then the "authentic" meaning (*Meinung*) of the logical Ideas, their general essence, has to be brought to Evidence. The necessities of communication bring it about that the phenomenologist makes use of objectifying expressions, as when he says: "We find" in direct 'lived experience' this and that. In fact, such talk refers to its objects only indirectly; all scientific or metaphysical objectifications remain completely excluded. Phenomenology therefore must not be designated as "descriptive psychology" without some further qualification. In the rigorous and true sense it is not descriptive psychology at all. Its descriptions do not concern lived experiences, or classes thereof, of empirical persons; for of persons – of myself and of others, of [207]
lived experiences which are "mine" and "thine" – it knows nothing, assumes nothing. Concerning such matters it poses no questions, attempts no definitions, makes no hypotheses. In phenomenological description one views that which, in the strongest of senses, is given: lived experience, just as it is in itself. One analyzes, for example, the thing appearance, not that which appears *via* the appearance; and one refuses to consider the apperceptions in virtue of which appearance and that which appears come into correlation with the ego to which the appearing thing appears. That critical illumination of knowledge which is built upon analysis of *this* sort is nothing other than *adequate abstraction based on intuition*. Within what is fixed upon phenomenologically, it brings to Evident consciousness, and therewith to "clear and distinct" understanding, *the general essence*, the "true and authentic content," of the concepts and laws of logic. The confused thoughts prior to critique, carried out in the inauthenticity of symbols and images, then are transformed into the corresponding clear, fully realized thoughts, whose possibility and validity are directly viewed and grasped. Obviously the psychological appercep-

tion here excluded can be fitted in at any time, and in such a way as to make the results of phenomenology and the critique of knowledge useful for psychology. Phenomenological analysis *then* takes on the character of descriptive psychological analysis; it *then* functions as the foundation for the theoretical explanations of psychology, the natural science of mental phenomena. Likewise, the illumination of the laws of logic supplies knowledge which claims fundamental significance in application to psychology. To the Ideal-law connections of those essences which are concepts there correspond, in the spheres of the corresponding particulars (of the corresponding psychic contents), rigorous uniformities of co-existence and lack of co-existence. But it is absolutely imperative for any self-consistent epistemology, or philosophy in general, that one effect with finality the separation in principle between purely immanent phenomenology or critique of knowledge, which holds itself free from all suppositions extending beyond the content of the given, and empirical psychology, which, even where it merely describes, makes such suppositions; and that one accordingly does not confuse, as is commonly done, questions about "origin" in the critique of knowledge with those about [208] psychological origin. It is in consequence of such confusion that the critique of knowledge has again and again slipped over into psychology and biology, and has become embroiled in the absurdities of skeptical empiricism and relativism. – Unfortunately, *Elsenhans* has also fallen into these errors, as clearly follows from the content of the above report, and especially from his placing the feeling of Evidence on a level with the feeling for language, as well as from his reduction of logic to the psychology of Evidence. This reduction – already suggested by *Mill* and carried out by *Höfler-Meinong* in their admirable *Logik* – has no more value than a reduction of arithmetic to a psychology of "correct" enumeration. Every mathematical law has a sense that can be given a psychological interpretation, and yet none has itself a psychological sense. For the more detailed analysis of the theory of Evidence, and of the psychologistic misunderstandings of Ideal laws (essence laws) in general, I will perhaps be permitted to refer to my *Logical Investigations*, First Part (*Prolegomena to Pure Logic*), Halle a. S., 1900, pp. 180ff. I have there also (pp. 89ff.) sought to show that the idea of the normal, mentally sound human being already presupposes the Idea of reason, and that it thus is totally unsuited to be first introduced in the delimitation of reason or the sphere of the logical.

R. Wrzecionko, *Das Wesen des Denkens. Beiträge zu einer Grundlegung der Logik*, Vienna and Leipzig, 1896.

In a style more aphoristic than expository, the author presents sundry mythical ideas toward a philosophy of feeling. Feeling is consistent experience (*konsequente Erfahrung*). It cannot be contradicted by experience because it is the basis of all experience. (p. 19) To be fact and to be feeling are the same. It is the British philosophy that comes closest to this truth. In *Hume*'s philosophy the analysis terminates in the invocation of feeling. (p. 20) "Only sense your way into the heart of humanity and you will see how, from out of the dark abyss of feeling, one phenomena after another rises up, how out of the receding tide of feeling there emerges island upon island . . . ." (p. 24) In the concluding sentences of the treatise (p. 38) we are told the following about the essence of thought: The symbol of thinking is ab [209]
$= b$, which consists in setting the whole to be identical with its part. Thus thought, if one considers it in itself, is a self-contradiction. But considered from the standpoint of reality, thought is rational because it is real.

H. Gomperz, *Zur Psychologie der logischen Grundthatsachen*, Leipzig and Vienna, 1897.

In the "Preface" the author characterizes his basic idea as "the dependency of discursive or conceptual thinking upon language." But "dependency" is not the correct word for the actual view of the author, which expressly *identifies* "concepts" with "general terms," and "thinking in concepts" with "using words in the place of things." (pp. 94, 103, etc.) In fact, he represents an extreme form of nominalism and sensualism. The vague concessions which he makes to conceptualism do not justify his location of himself at the mid-point between the two positions. But this does not adequately characterize the theme of his work, which, I would say, concerns the clarification of the psychological relationships between intuiting and thinking. To be sure, this is not a mere descriptive analysis and differentiation of the phenomena to be found in the two areas, but rather, at the same time, is a genetic and empirico-teleological analysis that exposes the biological effects of these two functions, and their separate development, as well as their progressive differentiation in response to extant needs and favorable circumstances. In any case, the individual observations – interrelated (as the author himself recognizes) in the fashion of a "mosaic" – fall under this general characterization. The entire manner of treatment of the subject, the forms in which the problems are set and resolved, correspond to the standpoint of the author already characterized above, that of the most radical of sensualisms. Nothing here betrays any awareness of the great difficulties which such a standpoint on the psychology of knowledge – and subsequently on the theory of knowledge – entails: difficulties of which older empiricists, such as *Hume* and *Mill*, were still painfully aware.

Now I do not mean by this to suggest that the writing before us [210] (possibly the author's first) is not a serious piece of work. To the contrary, it gives evidence of talent as well as of serious scientific aspirations. Thus, I cannot believe he will remain satisfied with this approach to the "fundamental facts of logic."

In the Introduction to the first Section (pp. 7ff.), he marks out the intellectual ideal, the ideal of "godlike" thought. It is, according to him (as also with Rickert), a thought which can comprehend the world of appearance in its totality, distinguishing each of the particulars therein and grasping the relationships, infinite in number, that obtain between them. Animal as well as human thought deviates from this ideal, being restricted in a two-fold manner: through the lack of power to distinguish things, and through the lack of sufficient scope of consciousness. The animal remains confined in this condition, its thought never reaching beyond the purely intuitive. Man, by nature a social being which develops language, creates instruments – general concepts (= general terms) on the one hand, and judgments and inferences on the other – which, if they do not altogether remedy those two imperfections, yet act as surrogates to mitigate them in practice. Conceptual thinking is not a matter of something inherently higher and of greater value, but rather only has the function of facilitating intuitive thinking, of abridging it or taking its place. (cf. p. 102) – So says the author. But now it seems to me that in sketching the intellectual ideal a small matter has been overlooked: namely, the intellect. This "godlike thought" is no longer *thought* at all, but an all encompassing, all distinguishing intuition, to the extent that *Kant* is right in saying that intuition without concepts is "blind." Thus, not a single *law* is discerned by this "divine" thinking: neither a logical law nor a natural law. Consequently, it also fails to apprehend a single thing as falling under a law. According to the author, it is only the finite human consciousness that requires thought in terms of laws, or conceptual thought in general, as a makeshift linguistic device. But would not the finite consciousness then possess – through its knowledge of the eternal necessities and possibilities, as well as through the attachment of singular facts to law – a towering superiority over the divine consciousness? Is that ideal of infinite wisdom indicated not rather one of an infinite stupidity, which *sees* everything and yet *comprehends* or cognizes nothing? These errors are only the all [211] too predictable consequences of that inadequate descriptive analysis of intellectual lived experiences which fails to distinguish the two pairs of oppositions, so radically different: that between the intuitive function and the symbolic, on the one hand, and that between the sensuous function and the categorial on the other. It therefore is incapable of dissipating the old equivocations that infect the concept of intuition (sensible intuition over against intellectual intuition = insight, Evidence) and the concept of thought (thinking in symbolic surrogates, over against thinking in categorial functions).[2] Certainly it is true that the author here only succumbs to errors that are almost universally dominant.

In the first Section ("Thought without Language") the author attempts to show that beings not endowed with language carry out their thinking only in images of particular things. It is only under the influence of language that the "basic facts of logic" develop, and concepts are the first of all. Concepts are dealt with in the second Section ("Word and Concept"). The author joins with *Wundt* in distinguishing (pp. 21ff.) "logical" concepts in the narrower sense, and "scientific" concepts. These latter (e.g., *gravitation*, *mammal*) are not elements, but rather objects of thought; not psychological experiences, but rather laws under which qualities or phenomena fall. Thoughts of them have the character of relational judgments: the α, β, γ, ... co-exist in conformity with law. Already in the exposition of these vague half-truths, and all the more in his treatment of the "logical" concept, the author evidences no suspicion of the number and difficulty of the descriptive-psychological relationships which must be illumined in the theory of the representation and the concept. According to him the psychological element in the concept reduces

[2] Cf. my *Logical Investigations*, second Part, "Investigations toward a Phenomenology and Theory of Knowledge," Halle a. S., 1901, VIth Investigation, Chapter 6 and following.

to the associative relationship between the sensible word-sign and the individual image from a similarity group. In general, this Section presents us with nothing that could materially advance us toward understanding of the "basic logical facts about the concept." Supposedly we are going to see that the mental life must be modified upon the entry of language. (p. 23) We are told that language is a spontaneous reaction, that it is in the first instance a tool used to [212] achieve understanding. Its origination means the formation of fixed associations between words and images. This does not come about in such a way that a proper name is formed for every individual image, nor in such a way that the common parts of resembling images are held together by one special term. (p. 24) The explanation for how any word obtains a general signification is found, rather, in *Taine*'s "tendencies of reaction." Each sense impression has the tendency to call forth a reaction, and under one term would be comprised all of those individual images which at the time when language arose were suited to evoke the same reactions. (p. 26) The author would be prepared, in his own way, for the objection that the particulars seized upon are after all only similar to each other, and that this similarity is the primary fact which first grounds the presumed identity of those reactions – moreover a fictitious identity, given such wide-ranging variations. Namely, he takes the strange tack of defining similarity in general by means of partial identity of reactions: whatever evokes identical reactions is identical, what evokes partially identical reactions is similar, and whatever evokes different reactions is different. (pp. 97ff.) There surely is no need of proof of the circularity and of the infinite regress involved in such a view. The author certainly makes every effort to provide analyses that are rigorously and systematically worked out. He distinguishes the standpoint of the hearer from that of the speaker and endeavors to describe their lived experiences. Thus, in the case of the speaker he distinguishes, as a matter of great importance, five psychical Moments: the representation which occasions the utterance, the feeling linked with that representation, the nerve stimulus, the verbal sound, and the written symbol. (pp. 28ff.) But on such matters as the following we learn nothing: the phenomenological significance of these distinctions for the "signification" and the "object"; in which psychical Moments the reference to the object (the conscious meaning of it) consists; the types of this reference to be distinguished; what phenomenological distinctions

are to be exhibited in those types; where to locate those "forms" which are not only dealt with in logic but totally permeate all actual thinking. Instead of analyzing the phenomena the author operates, as nominalists do in general, within vague phraseology: The individual representation is to "take the place of" the abstract concept, e.g., the spatial image that of the scientific concept of number. What does this "replacement" mean? The mere *association* as a dispositional fact? [213]
Or is it something manifesting itself in consciousness itself? Do we perceive in the word or image itself whether it replaces something, and what it is that it replaces? How could that possibly be, since the thing replaced does not itself fall within consciousness and, where we are dealing with scientific concepts, absolutely cannot fall within consciousness, according to the author? He just does not see the essential problems and difficulties. This is also shown, for example, where he distorts the principle difficulty of nominalism – *viz.*, to explain how thought which plays itself out as a flow of mere singular representations can be the bearer of general signification and validity – over into the question (certainly one that is indifferent to the questions of principle involved here) of how one can think in terms of singular representations without being led astray by the specific characteristics peculiar to them. (p. 37)

The fourth Section deals with "Sentence and Judgment." The author there carries on a critical discussion with *Wundt* and *Jerusalem*, and brings a devastating critique to bear upon Jerusalem in particular. The tendency of his own views are not determined on the basis of a serious phenomenological analysis of the lived experiences present in the meaningful utterance of a sentence, but on the basis, rather, of the languages of primitive peoples, the language of children, and the like. The results correspond: "Such a competent judge as *Sayce* holds that the division of the sentence into two parts, the subject and the predicate, is a purely accidental matter .... Had *Aristotle* been a Mexican, his logical system would have assumed a totally different form." (p. 53) According to the author, the judgment is, "... functionally considered, an instruction from the speaker to the hearer to form a total representation out of given elements." (p. 60) He considers, therefore, the analysis of the total representation into its elements. (pp. 64ff.) This analysis is – without being bound to our subject/predicate division – to be realized in conformity with three schemata: modification, co-ordination, and fusion. The sentence

expresses a mere enumeration of the particular partial representations and so forth. Of the categorial forms, without which the partial representations would form no unity of thought, nothing is said. The lack of *expression* of these forms in utterances of children, and in syntactically undeveloped languages, is confused by the author with the lack of these forms themselves, which also govern the lowest levels of thought because inherent in its very sense. I object no less to the author's explanations concerning affirmation and negation, as [214]
well as those concerning the theory of *belief*. He does not grasp the sound core of truth in the analyses of the judgment by *Hume*, *Mill* and *Brentano*, because he overlooks the distinction between primitive or simple *belief* and consent to a representation presented to the mind (possibly a propositional thought entertained) – a distinction which certainly is commonly overlooked by others also, even by advocates of the *belief* theory [of the judgment]. Simple consciousness of being, as present in any perception, memory, or expectation, as well as in any normal assertion (in contrast to the representing that goes on in free fiction or in the assuming of a propositional thought in reflections that are still completely undecided) is totally lost to the author. Moreover, we note at all points that the author has no sense for the task of purely descriptive analysis of the judgment (the foundation of a theory of knowledge), and much less, then, would he distinguish between that analysis and genetic analysis (which is irrelevant to the theory of knowledge). Accordingly, he also misconceives the meaning of talk about the irreducibility of ultimate phenomenological facts. Wherever he has laid out an account of a genetic process in which such a fact emerges, he believes he has already proven its reducibility. Thus in the discussion of his (in my opinion absurd) classification of judgments into judgments of conviction and judgments of communication, he "reduces" the fact of being convinced to an "adherence and repulsion of ideas," an effect of the "pushing and shoving of the ideas themselves." And he concludes that being convinced "is no specific capacity of the soul over and above representations, but rather is merely a manner in which representations succeed one another." (p. 69) What indeed the author understands by capacity of the soul, representation, and manner of succession among representations, is something I would hardly know how to say.

In the fourth Section the author deals in the same style with the "combination of sentences and inference." Inference is to be defined as the linguistic expression, occurring in two steps, for a pair of representations united by association, in which the second representation is newly thought, and now as a conviction. (p. 78) Even in the syllogism of traditional logic, the authentic process of inference (or, as we also read: the inference when it empirically occurs) develops through two terms. The *major* is said to be unessential to the psychological activity of inferring. In this there is much that is unintelligible to me, but above all how the author could write down such sentences with [215] such certainty. The vast, essential problems of a phenomenology and theory of demonstration have not even been seen by him. For example: Under what circumstances does the lived experience of [rational] motivation occur? What does it necessarily presuppose? How is the distinction between an alleged and an actual demonstration – or the further distinction between ground and consequent, or that between empirical and rational grounds – to be cleared up? We can, then, leave his further discoveries aside: for example, that there is just as little a distinction between deductive and inductive inferences as there is between analytic and synthetic judgments,or that to understand an inference means the same as associatively to derive one of two communicated representations from the other, and the like.

On the basis of the foregoing investigations, the relationship between intuitive and conceptual thinking is then to be made clear in the concluding Section. The ideas indicated above, in our discussion of the Introduction, are developed in more detail.

A REPORT ON GERMAN WRITINGS IN LOGIC FROM THE YEARS 1895–1899

Fourth Article[1]

Wilhelm Jerusalem, *Die Urteilsfunktion. Eine psychologische und erkenntnistheoretische Untersuchung*, Vienna and Leipzig, 1895.

The intentions of the author go far beyond a psychology of the judgment. Such a psychology is regarded by him as the foundation and presupposition for the whole of theoretical philosophy. (p. 2) Up to now the problem of the judgment has been dealt with only for the sake of logical interests; but it has never before been said, as he holds, that our whole world-view depends upon a happy resolution of the question about what it is we do when we judge, and about what in this function is of our own production and what is of a foreign origin. (p. 34) In connection with this he announces an attempt to solve the epistemological question by means of an investigation into the *form* of the judgment; and he adds: "Should we succeed in showing that the judgment form is the form which develops, conforming to psychological laws, in every human individual, and that this form must take in ... all that is given in consciousness in order for such matters to become ... actual contents of consciousness, then the solution to the ultimate questions of metaphysics will also be nearer. The concepts of God and the soul may be set in a new light. And also the question of whether we can prove that there is something extra-mental, occurring independently of us, becomes easier to deal with." In fact, the extensive concluding section of the book is devoted to just such questions. The author believes that, on the basis of his theory of the judgment, he has "overcome" Materialism, Critical Idealism and Empirico-criticism, and found "a harmoniously completed world

[1] First appearing in *Archiv für systematische Philosophie*, Vol. 9, 1903, pp. 523–543. Republished in *Husserliana XXII*, pp. 216–235. [D.W.]

view." He is also not sparing of "promises" in other respects. He believes himself to have opened the way to important insights in general grammar, logic and the theory of knowledge. Among such [217] insights he counts the proof that we cannot maintain the Evidence of inner perception, that sense perception is the ultimate and most certain source of knowledge, as well as the proof that the categories of substance and causality are not, "as *Kant* supposed," innate and original concepts of the understanding, present prior to all experience, but rather are thought forms that are jointly conditioned physically and acquired through experience. And so forth. (Cf. p. 5 of the "Introduction" and the retrospective survey at the conclusion of the book.)

Whoever reads the announcement of such goals, and at the same time considers the instrumentalities by which they are supposed to be attained, will not exactly enter upon the closer reading of the book with high hopes. However widespread the tendency to assign to psychology a fundamental importance for philosophy as a whole, very few will expect much from a metaphysics that presumes to find the key for the solution to the ultimate questions about God and the world in a genetico-biological theory of the judgment. Yet greater offense will be taken, at the claim to solve epistemological problems by such means, by all of those who have gotten clear upon the relationship between the *psychology* of knowledge and the *critique* of knowledge – and consequently have grasped the basic distinction of essence that separates the *explanation* (*Erklärung*) of thought processes (objective and theoretical analysis) from the *illumination* (*Aufklärung*) of the units of thought (analysis of signification and validity). The former locates us in the context of objective things and events: speaking of men as psycho-physical individuals, of their life circumstances, of their social unities and the psychical formations which develop therein. It delves into the elementary psychical states of those individuals, investigating their causal connections and their natural laws. Critical illumination, by contrast, does not theorize. It asks no questions about mental or corporeal *things*, formulates no natural laws for their properties, and states and explains nothing upon the basis of natural laws. It does not explain (*erklärt*), but clarifies (*klärt auf*). It inquires from whence, in general, the Idea of theory, as a unity of truths, is to be derived – along with all of the Ideas and Ideal laws constituting it. It asks what is really meant by "validity" and "being," and what can make their meanings intelligible.

That we have here distinctions and associated difficulties and problems whose clarification and resolution is the work of millenia, is something of which the author cannot possibly have even the most remote of conceptions – as is already clear from his "promises" and from all of his further expositions. [218]

But if, now, we consider the basic ideas in his theory of the judgment itself, leaving aside their epistemological and metaphysical applications, we then find that they have about them a naïveté which is hardly comprehensible. The author advances a theory of the judgment without any serious attempt at a description of the judgment – within which there obviously would also belong a purely descriptive analysis of all those lived experiences from which judgments (of the essentially different descriptive types) are built up. At bottom his theory consists of an idle fancy hounded to death over a few hundred pages. The author is guided by the conviction that every judgment is a sentence, expressed or unexpressed (p. 29), and that if you think away the linguistic expression of the act of judgment, then nothing is left over which one could call a judgment.[2] (p. 17) One would, accordingly, expect that the author would begin with an exact analysis of lived experiences of language, and would bring his investigation to bear upon the serious problems which have to do with the act of signifying and with its relationship to what is signified. That investigation would of itself have led him over into the more general question concerning what confers a reference to "objects" upon certain lived experiences – those which we designate as "representations," and which we sometimes conceive of so broadly as to include judgmental meanings, while at other times so narrowly that we actually contrast them with judgments in the full sense. It would also have led him to ask about what essentially different types of such references are to be distinguished; about which functional relationships, which possible forms of complication, are grounded in the essences of these differing types of representation; about which

[2] Other statements by the author are scarcely reconcilable with this: The judgment function is said to be already at work before the emergence of language, and already to have influence upon sense perception as an unconscious process. Sense perception too is, in "a certain sense," a judgment, but every "complete" judgment is a sentence. (p. 30) And the like. The closer exposition of all this does not get beyond incoherent and vague half-truths.

descriptive Moments can be confirmed in comparative analysis as present in the one or the other type; about how in such representations the fundamental logical and noetic concepts originate and are cleared of confusions and ambiguities. Is there indeed a systematic theory [219]
of judgment (and it is such a theory that the author intends to give us) without these investigations? Can a psychology of the judgment which is to be taken seriously begin otherwise than with them? But of all such investigations there is nothing to be found in the present book, however much it speaks of an analysis of the judgment. The ever-recurring clichés about the means of thought, or about the formations, the articulations and the objectifications whereby the events of the external world are brought to a condition suitable for consciousness, or about the biological function of concepts, judgments and so forth, can be of little assistance to us. All of those concepts whose illumination is no less presupposed by the psychology of the judgment than by the critique of logic – the concepts, namely, of *sign*, *word*, *expression*, *signification*, *concept*, *proposition*, *object*, *state of affairs*, as also *perception*, *representation*, *intuition*, and the like – remain completely unclarified, and are used in a confused and equivocal manner. The concept – or, as I have better named it above, the idle fancy – by means of which the author thinks himself to have resolved "the problem of the judgment" emerges, as if shot out of a pistol, at the beginning of the third Section (where the systematic analysis sets in). The representation of a tree in bloom, for example, can be realized without a judgment occurring. What is it in the representation, then, that is modified by means of the judgment *that the tree is blooming*?[3] (pp. 79–80) The answer, "tested over and over and found to be correct," runs as follows: "By means of the judgment the total representational complex, the unanalyzed process, is formed and articulated, in that the tree is set forth as a unitary entity imbued with powers, its expression of power currently coming to realization being, precisely, the blooming." (p. 82) In judging we *project into* the tree the power of blooming. Thereby the interconnection of the "process" with the other representations that I have in store is broken, and

[3] A "representation" is explained by the author to be a psychical process to whose essence it belongs to have a content that appears, to the one doing the representing, as an object or process of the external world. (p. 79) That is pretty well all the author produces as an analysis of that concept – or, rather, that bundle of concepts – which is most basic for the theory of judgment and travels under the title "representation."

it is as it were made ready for consciousness. Simultaneously with this "articulation" and "formation," *belief* (*ibid.*), or objectification (pp. 83, 85, *passim*), is realized in the judgment. The mere sensuous perception has, with the intensity of the feelings involved, something [220]
chaotically confusing about it, into which order is first brought by the judgment. So now trees, birds, etc., are centers of force which deploy their activity independently of us. (pp. 83ff.) The judgment is the most primitive and frequent type of apperception (p. 94), and the most comprehensive and primitive type of orientation toward our environment. (p. 84) That anthropomorphism which one encounters everywhere in the apprehension of the external world implants its mark upon the judgment function. In spite of its "meritorious falsity," we never can stand free of it. In every judgment we stand under its spell. (pp. 108ff.) The original anthropomorphism becomes, to be sure, ever more abstract. However, it never allows itself to be wholly eliminated. (p. 163) –

But enough of citations! The theory is amazing enough. No one will deny the anthropomorphism that dominates the world view of the child, and which also we more mature human beings "never wholly shake off." Certainly language (whose testimony the author loves to invoke) shows everywhere the traces of such anthropomorphism; and certainly an exposition of the human intellect along biological lines, or from the viewpoint of cultural history, will set out from these well-known facts. But to make them into the foundation and point of departure for an analysis of psychological elements, or to wish to found upon them even a theory of knowledge, a system of philosophy: *that* is an absurdity which is almost inconceivable. Need it be said (and, unfortunately, it *does* appear necessary to say, and not merely with reference to this author) what a πρῶτον ψεῦδος lies in a theory that undertakes to illuminate the essence of apperception *in general* by means of a genetic reduction of all materially determinate apperceptions to a *single one* from among them – one which develops naturally under the factual biological conditions of mankind? Is it not obvious that the problem – even should such a reduction succeed – would still stand there untouched? Very well. Suppose that all that is experienced is originally apperceived as a living subject, as a "center of force," or as activity of a center of force, and that in all things "a will is introjected" anthropomorphically, and so on. This apperceiving, apprehending, interpreting as such-and-such a specific thing – as center of force, activity of will, and the like – still does not illumine what in general apperceiving, apprehending, interpreting *consists in*, or then, further, what makes possible and intelligible its validity or non-validity, its true or false reference to objects. [221]

Should the author wish to object that apperceiving is not identical with judging, and that the latter is a special type of apperceiving whose peculiarity lies precisely in the "projection of the individual's will into the extraneous thing," then we reply as follows: Certainly a distinction is to be made between apperceiving in general and judging. For example, an act of mere imagining is an act of apprehending in the wider sense (namely, an apprehending of certain phantasized contents, which are lived through but not intended, and thus of phantasized objects, which are intended but not lived through); but it is no act of judging or presuming that the thing imagined *exists*, or that the event imagined *occurs*. Apperceiving in this sense is, then, obviously one and the same as what we normally intend under the vague title "representing." But to any representing there corresponds a possible judging with the *same* content (the same matter, the same significational content). It can thus never be a determinate content that distinguishes representing and judging in general. The anthropomorphic apperception is one determinate in content. As with any other, *it* can be carried out as a simple representing (thus in mere imagination) or as the positing in a judgment. How therefore could the elemental psychological essence of the genus *judgment* be characterized by saying: "All judgments apperceive phenomenal processes as a processes of volition," and such like? And further: How could the anthropological finding that all human apperceptions or judgments are originally introjections of the will pass for a genetic analysis of the judgment? What is it we want to do with such an analysis? Well, we certainly are not to be satisfied with a mere anatomy of the judgment – with a purely descriptive phenomenology of the essential, general characteristics of the judgment, of the elements and Moments which make it up, or with a purely descriptive formulation of its essential variations and of the essentially distinct forms in which it unites itself to more inclusive psychical formations. We will also want a physiology of the judgment: to find out about the elemental laws under which the entrance and exit of this descriptively fixed type of lived experience stands, under which its formation and transformation are carried out.

Let us consider a parallel case. Will one have solved the problems related to the anatomy and the physiology of man – indeed, will [222] one have even made a start on them – by establishing that all men derive from Adam, or from a concretely determinate race of human beings, or from a species familiar to natural history (the ape or the like)? We distinguish between natural science which is abstract and nomological, oriented in terms of elemental laws, and natural science which is concrete and ontological, directed toward the explanation of concrete natural formations. The latter is grounded in the former. Its theoretical explanations are carried through by application of the elementary laws arrived at in the abstract sciences. Would it not be an absurd undertaking to reverse this order and ground the abstract sciences upon the concrete ones? Precisely so, it would be absurd to ground genetic psychology, as an elemental and nomological science, upon child psychology, folk psychology, cultural psychology, or upon anthropology and concrete biology in general. Naturally this does not rule out the reception, from these concrete spheres, of motives that stimulate inquiry; nor does it rule out indirect testimony for what can, however, be realized in a definitive and decisive manner only through direct and analytic investigation into the elements. Now as to the present book, this perversion of principle constitutes its pervading character. The entire thinking of the author moves within anthropological considerations, within those of concrete linguistics or similar matters. Authentically elemental investigation, whether descriptive or genetic, is wholly omitted. And yet he believes he can pluck the fruits of such an elemental investigation while following his pathway, believing himself to have arrived at a completely adequate psychology of the judgment and – far more than that – the most profound of insights into pure grammar, logic and the critique of knowledge. In his presumed insights we can, unfortunately, only discover mistakes. Just consider: Through the peculiarities of that animistic apperception described above, the essence of the basic thought forms – such, for example, as *subject*, *predicate*, *copula*, *individual representation*, *concept*, and so forth – is supposed to be brought to analytical understanding. – I cannot enter into criticism of details; but it also is superfluous, in the light of results such as the following: The predicate expresses what the thing wills or can will, or what can be done with it without resistance. (p. 120) The subject "tree" (taken as an example in the illumination of the essence and origin of the subject function) signifies the bearer of those forces [223] which are actual in all trees in the same manner. (p. 109) Any assertion concerning a future event is a judgment about a tendency of will residing in present objects. (p. 136) By the introjection of will the bond of unity between the substance and what inheres in it is *created* once and for all. Crude anthropomorphism undergoes essential modifications in the course of the development of knowledge. When the judgment is fully expressed in language, the subject term becomes the bearer of potential influences, the bearer of capabilities. The will becomes force. (pp. 140ff.) In the judgment set in concepts (e.g., The dog is a domestic animal), what is expressed is that such-and-such potential influence resides in the existent center of force. (p. 141) Mathematical judgments are judgments concerning relationships, which are themselves operant forces. Such a judgment states: The relationship in question will prove itself to be in effect in all subsequent operations. (p. 157) And so forth.

Now there are books of a psychologistic tendency which, in spite of extreme errors, have rendered service – even an undying service – to psychology and philosophy. I call to mind *Hume*'s *Treatise*, a work flawed in principle, and yet of inestimable value for phenomenology and critique of knowledge. Unfortunately, the present book does not fall within this class. It deals with an area whose essential problems it consistently fails to grasp; and it treats of this area in a fashion that at no time has the character of that kind of work, so serious in its efforts and so penetrating, which brings lasting advancement to science independently of the viewpoints and prejudices of the authors. The, to some extent, very favorable reception which the book has enjoyed from other quarters may make amends to the author for this harsh judgment. But it must be strongly emphasized, and expressed in the surest of terms, that there also are professionals who cannot allow this book to stand as Representant (*Repräsentanten*) for contemporary efforts toward a theory of the judgment.

Nothing essentially new or worthy of notice relevant to this book is offered by the author's essay, "Ueber psychologische und logische Urteilstheorien," *Vierteljahrsschrift für wissenschaftliche Philosophie*, Vol. 30, 1897, pp. 157–190. It is in essence a defense against [224] O. *Kulpe*'s criticisms in the *Literarisches Zentralblatt* of 1896.

W. Kinkel, *Beiträge zur Theorie des Urteils und des Schlusses*, Giessen, 1898.

The author says it is the task of his book "... to discuss once more the essence of the judgment – yet without intending to give a theory of the judgment worked out in all its details – and, on the other hand, also to broaden the theory of deductive inference in one respect." In the author's presentations I have found nothing in the way of new ideas, and just as little independent work in his treatment of older views. Also, his critiques offer nothing worthy of notice. In what respect the author believes himself to have broadened the theory of deductive inference I am unable to say, in spite of an attentive reading of the book.

J. von Kries, "Zur Psychologie der Urteile," *Vierteljahrsschrift für wissenschaftliche Philosophie*, Vol. 23, 1899, pp. 1–48.

This extraordinarily stimulating and subtle work follows upon the author's well-known essay concerning judgments about the real and judgments about relations. (*Vierteljahrsschrift für wissenschaftliche Philosophie*, Vol. 16, 1892, pp. 253–288) In this latter work *v. Kries* treated that division of judgments which is designated by its title, and which in essentials coincides with *Hume*'s distinction between *matters of fact* and *relations of ideas*; and then he indicated, as subclasses for judgments about the real, the ontological judgment and the nomological, while as classes of judgments of relation we have analytical judgments, judgments of subsumption, judgments of logical relation, and judgments of pure mathematics. According to the author, this analysis – as, in general, *any* "analysis undertaken from the logical point of view" – requires supplementation. The results of logical reflection "... relate mainly to ideal cases of a scientifically clarified thinking, whereby there is at first wholly omitted from consideration very diverse formations of psychical occurrences that nonetheless are also called 'judgments'." It is of great interest, and also for logic, to give a *general* treatment of the actual structure of the processes of thought, and especially to assign to the special cases which serve as the starting point for logic their position in the total range of those formations. (pp. 1–3) [225]

In Section I (pp. 3ff.) the author wishes at the outset "... to survey and confirm from a more psychological point of view" his differentiation of the categories of judgments (stated above). He sets out from the guiding perception that in the judgment a number of

general representations[4] (concepts) are thought together under the supervention of an element that he designates as the consciousness of validity, or the feeling of validity or even of justification. This element is, according to him, the characteristic feature of the judgment. Now in each case where we are dealing with *typical* simple cases of judgments about reality or about relations, the difference in the various types of those judgments can be verified in a psychological difference in the feeling of validity. The validity of a relational judgment, such as that two numbers cannot be equal as well as unequal, is something immediately Evident, which is not the case for judgments about reality (such as that Constance is situated upon the *Bodensee* [Lake Constance]). In these latter the conviction of truth adjoins itself as, to a certain extent, something foreign, whereas with the relational judgments the validity finds its grounds directly in the content of the judgment, in the nature and signification of the representations conjoined, and in their reciprocal interconnection. The author characterizes this "psychological opposition" as that between an idiodetic and a heterodetic feeling of validity. (p. 5) But, as he further explains, not only does there correspond to the main differentiation of judgments, into the real and the relational, a distinction in the feeling of validity. Rather, he holds, there belongs to *each* essentially distinct type of relational judgment, unimpaired by their unifying similarity, a specific and peculiar feeling of validity. The feeling of our insight into a logical relationship is characteristically different from that for [226]
a mathematical relation. (p. 5)

This exposition by the author certainly occasions some misgivings. He speaks here of the different manner in which we have *insight* into

[4] On p. 7 the author adds: "Representations *or* attitudinal dispositions," since he assumes that frequently "... the substrate of the understanding of a word is not to be sought for in consciousness" but rather consists "... in certain physiological states, certain cerebral dispositions." I shall give no further consideration to this transition into the psycho-physiological domain in what follows. The "understanding of words," in those cases where there is no intuitive or authentic significational representation that brings the understanding about, is the so-called *symbolic* representation. It is the "more" in contrast to the empty verbal sound, by which the latter *signifies* something to us, or we *mean* something with it. For considerations of logic, or for those of epistemology or descriptive psychology, only this *lived experience*, the symbolic representation, matters; and its physiological substrates are utterly beside the point – as are also, of course, such substrates as may be associated with the consciousness of validity. This addition by the author, disregarding the fact that it is in error, is also superfluous, if the term "representation" is taken with that extension which is absolutely essential to every logical investigation.

the respective types of judgments, and discusses the *grounding of their validity*, whether in the judgment content itself or in something other than that content.

However, the classifications provided by the theory of logical form, and the *von Kriesian* among them, do not have reference to judgments *as* given with insight or as grounded or even as merely true, but rather to judgments in general – or, more correctly still, to *propositions*. Even an erroneous mathematical judgment is a relational judgment, and any false assertion about facts is a judgment about the real. So how can we give the general classification of judgment types its "psychological basis" (p. 4) through recourse to the distinguishing peculiarities of feelings of validity associated with Evident judgments of those types? The differences of judgment types *cannot* be "... verified as differences in the feeling of validity." Rather, there *corresponds* to the distinction between relational and real propositions, which lies purely in the specific sense of such propositions, a distinction in the possible ways they may become Evident – and one grounded, as an Ideal law, precisely in their senses. Here we are not dealing with distinctions which belong in general to the feelings of validity in real and relational judgments. On the other hand, we are also not dealing with such distinctions as may accrue to those feelings under contingent circumstances and from contingent psychological grounds. The distinctions in question are grounded in the Ideally identical senses of judgment forms of the two types. They are possibilities governed by Ideal law, and inherent in the genera of the propositions in question: possibilities concerning the determinately distinct manners in which such propositions become Evident, if it is possible for them to do so at all. But the difference amounts to this: that in the one case precisely a "fact," and in the other case a "relation," is intuited, and, indeed, is simultaneously intuited – is simultaneously "itself" given – exactly as it is meant in the judgment. To all categorial distinctions there correspond, in this manner, self-evident distinctions in the manner of becoming Evident. Each categorial form has a type of intuitive realization which is just its own. But these differences effect the "validity feeling," the *belief*, only insofar as it is belief with a content of this or that form. They belong primarily to the content: more precisely, to its logical forms.

I cannot concede that *v. Kries*' expositions provide us with a

natural starting point for an investigation of the "psychological basis" [227]
of logical distinctions, or for a phenomenological illumination of the logical in general. If one finds the characteristic element of the judgment in the "feeling of validity," then that is done, after all, only in virtue of a contrast with the "mere representation." But what the judgment *is* is no feeling of validity. Rather, it is inherent in the essence of the judgment, and ineluctibly so, to feel this or that as being the case – or better, to hold it to be the case. In this content held true, in the judged *what*, in the "representational content" of the judgment (which, under *v. Kries'* description, appears to be linked up in a quite superficial manner with the feeling of validity), there resides all of the distinctions of formal logic – although just this very judgmental content, and exactly as it stands, could also function as the content of a minding that is *merely* representational. To concentrate all logical clarification in the theory of judgment upon the feeling of validity, upon the *belief* quality, on the grounds that it characterizes the judgment in its contrast with the "mere" representation: that, it would seem to me, is to lapse into a pernicious partiality. It leads to the tendency to seek in the "validity feeling" for all of that in the judgment which is considered of importance to logic, and hence to misconstrue a whole sequence of the greatest problems in the theory of the judgment, which collectively have to do with the content, the underlying representation, and the sense of the judgment and its forms. The phenomenology of the "validity feeling" is not yet a phenomenology of the judgment. This latter is not something that is in any way separable from the phenomenology of the "representation," nor can it be isolated therefrom and closed off to itself, even if only in the manner of higher levels over against the lower. Recent efforts toward a theory of the judgment suffer from a pervasive lack of comprehensive phenomenological research into "representation" in that sense which reference to the judgment presupposes. However, closely considered, there are manifold significations assumed by the word "representation" in logical contexts, and these significations must be distinguished and clarified.

The author is, however, ahead of very many professional logicians, in that he vividly senses the necessity for a deeper and more subtle penetration into the differentiations within the phenomena of thinking. Two points impress him above all. First, that in many cases it cannot be stated immediately "... of what type the consciousness of validity is which is proper to the judgment" – or more simply [228 (since, according to our above critique, this manner of speaking is not correct), of what genus the judgment in question is, whether real or relational. *Von Kries* recalls, for example, the dispute over the logical nature of mathematical propositions. And second, he is struck by the fact that a feeling of validity is associated with extremely vague significations of words, indeterminately passing over into each other, and not merely with rigorously determinate combinations thereof; and this, according to him, occasions judgments "... which can hardly be brought into any close relationship with the logical types." Now let us take a look at the psychological analyses through which *v. Kries* hopes to solve such problems. In his concluding statement he characterizes the main task as that of "... gaining a certain overview, in connection with which we must broaden and generalize our ideal cases of the judgment in order to make them cover the full psychological polymorphism of our thought processes." (p. 7) In this formulation there lies one point to be emphasized. *V. Kries* supposes himself to have found a "foundation" for the treatment of his subject in his hypothesis of cerebral dispositions. In fact, that hypothesis plays no significant role at all as his work proceeds, unless it is in connection with the translation of the symbolic understanding of words – which is no hypothesis, but rather is a lived experience, and only comes into consideration for descriptive psychology – into the language of physiological substructures.

In Section II (pp. 7ff.) *v. Kries* chiefly deals with the variable degrees of strength and certainty in the feeling of validity. Here he is especially concerned with the judgment about reality. One should certainly suppose that exactly the same distinctions between a fixed, a more or less certain, and a hesitant persuasion, would occur in all judgments. The author does not touch upon this point, but rather (Section III, p. 8) lays a special emphasis upon the fact that an essentially distinct type of uncertainty shows up in the relational judgment. In the case of the judgment of reality, according to him, it is only our ignorance which serves as the source of uncertainty. In itself it is determinate whether the judgment is true or false. Not so with uncertain judgments concerning relationships or subsumptions. Whether a color presented to us *is* red is not at all something determined by the nature of the case, and is not even something subject to discussion. Only in certain cases of subsumption are we given a secure feeling about how things belong together, and have thereby something of the typical and of the rigorously fixed. Other cases, by contrast, are marked with something of the individual, the uncertain, the indeterminate, the atypical. And a similar point holds of all species of relational judgments. After this, *v. Kries* distinguishes between typic and atypic relational judgments. The latter, thus, would be judgments which are neither true nor false, and therefore admit of no discussion. [229]

Especially important are the atypic judgments of "psychological comparison," in their contrast with the typic judgments of mathematical equality and inequality. In the former (e.g., the similarity between a greenish and a pure gray, or the tone of *C* as sung and as played on an instrument), the relational feeling from case to case is said to be an individual: it is not exactly the same feeling, but rather only an "analogous" one. It is wholly different in cases where we assert that one real thing is here and another there, or assert a mathematical equation, a necessary logical relation, and so forth. The sense of these assertions about realities and the rest is everywhere the same. In a detailed polemic against *v. Meinong* (pp. 15ff.), the author states that he can assign *no sense* to the assertion of "genuine sameness" in the atypic sphere – as, for example, with the assertion that two degrees within a scale of intensities not only appear the same, but rather really are the same.

Although such solid conviction expresses itself from within the always thoughtful expositions of the author, we cannot follow him. Judgments which admit of no discussion? – Well, possibly *that* can be discussed. But judgments, a class of "atypic" judgments, which are neither true nor false, and *therefore* admit of no discussion? Let us take the atypic judgment of sameness. Does it make *no sense* to think of the red which we now see here as duplicating itself in several exemplars? This in such a manner that the several exemplars would be identical in their shade of red? Is not the meaningful possibility of absolutely perfect sameness conceded, if one answers this question affirmatively? The possibility that certain red contents which stand under a direct comparison genuinely are the same, and that, if we discover no differences and they therefore appear to us as identical, it is also possible that objective sameness could be the basis of that fact: – All of this certainly *is not* nonsense. Therefore it should not be denied that a grave problem remains here. Can anyone lay claim *in concreto* to Evidence for a rigorous assertion of identity in the sphere [230] of continua? But then from whence do we obtain the Evidence for the existence of any sameness at all within this sphere of continuous graduations, the Evidence for the indifference point marked by the sign "=" in the transition between levels from ">" to "<" ? It is curious that the author's scepticism stops short of pure mathematics. He supposes that we do possess such Evidence not merely in arithmetic, and thus with the discrete, but also in the geometrical sphere – and, indeed, even in the kinematic. These areas are considered by him as within the domain of "typic" comparisons. But is not geometric space an "idealization" of empirical space? And is not geometric sameness an idealization of the sameness of empirical space? Surely, therefore, one could never grasp a geometrical identity with Evidence, as one certainly can grasp a primitive identity of number – perhaps that of a two and a two once again. It is Evidence which here, as elsewhere, is the greatest problem, the genuine aim in the illumination of knowledge. Only those who take Evidence psychologistically, as a "feeling," and who make no attempt at its further analysis, just accepting it as a psychological fact that (given our psychical organization) this feeling actually sets in where it does set in, and that it therefore requires further investigations only in the manner of genetic psychology and biology: – Only such a one will take no offense at all this.

We pass over the remaining cases of atypic relational judgments which *v. Kries* proceeds to deal with, and thus the judgment of analysis (fusion) and, in the logical sphere, the judgment of probability. We can no longer regard it as established that all of these judgments "... do not fall under the main types in logic." Even here we are also reminded – and in an especially sharp manner with regard to probabilities – of the enormous problem about Evidence, whose solution is supposed to be provided by the (still wholly lacking) illumination of thinking in the empirical sciences.

In the IVth Section (pp. 26ff.) the author believes it is possible to identify psychological transitions between the "idiodetic validity" of judgments about the real and the "heterodetic" validity of relational judgments, by pointing out cases in which relational judgments – in virtue of their conceptual intricacy, which often leads through laborious definitional chains back to elementary concepts – lose the obviousness of their validity. This means that their "validity [231]
feeling," considered psychologically, becomes similar to that of the

judgment about reality. Even the sense of the assertion could be so far obscured, in the manner described, that its conceptual (analytical) validity would not be noticed. Yet such a judgment exercises its psychological function and retains its importance within the context of actual thinking. In place of the "typical" or "definitive" significations that are proper to "clear" analytical judgments, whether mathematical or real, now unclear and polymorphic feelings of justification, and significations which are merely "efficacious," come into play. In the place of insight we have an *expectation*, grounded in habituation, of ultimately arriving at correct results. With such judgments the domain of "logical reflection" is surpassed. Logic presupposes concepts which are absolutely and completely clear, sharply and rigorously defined: a presupposition which quite certainly is not fulfilled throughout the broad domains of our ordinary as well as scientific thinking. The author seeks to show by means of examples that thought is possible which is useful in practice and leads to correct results, "... in spite of a wide-ranging mixture of types of validity." And he also seeks to show that, by keeping in view the wider psychological situation of the judgment, there is also opened before one, on the other hand, an insight into the diverse possibilities of erroneous and wrong-headed thinking (through equivocation, through the fluctuating vaguries of concepts) – indeed, even the possibility of "pseudo-judgments," which seem to signify something, whereas they are, more closely considered, totally empty of content. (p. 37) The details of all this can be studied in the author's expositions, which are always stimulating, and whose importance he himself sees as follows: They impart to us a vision of the position assumed by the judgments commonly made the subject of logical investigation (with their validity feeling of a "definitive character") in the totality of our thinking, in the widest psychological sense. Logical reflection can disregard the multiplicity of psychological formations, since its most important issue – the question about the logical interrelationship between several judgments – can be posed *only* for judgments in the narrower sense, with their precisely fixed validity character. Such reflection thus only deals with select cases which, capable of no further reduction or explanation, present themselves as something [232]
definitively clear and definitively meaningful. (pp. 40ff.)

It need not be said, after the critical remarks above, that I am unable to confirm these views of the author. I would prefer to think

that logic does not have to do with idealizing fictions, and also not merely with the select cases from actual thinking, not with judgments of a preferred "validity character." The techniques of logic aim to place norms upon actual thinking as it stands. But pure logic, from which are drawn the principles for such norming, has nothing whatsoever to do with actual thinking, not even with select cases or idealizations of it. Therefore it also does not presuppose clarity, rigor, determinateness of concepts (verbal significations), or Evidence of the judgments. It presupposes absolutely nothing at all. Clarity and determinateness (or rather, clarifiability and determinability) and similar matters are requirements of *practical* logic, not presuppositions of *pure* logic. They are requirements derived from pure logic, not ones upon which it is founded. Pure logic poses, in my opinion, no requirements at all, just as little as does pure arithmetic. Acts or dispositions of representation and judgment are what are logically normed, while acts of enumeration and calculation, together with the corresponding dispositions, are arithmetically normed. But neither pure logic nor pure arithmetic have to do with such psychical matters. Rather, the one deals with concepts, propositions, objects, states of affairs, as to validity and non-validity, being and non-being – and this in "formal" generality. The other deals with pure numbers and number relations. A concept, a proposition (such as, for example, the concept of *angle*, or the theorem of the sum of the angles) is no psychical experience (*Erlebnis*), no psychical disposition. – Now, what determines the author's line of thought is always, as we have observed, the contrast between clear and unclear, evident and symbolical, conceptually fixed and vague or fluctuating thought: these, as well as the closely associated dispositional distinctions, the ability or inability to transform unclear representations into clear, vague into rigorous, and so forth. Of course the knowledge of logical distinctions and laws originates by means of generalization on the basis of "typical" cases, with an eye to lived experiences of representation and judgment that have complete clarity and Evidence, and an iden- [233]
tified determinateness. But "atypic" acts of thought – or, in order to designate the only sphere which really comes into question here, the acts which are vague and fluctuating in their significational content, as well as those in relation to which there is lacking the dispositional capacity for transformation into corresponding acts that are univocally determinate and clarified: – these atypic acts, I say, have in spite

of everything their significations. They mean something. And if the signification content is vague and fluctuating, then there precisely corresponds *idealiter* to each change, to each phase, a signification of its own, which could function as the possible sense in separate typic acts. If there is lacking the dispositions just designated – those toward clarification, determination, and rigor – that may cause trouble in practice. But it does not effect or restrict the spheres in which the laws of logic and their normative applications are valid. They extend to all significations, and hence to these also. That *we* are not able to carry out the norming in the conditions assumed changes nothing in all of this.

One will perhaps find a huge difficulty remaining here, analogous to the one before in the question about Evidence for the Ideal existence of an objective identity in spite of the fact of the threshold of distinction. Where do we get the Evidence that to *every* vague act of thought there *must* correspond a possible act of thought (or a determinate group of thought acts) with a signification that is clear and firmly delimited, and this in such a manner that in the latter act there is segregated out exactly what was confused in the meaning of the former? However is it not Evident, or something to be seen as *generally* true from arbitrarily selected cases, that every thought has a content,[5] a sense, and that it means something which is what it is whether or not we are factually in position, in subsequent thinking, to hold fast to its identity, analyze it, distinguish it from other senses? Even what the author calls a "pseudo-judgment" – insofar as it is not a mere sound, but rather is an expression – has its content, though it may turn out, when the concepts are clarified, to be absurd or ridiculously tautological. Accordingly, expressions and utterances of the most vague and "thoughtless" character also fall, in accord with the forms of their contents, under the "laws of logical relations." Only [234] *we* cannot apply those laws as norms in such cases, since the subjective preconditions of carrying out subsumption under them are, precisely, not satisfied. Where we cannot clearly distinguish things, and therefore cannot enumerate them, we likewise will not be able to use arithmetical propositions. What is lacking in *v. Kries*, as with most psychologistic logicians, is the distinction between pure logic and normative or practical logic; and the difficulties which are

[5] "Content" here does not signify *object*, but rather that which gives the act a sense, whether the object of the meaning may be impossible or even "nonsense."

forced into the open by his genuine concern for fundamentals seem to illustrate anew the extraordinary importance of this distinction. In other respects, I have only yet to mention that the unclarity and indetermination of the judgments under discussion have absolutely nothing to do with the "consciousness of validity" which characterises them as judgments, but rather with that which is in the respective case held to be true or is judged: the judgment content, the concepts and their forms of union, or, more correctly still, the peculiar manner in which the judgment content is phenomenologically lived through (*erlebt*). That "representations" which differ so greatly psychologically can possess the same significational content, that there is such a difference as that between representations [of the same object] which signify clearly and unclearly, is the important and yet ultimate fact.

In the last Section (V, pp. 42ff.) *v. Kries* gives a critique, well justified in part, of recent efforts to define "the essence of the judgment." The worthwhile result of a psychology of the judgment is said not to lie in the acquisition of a general formula – and, as he supposes, one highly indeterminate precisely because of its all-inclusive generality – which would do justice to all forms of the judgment. Rather, he demands a classification, an analysis of the various judgment formations and of their reciprocal relationships. – If I may here express my view, I would say that the *general* analysis of the judgment has very much more work to complete than current logicians, *v. Kries* among them, are apt to suppose. The current "formulae" are certainly very vague, and do not exactly yield exorbitant advances in insight. But that is not, I think, because of their generality – which is indeed very well compatible with clarity and exact determination. To establish with determinateness what gives unity to the many kinds of intentions of thought (and specifically those of the judgment), and what belongs [235
to their general essence, is something that requires the most comprehensive of phenomenological investigations, whose results are to be comprehended in no formula. Matters in the sphere of thought are not as simple as customarily assumed up to now. (At least *this* is something which my *Logical Investigations* may possibly have made clear.) I believe, further, that analytical research into the especially prominent specific formations must lead the way. Generality must arise out of specificity, if we are not to end up with those vague and elastic formulae upon which modern logic, as *v. Kries* very correctly observes, places far too great a value. The preliminary general

discussion and formulation must extend only to the point where the domain to be investigated appears sufficiently delimited for practical purposes.

A REPORT ON GERMAN WRITINGS IN LOGIC FROM THE YEARS 1895–1899

Fifth Article[1]

A. Marty, "Ueber subjectlose Sätze und das Verhältnis der Grammatik zur Logik und Psychologie," VIth and VIIth Articles, *Vierteljahrsschrift für wissenschaftliche Philosophie*, Vol. 19, 1895, pp. 19–87 & 263–334.

The series of articles, continuing through many years, which *Marty* has devoted to those fundamental questions in the theory of judgment bearing upon the elucidation of propositions without subjects, comes to a conclusion with the present articles. In the light of the exceptional importance of these works – instructive to the highest degree for logicians and psychologists of cognition – I shall preface my review with a brief summary of the previous articles. The first three, appearing in the year 1884 (this Journal, Vol. 8, 1884, pp. 56–94, 161–192, 292–340) are based upon the simpler form of *Brentano*'s theory of judgment, as that was presented in his *Psychologie* of the year 1874. In them the distinction between "simple" affirmations or denials and categorical judgments, which are "bi-level judgments" ("*Doppelurteilen*"), is still not allowed for. *Marty* divides his subject matter into three sections. In the *first* he deals with "... the possibility and the advisability of examining the judgment independently of the assertion." (p. 58) Here the author takes his stand, under point (A), against *Prantl*'s theory of the essential unity of thinking and

[1] First appearing in *Archiv für systematische Philosophie*, Vol. 10, 1904, pp. 101–125. Republished in *Husserliana XXII*, pp. 236–258. [D.W.] Cf. *G. Frege*'s references to this article in his letter of October 30/November 1, 1906 to Husserl. (In G. Frege, *Nachgelassene Schriften und wissenschaftlicher Briefwechsel*, edd. H. Hermes, F. Kambartel, F. Kaulbach: Vol. II, *Wissenschaftlicher Briefwechsel*, edd. G. Gabriel, H. Hermes, F. Kambartel, Ch. Thiel, A. Veraart, [Hamburg, 1976], pp. 101–105.) [This note by Bernhard Rang.]

speaking, and under point (B), against theories which exaggerate the significance of symbols in making thought possible. And he then shows, under point (C), that the grammatical categories are no true image of the logical ones, and that the analysis of sentence forms therefore cannot substitute for analysis of the forms of judgment. [237] *Marty* devotes his *second* section to "... description of the thoughts that underlie 'impersonal' sentences." (p. 75) He begins (under (A)) with an excellent summary of those various interpretations of the impersonals that assume the "usual view" of the judgment as a conjunction or relationship of two concepts. After the rejection of these interpretations there follows (under (B) of the second article) a proof of the view which *Marty* takes over from *Brentano*. The theory of the ambiguity of the so-called copula (*Mill*, *Sigwart*, *Jordan*) is extensively criticized (during which, on pp. 171ff, the relationship between the concepts of existence and of reality is discussed), and a proof is sought that in the categorical, as well as in the ordinary existential assertion, the object of the judgment – which is designated by a name composed of the subject together with the predicate – is acknowledged by means of the "is" and rejected by means of the "is not." In each *judgment* the form (acknowledgement/rejection) is to be distinguished from the matter (the acknowledged or rejected content). These correspond to the two essential constituents of any complete *assertion*: a name, whether simple or complex (*S*, *SP*, and the like), to serve as a sign for the underlying representation, and a sign ("is", "is not") which indicates that what is represented is to be "acknowledged" or to be "rejected." The existential judgment, accordingly, is said to be the standard type for any simple judgment, and all other forms of assertion must be traced back to it. In conclusion there is a noteworthy discussion of *G. Frege*, and a defense of the equal standing of affirmative and negative judgments, contrary to the views of contemporary logicians.

In section III (Article 3, pp. 292ff.), the question is raised about the origin of merely *apparent* subjects and predicates in assertions which, like the impersonal and existential, lack corresponding distinctions in their signification (assuming *Brentano*'s interpretation). This leads to the question about the origin of the truly categorical assertion – its distinction from the existential assertion being indeed a "merely grammatical" one, but yet not merely one of the sounds or marks involved. The nature of this "inner" distinction therefore requires investigation. But we do not get to that in this article, which is devoted to a merely preliminary (though very important and instructive) investigation into the nature of the Etymon, or *W. v. Humboldt*'s "inner form of language." The inner form of language [238] includes collateral representations which do not form a part of the signification, although they are most often confused with it. These must serve as auxiliary representations, as a means to communication, as an associative bond between sound and signification. In this context there occurs the discussion, much utilized since it appeared, concerning what is signified and what is manifested in the name and the assertion (pp. 299ff.), and how they are related to the Etymon. Further, there are here detailed criticisms of the views of *Steinthal* and *Wundt*, who, according to the author, confuse signification and Etymon. –

A decade later there appear articles continuing the series, and not without essential modifications of the original theory, following indications given by *Brentano* in his book on the origin of moral knowledge. However, *Marty*'s discussions go much farther. The fourth and fifth articles (this Journal, Vol. 18, 1894, pp. 320–356, 421–471, subdivisions A–E) engage in critical discussions with various authors: with *Sigwart* (*Logik*, second edition, revised and expanded, Freiburg i. Br., 1889–93; and *Die Impersonalien, Eine logische Untersuchung*, Freiburg i. Br., 1888), with *H. Paul*, *Schuppe*, *Puls*, *B. Erdmann*, and *Wundt*. We find a genuine continuation of the investigation in the two articles now before us, which are exceptionally interesting and bring the series to completion. With critical reflections still continuing at first (under (F)), the sixth article deals with the theories of the existential proposition that were held by *Hume* and *Kant*, the former having been misunderstood by *Erdmann* and the latter by *Sigwart*. Kant is said to be no more than Hume a party to the theory that in the existential proposition also two concepts are linked up together.[2]

[2] This is clear with *Hume*. On the other hand, *Marty* overlooks the fact that *Kant*'s words do agree with *Sigwart*'s interpretation, insofar as it claims that "*Being* is a relational predicate bearing upon a concept." Kant supposes, as Marty himself states, that in the existential proposition "... an object is attributed to the concept, or is posited in relation to it," the categorical relationship obtaining between concept and object. (p. 25) But this means that the existential proposition in its subject term directly judges about the concept, instead of about the object of the concept, and in the predicate asserts of that concept that an object corresponds to it. (In any case it is therefore from Kant that *Bolzano*'s interpretation derives, according to which "*A* is"

With this view, which seems to me to be a misunderstanding, *Marty* [239] combines another, to the effect that for *Kant* the existential proposition does not have a concept as predicate at all: not the concept, but the object, is supposedly adjoined (and indeed synthetically) to the subject.[3]

Yet more interesting is *Marty*'s own discussion of the question: Is existence a predicate, and, if so, of what kind? (Under (G), p. 32) He distinguishes two kinds of existential propositions. In "simple and primitive" propositions of the form "*A* is," the concept of existence is not given in the predicate, and "*A* is" here signifies nothing other than the acknowledgement of *A*.[4] But then, in reflection upon such acknowledgement where it is correct, the concept of being correctly acknowledged – i.e., of existence – can be abstracted. And then this concept can serve as predicate in the predicative existential proposition, "*A* is something that exists," which is equivalent but not identical with the simple proposition, "*A* is."[5] *Marty* distinguishes predicates into three classes: into the real (e.g., red, two feet long),

= "The representation *A* has objectivity".) Thus we have here a categorical judgment which, as expressed, however, is "... no synthetic proposition in the usual sense." The existential proposition in *Kant*'s sense has a logical subject that is different from what it grammatically expresses: Not God, but rather the concept of God is the subject of the proposition, "God exists." Since now – this being so – not the concept of God, but rather the concept of the concept of God, is functioning as the subject concept, then, of course, the concept of God is itself also not synthetically enlarged through the predicate concept.

[3] The conclusion drawn in the previous note makes clear how one can avoid interpreting Kant's obscure manner of speaking as an absurdity.

[4] "Signifies" nothing other? Acknowledgement is, however, a contingent act, and presupposes someone doing the acknowledging. But "*A* is" cannot be taken as stating or as presupposing – and no matter in what manner – that someone acknowledges *A*. That also cannot be *Marty*'s view. But how does one escape this difficulty if one takes "is" to be the sign of acknowledgement?

[5] In a polemic against *Jerusalem* inserted here (pp. 35ff.), *Marty* touches upon the relation between the concepts of existence and Evidence. He adopts the view according to which the existence of something indeed requires that it can be acknowledged with truth, but does not require that it can be acknowledged with Evidence. In this Marty certainly goes further than he would have to or can, even on his own position. The possibility of true acknowledgement and that of Evident acknowledgement are equivalent – indeed, even identical – possibilities. Only they cannot be understood as real possibilities, but must be understood as Ideal ones. That "agreement with the object" which is designated as truth is precisely the Ideal relationship (of the Ideal unities: signification and object, or proposition and state of affairs), the actual being-given of which is the lived experience of Evidence. This means that, in fact, truth and the *Ideal* possibility of Evidence are one and the same thing.

the not real (e.g., not existing, having been, merely possible, merely represented), and the ἀοριστα (e.g., not red, not human). Real predicates can belong only to real entities, the not real (unless in some modified fashion) only to the non-real, and the ἀοριστα to the real as well as to the non-real. The predicate *existence* belongs to the latter class.[6] However, it enlarges, as does any predicate, the [240]
content of the subject concept, unless one is determined to understand only *objective* conceptual properties under the heading of conceptual content.

In an interchange with *Erdmann* and *Sigwart* (pp. 39ff.) concerning, precisely, the controversy about the extent to which "... being is a *constituent* of the represented content," the author explains (among other things) that the conceptual synthesis of the subject concept with the predicate concept is the total representation that immediately underlies the judgment: For example, in the judgment that this square is white it is the synthesis square-white; and, likewise, in the judgment that *A* is something existing it is the synthesis of *A* and existing. Thus, it is hard to see how one could avoid speaking of an enlargement of content.[7]

---

[6] This distinction would require a phenomenological investigation. Another distinction, more important for epistemology, which would come into question here, would be that between the "real" and the "categorial": the real as that which is intuitable in intuition plain and simple, the categorial as what is intuitable only in categorial (or founded) intuition. Purely categorial concepts are *being*, *identity*, *difference*, *relation*, *totality*, *number*, and so forth. Also existence is, in this sense, a concept which does not enlarge its subject by adding to it a "real property." No purely categorial ("purely logical") transformation of a concept or proposition introduces into it additional "real" content. Closely related to this point is the distinction (relative and absolute) between matter and form. Cf. Husserl, *Logical Investigations*, Second Part, "Investigations in Phenomenology and the Theory of Knowledge," Halle a. S., 1901, VIth Investigation, Chapter 6, p. 606 and elsewhere.

[7] What *kind* of "conceptual synthesis" do we have here? Square is still being square and white being white; and what unifies the two of them into a conceptual unity? Further, is the synthesis here in mind the same as what "founds" the judgment? The "total representation underlying" the judgment is, however, the one we name in answer to the question: What in this judgment is believed and what persists as "mere representation" where the "belief" falls away? But for *Marty*'s example there is no other answer than this: "That this square is white." One will not be able to evade the admission that the "mere" representation – in the sense of the contrast with a judgment of "the same content" – coincides in every detail, in all articulations and forms, in the "is" and the "is not," with the judgment. The two have precisely "the same content," that is, the same propositional signification content (again, one important concept of *the underlying representation*!), which is qualified in the one case in the mode of belief, in the other in that of mere representing. But of course

Summarizing the results of all his critical reflections, *Marty* draws the conclusion "... that the most recent efforts to detect subject and predicate in the impersonals and in existential propositions have also failed." With respect to the impersonals he seeks to prove that the [241]
series of attempts that have actually been made exhausts all the possibilities. We refer specifically to the table of such possibilities given on pp. 49f.

Now *Marty* turns to the problems raised in the third article (cf. above, p. [237–238]), but not yet solved. He recapitulates its IIIrd section and supplements it with a subsection (B), concerning inner form within the domain of syntax. "Syntax" covers the cases where the union of several elements of discourse has a signification that is not the simple sum of the significations of those elements, and where there emerges in signs a mode of signifying that is carried by no sign independently, but rather is only possible for signs signifying jointly. (p. 56) According to *Marty*, the opinion that the impersonals and the (simple) existential propositions have a subject and a predicate is to be traced back to a confusion of the inner form or of etymological after effects with the actual signification. The more detailed discussions of this matter are given in the IVth section. The interior speech form apprehends *being* as a predicate, whereas at the level of signification the predicate is lacking. In other words, the grammatical predicate undergoes a change of function. In order to clear this up, Marty sets out from a careful and penetrating investigation "... concerning the nature of categorical or '*bi-level*' *judgments* and the predicative synthesis of representations grounded thereupon" (under *A*, pp. 63ff.). The categorical or two-termed form of assertion has arisen as a suitable expression for "bi-level judgments." In propositions such as "This flower is red," or "Certain trees in the garden

---

one must not confuse the "mere total representation underlying the judgment" in *this* sense with mere representation in another sense, which belongs only to the judgment realized in intuitive authenticity. The unified total intuition on the basis of which a judgment of this latter type is realized is not yet the propositional representation, just as little as its object is already the judgment content, the state of affairs. It is, rather, the foundation for these and those categorial apprehensions and formations which first provide the "*S* is *P*" that, in the cases of intuitive judgment, bears the qualification of *belief*. Cf. My *Logical Investigations*, Part II – and, among other passages, the Vth and VIth "Investigations" – which make clear how much more numerous are the fundamental distinctions in the domains of "representation" and "judgment" than investigations up to now have assumed, but also how very much our knowledge of these domains is in only its bare beginnings.

have suffered from the frost," there lies a peculiar type of composite judgment that cannot be resolved into a sum of simple acknowledgements. As soon as we say "this" or "this flower" (also, "*the* king," "some of *the* members of the society," "*my* house," and the like), the acknowledgement of an object is carried out. In general, demonstratives and personal pronouns (p. 63, note) taken by themselves alone are complete expressions of judgments. Upon this sort of basis, then, a second acknowledging is built, which would be inconceivable by itself alone, so that we have only a unilateral separability between the two elemental judgments. Accordingly, one must distinguish the partial judgment in the subject and the partial judgment in the predicate within the categorical judgment as a whole. It is impossible to render the sense of the bi-level judgment by means of a formula that corresponds [242] to one simple judgment. One cannot say in place of "This flower is yellow" that "There exists a yellow flower." Moreover, the latter expression contains the composite concept *yellow flower* (i.e., a flower, which is yellow), which could not be formed without a prior and predicative bi-level judgment. – To affirmative bi-level judgments there also belong some ambiguous forms of expression, which under one interpretation only *seem* to express categorical or bi-level judgments, but under another interpretation actually do so. It is thus with formulae containing "any" or "some." "Some *S* are *P*" can mean, as a simple and purely existential judgment, "There are *S*'s which are *P*'s." But it can also mean: "Some of the *S*'s are *P*." For example, that some of the members of the society (that is, some among *the* members of the society) are ill. The judgment is then categorical. – Besides the affirmative bi-level judgments there are ones which are denials. Upon a judgment affirming the subject we have constructed, not an affirmative, but rather a negative predicative judgment. For example, "Socrates is not ill," or "Men, all men (= men all) are mortal," or "Man is mortal." In the latter proposition the class concept *man* is affirmed, and upon that affirmation is constructed the judgment that nothing which is not mortal belongs to that class. It is true of each of these assertions that in the existential form only an equivalent can be produced, or not even that. Thus, for the proposition: *All of the apostles were Jews*, the pair of propositions: *There are apostles* and *There are no non-Jewish apostles*, is equivalent but not identical. Something identical would be given in the formulation: "There are a number of apostles, and

none of them is a non-Jew," where in the "none of them, etc.," we then precisely would have, again, a synonym for "the apostles all, etc." There then follows (pp. 69–72) an investigation in which the author seeks to carry over to bi-level judgments the theorem enunciated for the "simple" judgment by *Brentano* – that with a judgment of indeterminate matter, it will always be particular if it is of affirmative quality, and always universal if the quality is negative – and in which he discusses the origin of the negative predicate. – I cannot here go into details on these points and only mention that *Marty*, in order to carry through with his general rule in the case of particular, negative propositions, locates the negation within the predicate concept, and thus, for example, interprets "Some of my trees are not productive" as saying "they are non-productive." If one assigns the "not" to the copula, then, he supposes, the sense of the sentence would be the [243]
denial that any of my trees are productive, and that would mean that none are productive. (p. 70)

The upshot of all these analyses is: There is besides the simple judgment a specific type of "relating" (*beziehendes*) judgment – and, beyond *simple* positing and rejection, a certain acknowledging and disallowing, granting and denying – which is to be affirmed as an ultimate element of our psychical life. The traditional theory erred in holding that in *every* judgment there was present such a relating of one thing to another – which relating was, moreover, misinterpreted by it as a mere relation of representations. But judging is a phenomenon *sui generis*. Only, within this genus there are, besides simple judgments, those phenomena intimately fused together in a characteristic fashion, which are just as little reducible to simple judgments as simple judgments are to mere acts of representing. It would be absurd, as *Marty* further explains, to want to define predication by recourse to the concept of identity, just as it would be absurd to define plain and simple acknowledging by the concept of existence. The relationship between the assertions "*A* is *B*" and "*A* is identical with *B*" is analogous to that between the assertions "*A* is *B*" and "*A* is something existing." To exist is to be that which can be justifiably acknowledged; and we have identity where one thing can justifiably be taken as the other, or, what is the same, where the one is the other in truth. (p. 77) The concept of identity is abstracted from judgments with the content *A is B*. (p 77) The *fundamentum in re* for the bi-level judgment and for the positing of identity resides in the compounded nature of the objects, which also brings it about that those objects are apprehended through repeated acts, step by step. We form a bi-level judgment on each occasion where, in an object already known under certain determinations, we discover a further determination, a Moment not attended to up to now – in short, a new part (in the broadest sense) of any kind.

The author then further discusses the origin of concepts resting upon predicative synthesis, in the course of which he distinguishes those cases where the corresponding part/whole relation is given in intuition from those where that is not the case. Among the latter he counts the concepts of the contrary and the contradictory, and concepts such as the four-dimensional continuum, but also syntheses of determinations that would be given to intuition as qualities from [244] different sense-fields – as, for example, in something which has a color and a sound. To such predicative syntheses we owe, according to him, nearly all of our ordinary representations of *things*. (pp. 78f.)[8]

The great importance which the bi-level judgments have for the structure and development of our thought makes it intelligible why, according to the author, the peculiar linguistic expression formed to suit it gained preponderance over every other form for the expression of judgment, and became the exclusive type for our assertions. (p. 87)

If I might be permitted a few critical observations on these important theories, I would wish to say, in the first place, that the essence of categorical judgments – even if we place ourselves within the theories of *Brentano* and *Marty* – is not adequately characterized by viewing them as bi-level judgments; for genuine existential judgments also can be double judgments, as when we say: There is an inn on the river Lahn. Here a plain and simple "acknowledgement" is carried out with the "matter," *an inn upon the Lahn*. But this matter already involves an "acknowledgement." A judgment matter can contain arbitrarily many of such "acknowledgements," however it may otherwise be structured. This also holds true for categorical

[8] This last point is not really intelligible to me. We have representations of things prior to all predication, which I, however, do distinguish from identification. It seems to me that one will have to hold a certain consciousness of unity prior to predication in distinction from that predicative consciousness of unity which is made possible by predication alone. Cf. my *Logical Investigations*, Part II, first edition, pp. 507f., 622f., *passim*.

judgments, and specifically for those that are simple. Their simplicity as categorical judgments is undisturbed by the number, or peculiar manners of insertion, of the suppositions showing up in the matters at the subject or predicate positions. (E.g., The king marched by his honor guard at the train station.) The essential point does not lie in the duality of the acknowledgement (in opposition to the simplicity or even the multiplicity thereof), but rather in the characteristic type of synthesis which confers the unity of *one* objectifying act upon the discontinuous, partial, objectifying acts – and, more precisely, in that judgmental unity which produces the unity of the matter in the categorical judgment *via* the *belief* quality that unifyingly compasses it. This is similar to what *Marty* himself has in view when he speaks of a "... peculiar type of complication and interweaving [245] of two judgments," which must be regarded "... as an elementary trait of our psychical life, not further reducible." (p. 67) In any case, therefore, *Marty*'s phraseology, "bi-level or categorical judgment," is inappropriate.

Other doubts go much deeper. By means of the doctrine of the bi-level judgment, *Brentano*'s theory of the judgment loses much of its original paradoxical air. It does not, as it seemed it must when first published, reduce *all* judgments to non-predicative, existential judgments, but rather assumes, in addition to "simple" positing and denial, that a predicative acknowledging and disallowing is an "ultimate element of the psychical life." (p. 72) The disagreement persisting, therefore, is that according to the dominant viewpoint every judgment is predicative, while according to *Brentano* and *Marty* this is not so. Even existential judgments are, under *Marty*'s expositions, in part predicative. But there also are supposed to be "plain and simple" ("*schlichte*") existential judgments, and here is where the argument becomes critical. Now in the first place, as concerns independent existential judgments or propositions, one will, in my opinion, have to allow of them as a group that they mean what they manifestly say: namely, that this or that exists. They predicate existence. But how about the "subject relevant" judgment within the bi-level judgment so-called? Now it is certainly an important point which *Brentano* and *Marty* make, that in the context of the judgment there can occur nominal representations borne upon a *belief* properly their own. If one sees in this "belief" or "validity consciousness," or whatever else one may call it, the essential thing about the judgment, then we will be able justifiably to claim such nominal acts as judgments, and take *Marty*'s examples of categorical judgments as bi-level or multiple judgments. But names are not propositions, and the significational content of the nominal act does not have the character of propositions in the logical sense. As long as one holds the concepts of *proposition* and *judgment* in essential relationship, and takes that which (as the identical sense) is judged in "judgment" to be a proposition – as, indeed, very weightly logical grounds indicate we should do – just so long will we have to exclude the nominal position from the sphere of the judgment. "The king," "this house," and the like, are no existential propositions, because they are not propositions *at all*, neither grammatically nor logically. They posit objects nominatively, not states of affairs assertively. To each positing name there corre- [246 sponds, in the logico-Ideal manner, a possible equivalent existential proposition – but one which unquestionably must contain the same or some other positing name.

An act of "belief" yet constitutes no judgment in the *logical* sense. Indeed, that act does in a certain sense intend an entity, but not in such a way that the entity would be contained in its content (sense). (Here the fundamental core of *Brentano*'s ideas comes into its own.) If the belief is correct, if it "corresponds," then, ideally speaking, a verification is possible, in which the correspondence – or the identity of the intended and the given – would be a lived experience. In virtue of such lived experiences the intended is intuited, conceptualized and designated as "being," as identical with *itself* (just as it is intended). In the plain and simple nominal posit the concept of *being* is lacking, which concept first shows up in the existential proposition in that that self-identity intuited in living through the Evident "correspondence" is assigned, as a categorial predicate, to the intended object. Correlatively, there corresponds to this predicate of the represented object a predicate of the representation, namely, that it is capable of being brought to adequation. Hence the proposition, *To the representation A there belongs an objectivity*, is not, as *Bolzano* supposed, synonymous with *A is or exists*, but is only its immediately Evident equivalent. – The affinity of this "is," the predicate in the existential proposition, with the "is" of the copula leaps to the eye. This latter also expresses an identity: not that identity involved in the apprehension of the thing by itself, but the identity between subject and predicate which enters into all predicating. Hence, neither in this case nor in any other does the copula express “belief.” How could it ever do so, since *everything* in what is expressed belongs to the *state of affairs*, which is no psychical experience (*Erlebnis*), but rather is an intentional unity? And so, then, the “is not” also expresses no disbelief, but rather a predicative opposition, an objective relationship. I also diverge from *Marty* and *Brentano* in that I, in agreement with *Hume*, do not assume there to be in general a negative opposite to *belief*. This “believing” is no affirmation, acknowledging, assenting – opposite to which there stands a denying, rejecting, refusing. The word misleads. If one sticks to the *Erlebnis* itself, to the plain and simple meaning of the assertion, then one indeed finds within the [247] higher genus of “objectifying act”[9] a phenomenon that is coordinate but not opposed to it: the act of “mere” representing. This also explains why a negative can never function in the bi-level judgment as the “subject judgment,” but a mere representation can quite well do so – which certainly is not in accord with *Marty*’s theories. Deeply absorbed in contemplation of an image, do we not make judgments concerning the objects represented in the image – without, however, seriously holding them actually to exist? Or, indulging in the play of phantasy, do we not do the same with regard to the forms which it magically sets before us? Do we want to explain away all of these modified judgments, judgments on the basis of mere representations, by reinterpreting them? The subject representations in those judgments do not function as positings (*Setzungen*), but rather as what is prior to positings (*Voraussetzungen*). In the sphere of the conditioned judgment exactly the parallel role is played by genuine (propositional) judgments at one time, and by propositional *assumptions* at another, as foundations for the judgments built upon them – of different characters depending on the circumstances.

With full acknowledgement of the irreducibility of categorical judgments to simple existential judgments or mere sums of such, there persists even in the present form of the *Brentanian* theory an excessive favoritism toward the existential judgment. The radicalism with which it sought to interpret away a whole sequence of fundamental and ultimate logical distinctions is only slightly moderated. Certainly this follows from the basic conception that dominates the theory, which sees in *Hume’s belief* the essence of the judgment and

---

[9] Cf. my *Logical Investigations*, Part II, pp. 445ff. *passim* of the first edition.

which correlates the "is" and "is not" of the sentence to the Moment of belief or certainty in the act of judgment. It is striking that *Marty* has overlooked the difficulty into which the theory falls because of its interpretation of categorical judgments. Such a judgment is not only a "bi-level" judgment: i.e., a peculiar sort of interweaving of elementary judgments. It is quite certainly at the same time *one* judgment. As a whole, it "believes"; and it believes something. Thus the distinction of quality from matter, of acknowledgement from content acknowledged, must therefore find its place here as well as anywhere else. It accordingly is not clear why the general schema "*A* is" will not do here, and thus why the bi-level judgment, without the slightest change of sense, should not be construable as one unified existen- [248]
tial judgment. Now certainly – in judging "This *S* is *P*," say – we can distinguish the believing and the believed. But to the question about what therein is believed, we obtain only one correct answer: precisely that this *S* is *P*. If, now, we formulate the existential proposition according to Brentano's principle, and get "That *S* is *P* exists," then we have obviously obtained a modified thought content and the prospect of an infinite regress. The Evidently inadmissible correlation of the "is" in the sentence (the copula) with *belief* indeed leads *Marty* to draw the "*SP*," or the "this *SP*," together as *matter*, in the traditional sense of that concept. But, unfortunately, that yields no unity, without the "is" Moment, not even in mere representations. "This red round" is not a thought. We have to say, perhaps, "This red round-thing"; and there, in the adjectival inflection (*"rote Runde"*), we have the idea of being. The distinction between two fundamentally different concepts of matter is lost to this theory: (i) Matter as correlate of the *belief*, as the unity of what is believed (e.g., as sense of the asserted sentence, which, instead of in the believing, could also function identically as sense in a mere representation); and on the other hand, (ii) Matter as the totality of the "terms," in contrast to the categorial forms giving unity to the sentence ("a," "some," "all," "is," "not," and so forth).[10] And likewise with the distinction between the two fundamentally distinct concepts of quality: quality in the traditional sense of formal logic, related to the proposition, and quality in the sense of descriptive psychology, related to the judgment (namely, as the Moment of *belief*). The reinterpretations

[10] Cf. *Logical Investigations*, Part II, pp. 608, 654, *passim*, first edition.

to which *Brentano*'s theory finds itself pushed are often interchanges of equivalents, but often not even that. The latter holds for the interpretations of the hypothetical propositional forms and others closely related to them, and the former for the interpretations of the universal, general and particular forms. Only these more simple forms can for the time being be considered. In their case *B. Erdmann* had the fortitude simply to follow the natural sense of the words, and in so doing he hit upon the right path, in my opinion. With him I take it for certain that "All *S* are *P*" is a purely affirmative judgment. Being-*P* is asserted of all of *S*, and the "all" belongs to the subject. *Brentano* [249] unfortunately did not profit from *Sigwart*'s important discovery of the formal irreducibility of *plural* predication. The judgment indicated is a plural one, and the plural subject by itself contains a two-fold negation, the representation of a collectivity which functions as the foundation (substrate) for a distributive predication. It is not just the exceptionlessness with which the *S*'s are *P* that the "all" expresses; for otherwise "the *S*" would again bear collectivity in itself, making *every* universal proposition require not merely two, but rather four negations. We certainly do need them *if* we wish to *assert* that exceptionlessness. But in the normal case the mere indeterminate plurality representation, "of *S*," is determined in subject position in such a way that no *S* is to be not contained in it; and the representation so determined serves as the complete subject term for a distributive or, what means the same, a plural predication of *P*. –

I also hold the general judgment to be an affirmative categorical judgment. I have to reject *Marty*'s interpretation of *Man is mortal* as identical with *There is a class concept of man, and nothing which falls under it is non-mortal.* Of *the* human being *in specie* mortality is here affirmed in generality. The "is," the entire sense of predication, is here otherwise affected than in judgments concerning particulars. Predication is otherwise affected once again within the form *An S is (in general) a P*, the form of the indeterminately general proposition, which contrasts with the indeterminately particular proposition, *An S is P.* And similarly with many ultimate distinctions here – distinctions such as affirmative/negative, singular/plural, determinate/indeterminate, individual/general, particular/ universal – whose description should be organized around fundamentally different points of view, while the directly Evident *equivalences* that exist here in many cases, and are customarily confused with *identi-*

ties, already belong within the theory of the laws of logic. But, as opposed to *Marty*, I must yet observe that particular judgments (more specifically, the indeterminately singular or plural ones), appear to me to be categorical, not existential, and exactly in the same manner as the determinate singular and plural judgments closely related to them (e.g., S_0 and S_1 are P). As to the negative judgments of these forms, they pose no difficulty if one will only distinguish between a negative predicate concept and negative predication. Negation belongs to the [250] predicate, of course. Of *all*, or of *some*, or of *these* it is said that they are not P. But it is in nowise said that they are non-*P*. To say of people that they are not smokers does not mean to say that they are non-smokers. (To *reject* smoking is not the same as to *acknowledge* non-smoking.) For the former the negative concept, which may not have yet been formed by the one judging, is not required. And obviously the origin of the negative concept presupposes negative predication. It amazes me that *Marty* denies this right where it seems corroborated precisely in the content of his counter-argument. (p. 71) He also supposes that if the "not" belongs to the copula in the particular proposition, and thus to the predication, then the entire judgment would be negative: *Some of the S's are not P = Not one of the S's is P*? I do not understand how this interpretation is possible. I refuse to acknowledge of some *S*'s that they are *P*, and I do not assign not-being-*P* to some *S*'s. The act of predication is realized upon the basis of the act of taking something as subject, and this latter remains unchanged in the acknowledgement or refusal of the *P*. – However, I cannot here present all of the difficulties, and so return to the review.

In the VIIth (and concluding) article Marty treats (under (B)) "... of the internal form of the categorical assertion." The formula for the categorical assertion is the characteristic and adequate expression of the "bi-level" judgment. The peculiarity of the subject-predicate syntax, which is of such a kind that the positions cannot be interchanged without a modification of sense, is rooted in the peculiar construction of the bi-level judgment from its two constituents, as described (pp. 263ff.), but not in the fact that subject and predicate express the relationship of subsistence and inherence between two contents – as modern logic, following *Kant*, usually views the matter. There are assertions which bring such relationships to expression but are not truly categorical (e.g., "There are yellow flowers"), and conversely, truly categorical assertions to whose signification there cer-

tainly belongs no such relationship (e.g., "This color is red"). (pp. 264f.) It is not sufficient to recognize that an inherence relationship *does not always* form the signification of the categorical assertion (*Lotze*), and it also is not correct that the assertion's copula is equivocal in character (*Laas*, *Sigwart*). The signification of the subject-predicate syntax is never that indicated by *Kant*, and it is always the same signification. The categorical copula is univocal, as is its closely [251]
related judgment sign in the existential sentence. In this latter – depending upon whether it is something real or non-real, simple or composite, an absolute determination or a relation that is acknowledged – *being* undergoes, not a modification, but rather a supplementation. And likewise for the copula in the categorical sentence. In all cases it signifies acknowledgement or rejection, and all distinctions belong to the matter.[11] (pp. 267ff.) *Kant*'s theory is to be explained by the confusion of inner speech form and signification. (p. 270) The thing representation and the representation of a property inhering in the thing (more specifically, of something it is inherently doing or undergoing) are linked up with our categorical *propositions* as their "internal form," whereas they do not necessarily have anything at all to do with categorical *judgments* as such. *Marty* traces out a detailed and very worthwhile line of inquiry into the origination of these "internal forms." He first discusses the motives which determine us to favor reception of a concept into the subject of predication, then applies the result to the concepts which define the relationship between *thing* and property, and indicates how this relationship surpassed all others in frequency, and, therefore, gave the most frequent occasion to the bi-level judgment. The thing, the substance, thus came to function most frequently as subject, and the accident as predicate. "The consequence of this was that, to the language-forming consciousness, the representation of the *substance* could come to stand as the type for the *subject*, and the name of the subject could become something like our "substantive" – a form of language which, even where it does not really designate a thing, nevertheless thrusts whatever is designated under the linguistic *model* of a substance." (p. 272) In an analogous fashion the accident became the type of the predicate; and furthermore the grammatical categories of the active and the passive,

[11] Cf., by contrast, p. [248] above, concerning the concept of matter. My scepticism concerning particular points do not, of course, concern the main elements in this presentation of Marty's, which are as illuminating as they are instructive.

and of the *verbum finitum*, are to be explained as necessary elements of every predicate. (p. 273) The representations which originally belonged to the signification, and conforming to which the grammatical categories were originally formed, everywhere fall to the level of the mere interior form of language.

Marty devotes Section C to the "expression of simple judgments," and in particular to "pseudo-categorical assertions" and their internal [252] forms. (p. 277) According to *Marty*, impersonalia and the primitive existential propositions are not predicative, in spite of their linguistic form. Thus there are pseudo-categorical assertions in which the "being-predicated-of" is only an image, but is not signification. Among these *Marty* also numbers mathematical propositions, such as that all triangles have the sum of their angles equal to two right angles, or that the angle inscribed in a half-circle is a right angle. The first is identical in its sense with: There is no triangle that does not have the sum of its angles equal to two right angles. And the other one is likewise purely negative. In both cases *laws* are supposed to be expressed, and the existence of the apparent subjects is on no wise asserted in them. – All of which can certainly be doubted. A transcription of mathematics into negative existential propositions would be impossible not only for linguistic reasons, but also for reasons of the subject matter. The mathematician asserts no universal and general sentence without previously having demonstrated the existence of the (not apparent, but actual) subjects, or having axiomatically specified it at the start of the deductive system (through existential propositions or existential assumptions, depending upon the type of mathematics concerned). Mathematical existence is that of the lowest species: e.g., that of the determinate numbers in the number series, determinate spatial figures, and the like, of whatever identifying characteristic (falling in the conceptual domain specified by the axioms and definitions). These are the, so to speak, "mathematical individuals," whose existence, as so-and-so defined, is deductively proven with each step. No mathematical proposition refers to individuals in the literal sense of the term. The negative existential propositions of mathematics, which occur *in addition to* its categorical ones, do not exclude real existence, but mathematical. Moreover it is clear that *Marty*, in attributing the character of laws to mathematical theorems, would have to draw exactly the opposite conclusion from the one he in fact draws: namely, that they *cannot* express existential propositions in his sense. One sees this from the negations: The denial of a necessity that an *S* be *P* yields a possibility that an *S* is not *P*; but the denial of the proposition that there is no *S* which is not *P* yields the *fact* that there is an *S* which is not *P*. This objection altogether refutes *Marty*'s interpretation of the general proposition. I therefore hold his polemic against *Sigwart*'s distinction between narrative and explicatory [253] judgments, and against *Riehl*'s distinction between judgments about an existent and judgments about a mere relation of concepts (in which *Hume*'s distinction between *relations of ideas* and *matters of fact* is revived), to be ineffectual. However contestable the terminology, and however incomplete may be the relevant phenomenological analyses, underneath these distinctions lies the fundamental contrast between "fact" and "law." Laws as propositions which assert no "facts" are, therefore, in no wise propositions which deny facts of any kind. A proposition which speaks of pure conceptual beings, and of that which is grounded in them, can be completely affirmative. To reduce it to negative existential propositions, by recourse to the extensions of individuals under the pure concepts, is to surrender the character of *law* ("law" understood in the most rigorous sense).

Analogous objections arise against *Marty*'s interpretations of many types of *particular* propositions as pseudo-categorical, although he himself acknowledges that particular propositions no doubt more commonly express bi-level judgments. When, in order to meet *v. Meinong*'s objections, he (pp. 285ff.) interprets the proposition, *Some isosceles triangles are right-angled*, as containing two existential propositions (i. e., *There is a certain sub-class of isosceles triangles*, and *There is nothing which is both a member of this class and not right-angled*), then – provided we interpret the existence in question as mathematical – this indeed does provide us with an *equivalent* to the original proposition, but one so complicated and changed in its signification that it would be very laborious to use it as a substitute, even in the easiest of inferences.

In the developments which follow (pp. 287ff.), *Marty* has sought to show how, on his view, the transition from categorical formulae to simple judgments could come about. One will be able to study these developments profitably, and derive many worthwhile things from them, without sharing his standpoint. It is thus, for example, when he shows how, in the case of existential propositions involving subsistence, occurrence, having occurred, and the like, the original predicate fades out to the point where there emerges the signification of mere being – "in the sense of acknowledgeability," as Marty adds. (pp. 292ff.)

The Vth main Section (p. 298) aims to establish "a classification and delimitation of subjectless propositions or thetic assertions." The [254] domain of subjectless propositions coincides with that of simple judgments [in Marty's sense]. The above analysis of pseudo-categorical propositions has shown that simple judgments can exhibit in their matter complexities stemming from predications. Now, in supplement to that point, it must be noted that in yet another fashion the matters of simple judgments can be "reflective of judgment" – for example, matters of the form, *existence of an A* (that *A* exists) or *nonexistence of a B* (that *B* does not exist), which are representations that are not formed in reflection upon *predicative* judgments. Of greater interest than these composite forms and existential propositions (the questionable interpretation of which we omit to criticize) is *Marty*'s inclusion of copulative, remotive, and disjunctive propositions, which he interprets as existential propositions with the matters: *A with (without) B*, and *Union of A and B*. In combination with these complex matters, he teaches, simple judgments are not possible of every type: with the copulative, simple judgments are possible only in the negative, and with the disjunctive only in the affirmative sense.

As to this latter point, we judge affirmatively and simply: A union of *A* and *B* exists. But if we judge negatively, that there is nothing consisting of the union of *A* and *B*, then we express a number of negations, as if we said that neither *A* nor *B* exists. (p. 300) There is a number of negations here which are drawn together only in language. (p. 303) As to matters which are copulative and remotive, there is here the simple negative judgment: *A* with (without) *B* does not exist. If to the contrary we judge affirmatively, then instead of one simple judgment we would have a number of such judgments: There is *A* and *B*, i.e., *A* is and *B* is; or else *A* and not *B* is the case, i.e., *A* is the case and not *B* is the case.

I must hold all of this to be completely untenable. One of the most primitive and most important forms of thought unification is removed if one interprets, and interprets *away*, the copulative affirmation and negation as sums of the affirmations and negations merely logically implied in them. *Both* things mentioned are (or are so and so); "both" are not (or are not so). The thought of the *both* is lost, which nevertheless does belong to the unitary proposition, to the *sense* of the unitary thought, and not to the mere words (as sounds), or even to psychological incidentals – such as, perhaps, that the two partial judgments are made simultaneously and in the same ego. A plural affirmation and negation is absolutely never to be replaced [255] by a mere summation of singular affirmations and denials; and even a logical equivalence is obtained only through connecting up the separate sentences by means of the "and," to which corresponds a unitary consciousness linking the separate judgments in matter and character of belief. For example, "S_0 and S_1 are P" = "S_0 is P *and* S_1 is P." The "and" is circumscriptively expressed, let us say, by the words: "*Both* propositions are true." On the left hand of the equation stands a simple categorical judgment, and on the right a unitary judgment, indeed, but one composed out of two simple categorial judgments. Analogously in the parallel case. The author (and *Brentano* as well) believes himself to have reduced hypothetical and disjunctive judgments to existential judgments by means of those judgments about remotive and disjunctive matters. For example: If *A* then *B* = No *A* (or *A* is not valid) unless *B*. The congruence claimed can be contested on good grounds. There is not even equivalence (or equipollence) here, and the test of negation, which surely must yield propositions that once again are of the same truth value, rejects it. Negation of the left hand yields: *A can* be true without *B* being true; and of the right: *A is* true without *B* being true.[12]

If indeed categorical judgments do fit, if only apparently, within the framework of *Brentano*'s theory of the judgment, hypothetical judgments totally refuse to do so. Of course the schema of "quality" and "matter," of belief and the believed, is suitable here as everywhere. But only the prejudice that belief signifies "existential acknowledgement or rejection," and that in the existential "is" or "is not" it finds

[12] An interesting and significant logical paradox which I chanced upon about ten years ago, and which induced me to make thorough investigations into the hypothetical judgment, is here communicated: From the proposition, If *A* then *B*, there obviously follows, No *A* (or *A* is not true) unless *B*. It is also Evident that from the second proposition the first follows. Thus it is totally certain that the propositions are equivalent. But the test of negation on both sides (cf. above in the text) proves, again with Evidence, that the two propositions are *not* equivalent: a complete contradiction. Now *A* and *B* can signify propositions, but also concepts. Thus we perhaps more precisely read: If the proposition *A* is true, so is the proposition *B*; or, If something is *A*, then it is *B*; or, If *A* exists, then *B* exists, and so forth. The solution is by no means obvious.

its appropriate expression, permits it to seem obvious that also every hypothetical judgment, and especially every "simple" judgment of that type (bearing only one act of belief), must have the character of a plain and simple existential judgment, and consequently must admit [256] of expression as an existential proposition without modification of its sense. That prejudice explains why this theory is given no pause by interpretations that are so far removed from the actual sense of the propositions concerned that mere signification analysis – even itself misguided – could not lapse into them. Only in a few cases (as with particular propositions for example) is the transition into equipollent existential propositions so very enticing to natural thinking that the temptation to confuse *logical analysis* (which is directed upon what is *logically* contained in the sense of the proposition, what is to be logically deduced from it) with *genuine signification analysis*[13] (which seeks what is actually contained in the sense) proves favorable for the existential interpretations.[14]

There also are linked to this prejudice gratuitous problems. Since each simple judgment must have *one* "quality," and indeed (given the identification of "belief" with "acknowledgement or rejection"), *either* an affirmative *or* a negative quality, then there arises the question of whether simple hypothetical (and disjunctive) judgments can have both kinds of quality, or only one; and, if only one, which

[13] The "purely grammatical" in the sense of my concept of "pure grammar." Cf. *Logical Investigations*, Part II, IVth "Investigation."

[14] How the theory also forces us to put a wide distance between what is in fact closely related is shown by the interpretation of the so-called "causal" judgment. (p. 307) Whereas the hypothetical judgment, *If A is, then B is*, simply rejects the matter, "existence of *A* and non-existence of *B*," the causal judgment, *Because A is, B is*, is supposed to hold that *A* is, that *B* is, and that the one fact is grounded in the other. Should it not be the obvious course, on an unprejudiced analysis, to assume *the same* main idea in both cases: namely, that on the presupposition that *A* is, *B* is also to be posited – only that the presupposition is characterized as a mere "assumption," a mere "hypothesis," in the one case, while in the other case as the positing of an established fact. The function of a proposition as "presupposition," or as one which "grounds," actualizes itself phenomenologically in a determinate character of the respective propositional *act*. But this act in another aspect bears within itself either the character of "belief" or that of the "mere" representation; and thus results the distinct phenomena of the "If *A* is" and the "Because *A* is." Analogously for the apodosis. Quite similarly, a nominal representation can be a "mere" representation or else a positing representation, and simultaneously assume the characterization of the subject representation for a predication built upon it, which accordingly possesses the normal or else the modified character – of a predication founded on mere phantasy, and the like.

one. The interpretive existential proposition of course supplies the answer: The hypothetical judgment is always negative, and the disjunctive affirmative. (cf. pp. 303f.). These questions are pointless for the one who at all times assigns the "is" and the "is not" to the "matter," i.e., to the content of the belief, and thinks that even in each negative judgment (e.g., *S* does not exist) something is believed (but [257]
not "affirmed or negated!"): namely, *that S does not exist.* Such a one holds that the belief itself does not have different qualities, and that (if one takes affirmation and negation as qualities) in the matter of the belief there occur just so many qualities – now affirmative, now negative, now the main one, now subordinate – as there occur in it unitary categorical propositions, with subordinate propositions or attributions in their terms.

I have yet to mention, finally, that *Marty* everywhere carries through with the point of view of the theory with great consistency and clarity, and divides the assertions in which *simple* judgments find their expression into two classes: namely, 1) thetic assertions with a pseudo-categorical form, and 2) thetic assertions with a conjunctive form. For the former the acknowledging or rejecting character of the manifested judgment is said to be expressed by means of a semblance (*den Schein*) of acknowledgement or rejection: – thus with existential propositions, the impersonalia, and with propositions such as *There is no color in tone*, etc. In the latter, this character is expressed by means of a peculiar type of union between the elements of the proposition: thus in disjunctive and hypothetical assertions, insofar as they give expression to simple judgments (or are "thetical").

There then follows, under VI, a concluding word concerning the interrelationships between grammar, logic, and psychology. (pp. 326–334) *Marty* here draws some more general results from his investigations: the necessity of considering thought independently of its linguistic expression – indeed, yet more, the necessity of a *systematic* emancipation of logic from grammar. But of course "emancipation" does not mean simply to *ignore* the linguistic and the grammatical. To the contrary, to take them into consideration is, he holds, indispensible, if one aims to get free of the troublesome influence of linguistic representations. In particular it is necessary, among other things, to keep distinct that which constitutes the signification of our linguistic expressions and those accompanying representations which do not belong to the signification, and which have been called the inner form of language. On the other hand, grammar must have recourse to logic, in the sense that it has to do not merely with the external and internal form of linguistic expressions, but also with their signification, and thus with, so to speak, the logical aspect of language. In the discussion of the relationships between genetic [258 and descriptive psychology and grammar, *Marty* protests against the "psychological" manner of thinking about language, which has been energetically pursued in more recent times, and which would turn the grammarian away from taking meanings into consideration. In connection with this he attacks the strange theory that the signification of our linguistic expressions (the "logical" aspect) is something that in speaking and understanding does not itself enter into consciousness at all, but rather is wholly represented there by the representations and apprehensions of the internal form of language, which is called "psychological" thinking. For my part, I would certainly suppose that – in spite of numerous exaggerations, and the confusion between signification and internal form of language censured by *Marty* – a *part* of the truth also lies on the side of his opponents. A true solution can, in my opinion, be brought about only by a broadly extended and deeply penetrating *phenomenology* of the lived experiences of thought, and especially of those that are linguistic. Of such a phenomenology the beginnings alone are as of yet on hand. But it is certain that whoever today selects his field of work within the area of these problems – equally important for epistemology, psychology and logic – cannot ignore *Marty*'s weighty contributions.

COMMENTS IN A. LALANDE, *VOCABULAIRE TECHNIQUE ET CRITIQUE DE LA PHILOSOPHIE*[1]

1. On the Article, "Faculty"

Kant (*Anthropology*, subsection 7) opposes the concept of faculty (*Vermögen, facultas*) *in the strong sense* to that of receptivity (*Empfänglichkeit, receptivitas*): "Power to act/susceptibility to undergo." In the discourse on the faculty of knowledge (*Erkenntnisvermögen*) he employs, by contrast, the concept in the broader sense: sensibility, which is a "receptivity" (*Rezeptivität*) constituting the "inferior faculty of knowledge."

One must note that the concept of a psychical disposition belonging to each particular empirical subject – a concept included in that of the faculty of the soul – vanishes entirely in the usage of the word "faculty" for purposes of the critique of knowledge. The theories of the Critical Philosophy concerning the constitution and interconnection of the faculties (sensibility, understanding, etc.) are not psychological theories about the mutual relationships between psychical dispositions. They involve, under a form that is mythological and incompletely clarified, a doctrine concerning the necessary relations inherent in the *essences* of the various types of phenomena: phenomena of knowledge, of the will, etc.

[1] *Husserliana XXII*, pp. 259–260. Comments 1–3 first appeared in file #9 of the *Vocabulaire*, in *Bulletin de la société française de philosophie*, Vol. 6, 1906, pp. 293, 296, 299. Comment #4 first appears in file #12 of the *Vocabulaire*, in *Bulletin de la société française de philosophie*, Vol. 9, 1909, p. 263; and see p. 235 for the editorial note. [My thanks to Fabrice Pataut for his advice on the translation from the French. D.W.]

2. *On the Article "Fact"*

Naming the objective correlate of a valid judgment a *state of affairs* (*Sachverhalt*), we then call any *state of affairs* in which an individual [260] existence is involved a *fact* (*Tatsache*). A "thing" is not a fact. What constitutes a fact is *that* this "thing" *exists*, or *that it is* of such-and-such nature, and so forth. A mathematical *state of affairs* is not a "fact," because there is nothing of individual existence about it. Individual existence is *given* in perception, while facts are given in perceptual *judgments* – or again, when we are concerned with *re-*presentation, facts are given by memory in memory judgments. They are acknowledged from a foundation, but in a mediated fashion, in virtue of inferences that rest upon such judgments. These judgments and reasonings taken together constitute "experience." One therefore calls any *state of affairs* which can be given in experience, or founded upon experience, a fact.

3. *On the Article "Phantasy"*

Phantasy (in the sense where the word applies to an isolated act) designates the pure and simple representation of something individual (the fact that one has it purely and simply before one's eyes), but in the absence of the conscious sense of existence (*belief*) which would posit it as an object of perception or of memory. One has it in view, but without deciding whether it is believable, or even without believing it.

4. *On the Article "Intention"*

In Germany the words "intention" and "intentional" are still much used by philosophers who are associated with *Brentano*.

See among others *Husserl, Logical Investigations*, 1st ed., Vol. II, pp. 346 and following.

[Editorial notice on the article "Individual," not composed by Husserl]:

This article amended and completed in accordance with remarks from J. Lachelier, Rodier, Bernès, Brunschvicg, Dwelshauvers, Drouin, Husserl, Van Biéma, and conforming to the discussion which occurred in the meeting of July 1st.

REVIEW OF A. MARTY
UNTERSUCHUNGEN ZUR GRUNDLEGUNG DER ALLGEMEINEN GRAMMATIK UND SPRACHPHILOSOPHIE[1]

In this vast work, projected in two volumes, the author sums up the results of his indefatigable inquiries – nursed along through decades – in the domain of philosophy of language. His scientific career opens in the year 1875 with that excellent book, *Ueber den Ursprung der Sprache*; and since that time, as his many treatises bear witness – "Ueber subjectlose Sätze und das Verhältnis der Grammatik zu Logik und Psychologie,"[2] "Ueber Sprachreflex, Nativismus, und absichtliche Sprachbildung," among others – problems in the philosophy of language always stand at the center of his interests. This work is dedicated to *F. Brentano*, his brilliant teacher and friend. *Marty* is indebted to him for what is called the "philosophical point of view": thus, not only for a basic supply of philosophical convictions, but rather for the overall type of orientation toward the problems – the style in which philosophical problems are set up and the kind of methodology used on them. Only in extreme necessity, and only insofar as the general style permits it, does *Marty* commit himself to modifications of *Brentano*'s views. One can state that this constitutes the strength, but also the weakness, of his work. Its strength: for the flood of theories, up to the most specific conceptualizations, is illuminated by the unified standpoint, established with the greatest of care. Its weakness: for that illumination falls upon the facts (*die Sachen*) only insofar as, and in such a manner as, they are visible *from* the standpoint. With all serious endeavor, the author is still

[1] Vol. I, Halle a. S. 1908. This review first appeared in *Deutsche Literaturzeitung*, Vol. 3, 1910, columns 1106–1110. Republished in *Husserliana XXII*, pp. 261–265. [D.W.]

[2] Cf. Husserl's "Fifth Article" above. [D.W.]

unable to overcome the partiality of his position. And where it is a question of putting into effect that essentially new orientation which [262] phenomenology (in the sense of my *Logical Investigations*) presupposes, he is unable to follow through. But it is precisely with this phenomenological orientation, I am convinced, that there is for the first time disclosed the radical sense of the philosophical problems about "origins," and therewith the possibility for a definitive "laying of philosophical foundations." But that does not perhaps mean that we must assess *Marty*'s work as being of little value. Given the well-known carefulness and thoroughness of his researches, his acuteness with subtle conceptual distinctions and significant synonymies, and the rigorous objectivity of his (only all too extensive) critiques, it is, rather, obvious that favorable expectations of many useful results will be fulfilled. And this is so whatever position, in the philosophical confusions of our day, one may take – as logician and epistemologist, as psychologist and linguist – toward the author's theories.

From the content of the work I here can emphasize only the essentials. The first and introductory segment gives a detailed treatment of the concept and the task of the philosophy of language and general grammar. Corresponding to his view of psychology as the science basic to philosophy, *Marty* defines the philosophy of language as the unity of all of the problems bearing upon the general nature and lawlike order of linguistic phenomena: problems which either are of a directly psychological nature, or at least cannot be solved without major assistance from psychology. (p. 19) It divides into theoretical and practical philosophy of language. The latter (the "Glossonomie") deals with language as teleological formation, and, indeed, with respect to the aesthetic, logico-scientific, ethico-political and ordinary practical demands that can be placed upon it. (Cf. the fine appendix on pp. 756ff., concerning the problem of an international language.) The present work is not, however, devoted to this practical philosophy of language, but to the theoretical; and, more precisely, to the most important part of theoretical philosophy of language: the general descriptive theory of signification. The tasks of this part of the philosophy of language are carefully discussed in the concluding chapters of the first segment of the book. *Marty*'s advocacy of the Idea of a "general grammar," in the sense of his "Semiotic" ("*Semasiologie*"), is certainly very meritorious; only I must deplore the fact that in this he proves absolutely incapable of doing justice to *my* Idea of a "pure" grammar, as an apriori and self-contained discipline dealing with the forms of significations as Ideal unities. Now, the same as ever, I hold this to be the fundamental part (the sub-soil, so to speak) of pure logic, and, on the other hand, to be the core of general grammar. It is precisely the plain sense of the concept of "pure" signification which *Marty* is unable to understand, and throughout the entire work he is unable to liberate himself from the merely psychological and communicative aspects of signifying. The true fundamental concept of the theory of signification is rejected by him, in that he psychologistically interprets it away. He attacks my "Platonism" by reproducing the old Aristotelian arguments (pp. 337ff.), and does not notice that "Ideal" objects and the *Platonic* Ideas (as *Aristotle* understood them) are totally distinct things. Of course it would never occur to me to challenge in any way the right of the general psychological investigation of signifying; but I am indeed concerned not to allow the newly unearthed treasures of the logic of *Bolzano* to disappear again, and, on the other hand, I am concerned to secure a place in the scientific consciousness of our time for the basic problems of a phenomenological theory of reason, which are most intimately bound up with Ideal significations. [263]

The second segment of the book deals, in ordered sequence, with the main problems of semiotic, and, indeed, from the vantage point of the equivocal distinctions between "form" and "matter" in language. To mention only the most important points, *Marty* distinguishes between signification and "interior form" of language, the latter being regarded by him as mere *means* of expression. As the "exterior form of language" takes in the means of expression that are sense perceptible, so the "interior" form takes in those peculiar characteristics of the means of expression that are experienced only within.

In an important supplement to his older presentations, *Marty* distinguishes between a "metaphorical" and a "constructive" interior form of language. The *metaphorical* consists of certain representations that accompany the signification, mediating in a favorable manner the association between sound and signification, and yet more often assuming the function of evoking aesthetic enjoyment (e.g., I am deeply moved, edified, etc.). The *constructive* interior form of language refers to cohesive speech whose signification appertains to the discourse as a whole. But in such discourse the particular parts also evoke, through the manner in which they are constructively interconnected, pre-interpretations and expectations in relation to that which is meant by the discourse as a whole. Thus, understanding also is prepared for and mediated in this way, and in a way which yields distinguishing characteristics not only for the manner in which the individual speaks, but also for whole languages and families of language. – [264

Let us now turn immediately to the second part of this section of the book, which deals with the distinction between the formal and the material elements in the signification, and contains *Marty*'s own theory of signification. The starting point is formed by the crucial distinction between autosemantic and synsemantic expressions (independently significant, and significant solely in conjunction with other expressions), the analysis of which certainly leaves many unresolved difficulties. There follows the comprehensive treatment of autosemantic expressions, and first their division into the three basic classes: those which suggest a whole representation (and especially names), assertions, and "emotive" terms. These classes of autosemantica must, according to *Marty*, parallel the basic classes of psychical experiences, for the expression of which the autosemantica are appointed. *Brentano*'s well-known classification is taken over by *Marty*, and defended against objections (in particular against *Meinong*'s theory of assumptions, in which I surely can see only a flawed transformation of my theory of "propositional representations"). With respect to signifying in general, *Marty* distinguishes what an expression manifests or "expresses" – namely, the psychical life peculiar to the one speaking – and what it signifies – namely, certain psychical phenomena which the speaker intends to evoke in the hearer through the instrumentalities of speech, e.g., through the assertion of a judgment. There then follows the philosophically most important chapter of the book. For, in connection with a closer investigation into the basic classes of autosemantic expressions and their significations, the content and origin of the most fundamental concepts of logic, axiology and theory of knowledge are simultaneously dealt with: concepts such as representation content, judgment content (state of affairs), existence, subsistence, truth, Evidence, apriority, necessity, impossibility, real, non-real, value, etc. Further, within the theory of names there is tested a solution to the problem of the "immanent" and the real objects, as well as to the problem of the

Ideal *conformitas* between representation and represented, between consciousness and content of consciousness. The epistemological- [265]
ly fundamental distinctions between "authentic" and "inauthentic," intuitive and conceptual, clear and confused representing are also thoroughly investigated. All of these careful and profound analyses will be of great value to philosophers. Their utilization is facilitated by an excellent table of contents and index. Here the space is lacking for a nearer characterization of this theory, and especially for a specification of the serious misgivings which I would have to bring up on almost every point. A lengthy critical appendix deals with *Wundt*'s theory of signification *change* and with his critique of the teleological way of thinking about language.

The second volume of the work is to be devoted to a detailed discussion of synsemantic expressions, some intimations of which are given in the concluding chapter here.

A. Unpublished Treatises

INTUITION AND *REPRÄSENTATION*, INTENTION AND FULFILMENT[1]

I. Representation as Intuition

Subsection 1. Unity of the Perceptual Process, Two Concepts of Intuition

How does the unity of a long-continued process of change, a unity achieved successively or through a developmental process – e.g., the unity of a melody – come to representation? Only quite small segments of a temporal succession and elongation can be taken in at *one* glance, in an instantaneous intuition; and so, likewise, only quite small parts of a melody can be intuited in any one instant.

A melody as such also is not a *sum* of separate intuitions, since the tone formations (*Tongestalten*) belonging to it, succeeding or developing out of one another, unfold within *one* (temporally enduring) act. Though distinct acts may well correspond to the particular tones and structures, there must be present an act which comprehensively embraces the unified content, inasmuch as it is in each moment content of one act of perceiving. So long as this act endures, its unity is retained, although perhaps the content varies and is modified from moment to moment, the act thereby being also modified in a certain sense. There goes hand in hand with the temporal elongation a temporal foreshortening, in virtue of which a part of what was intuited earlier in the act is modified or disappears, while on the other hand

[1] Probably from 1893. First published in *Husserliana XXII*, pp. 269–302. It should be kept in mind that these are exploratory and occasionally tentative sketches. "Representation" and its cognates, with capital "R," are used to translate "*Repräsentation*" and its cognates, except where grammar otherwise requires a capital "R," in which case the German is used. [D.W.]

new elements emerge. Taken in the narrower sense, the "intuition" is thus a different one from moment to moment. And in each case it is but a fragment of the melody, with its tone formations situated in a certain stage of the melody's development.

Let us consider the case where a *familiar* melody is just beginning. It commences with an intuitive, characteristic tone formation that clearly stands out against the "background" of noises or other tones. Another tone formation then builds upon the first form, and [270]
so, in general, formations develop out of formations. In so doing the new ones do not simply exist alongside the ones remaining. The tone or tonal structure with which the melody begins remains in consciousness only for a small time *t*, irrespective of the modifications of content (the temporal recession and weakening) which it undergoes. Within the temporal order of the content it is, at each moment, the last segment, and in general what comes later, closer to the 'now', that dominates in clarity and fullness as compared to what came earlier. Once we move beyond the span *t*, the beginning *disappears*, and from there on ever new parts of what follows it do also. But this occurs without being especially taken note of, due to the obscurity of the disappearing parts, especially since one's interest attaches to the livelier and newer contents and is directed forward all along the way. The tone formations thus weaken and gradually disappear without that transformation being noticed. One notices only what is gained, not what is lost. Yet there is a lasting gain from what is lost. In so far as it was followed with a focused interest, the entire preceding development has an influence upon the aesthetic character or feeling of what happens to be present. The effect is to some degree retained, although that upon which it is based is lost to intuition. If the melody comes to an end, we have a characteristic limit as such, the consciousness of completion. The manner of the ending leaves me with nothing further to expect or require, like the period of a sentence. A half-written word, a dangling antecedent clause, or even a sentence fragment, a word (one which, owing to emphasis, does not function as a whole sentence), evokes an expectation that it does not satisfy – just as when, sitting down to lunch, nothing is served after the soup. A tonal movement which begins but does not completely play itself out possesses the character of the unfinished, the deficient. We "feel ourselves drawn onward," the movement begs continuation, or however we may express it. In any case, the content intuited at

the moment displays a peculiar character, a "*quasi*-quality." With it thoughts can be associated in virtue of which we bring *explicitly* to consciousness the fact that new tone formations must (or must not) follow and be linked in a typical manner to the contents just intuited or arise out of them. What *Kerry* and *Lipps* have called the "impulse toward continuation" is relevant here, although it is not the same thing. The habit of representing or acting along uniform pathways projects several steps ahead, or produces the "consciousness" of a possible continuation, which of course is not represented. A progression in a uniform direction, within a series, an order, a harmony of objective relationships, does indeed produce specific "sensations," and dispositions are established which, when linked to contents that call them into play, make the continuation or elaboration possible, along with the awareness (*Wissen*) that it is possible. [271]

Now the question is: What enables us to know, when we are at the end, that anything at all has gone before, that what is last present is not the entire melody? What lets us know, in the case where perception or phantasy (or both) of the initiated melody ceases, that anything at all *should* follow, that something is lacking in the totality of the melody?

If in perception or phantasy the melody smoothly runs its course, then the continued appearance of new tone formations recognized as proper lets us know that the melody is indeed not yet at its end. If the perception breaks off, the further course may be completed in phantasy; and inasmuch as what that offers us is sensed as only an imperfect substitute, we arrive at the knowledge which we express, for example, in the sentence: "The performer has stopped in the middle." Often such a continuation in phantasy does not occur, and we nevertheless are able to make the same sort of judgment. Perhaps some few measures quite indistinctly enter consciousness, or perhaps not even this. And how are we nevertheless aware, and so state, that the performance has broken off long before the end? What mediates this judgment in that case is the feeling of lack, of disappointment, of a more or less vivid blockage, perhaps simultaneously with the feeling of surprise, of being unsettled, and of cheated expectation. The same holds true when we ourselves break off the melody in actual performance, for perception, or in mere imagination. Except that in this case the feeling of surprise, of being unsettled, which otherwise probably sets in, falls away.

To this point we assumed the case of familiar melodies. The situation is analogous with *unfamiliar* ones. Our general experience with music allows us to understand very well what is a completed melodic whole, and what is not. It is analogy that guides us. [272]

Those feelings therefore imprint on what is momentarily intuited the character of incompleteness, defectiveness or imperfection, making it then serve as an inauthentic (*uneigentliche*) representation, as Representant (*Repräsentant*) for the whole and complete object. The psychological origination of these Representations does not differ from that of any others, and here as elsewhere what they accomplish is clear.

Should we wish to explicate these Representations, then the dispositional interlinkage of the intuitions that make up the well-ordered and objectively determinate process of intuiting the whole melody comes to our aid. Forward and backward we pass along the continuous series. We let the melody develop onward in phantasy; and we refer back to past tone formations. A simple *reversal*, passing backward from tone to tone, is in fact not possible. In the actual generation of the melody the individual tones were of interest to us only as foundations for the tone formations constructed upon them, and the composition of these latter into more inclusive or intertwining formations (*Gestalten*) brought about the coordinated unity of the total melody. Thus in our retrospective glance we can only grasp – or better, reproduce – the relatively independent fragments or formations which separate themselves off as units; and hence we do *not* get back to the concatenation of the individual tones, but rather to that of the tone formations, as our beginning. This is the fullest possible explication in the case of melodies that are well-known and freely available to our phantasy. Of course the explication can also be more or less fragmentary, or wholly inauthentic, and must be so where recollection is deficient or the given fragment belongs to a melody that has never been heard – in which case the judgment that it is a fragment of a melody is made only on the basis of certain indices. All such explications occur, involuntarily or voluntarily, upon reflection; and they can occur then because those feeling qualia that attach to the surrogates given at the moment, and that make them into Representations, stand to the explicating process in a dispositional, i.e., experiential, relationship.

It seems to me we cannot avoid giving "intuition" a *narrower* and a *broader* signification. Intuition in the narrower sense is the imma- [273] nent and primary content of an instantaneous act of representing – or better, of perceiving (*Bemerkens*). Intuition in the broader sense is the content of a unified, enduring perception. If during intuition in this latter sense the content remains unchanged, then, disregarding temporal elongation, it shows no distinction from instantaneous intuition. Things are quite different in the case where the content changes with continuity, or where instead of one content a constantly changing multiplicity appears in the unified act of intuiting. We then have a cohering sequence of instantaneous intuitions, encompassed by the one enduring act of perception. Here there are not inessential distinctions, depending on whether the content changes with continuity or discretely. In the latter case the cohesiveness of the total intuition is not disturbed by the fact that separate acts of perception may single out some individual part, just so long as the whole sequence of acts occurs *within one* overarching act. Moreover, even where there are discrete changes in contents there are always continuous ones as well: the natural changes in contents due to temporal displacement and elongation. Ideally, the unified act of the enduring intuition might likewise be analyzed into instantaneous *acts*, and we can speak of continuity in the "intuitional process" in that sense, regardless of how the content changes. However, it is better to understand by "intuitional process" the sequence of *discrete acts* of perceiving, within which the multiplicity of changes in content are contained: acts which all transpire within the one continuous act of perceiving.

Were one to admit only the instantaneous intuition as an "intuition," that would remove us all too far from the ordinary use of the term. We have one intuition of a spatial object when we observe it from all sides. The object, as conceived of in its objective existence, is a presumed, inauthentically represented total intuition that includes in itself everything which the intuitions received from various sides present as novel over against each other. We bring the object to intuition when we intuit all of that which is to be intuited in it, and thus observe "it" from all sides. For each part and every property of the thing there is a standpoint from which we can grasp it "best." That is, in the continuum of alterations which each intuitive [274] Moment undergoes with the change of standpoint, there is always a phase where the Moment most satisfies our interest. The pertinent standpoint is the "normal" one, and the Moment forms one part or, as a determination of the whole, one property in the Ideal synthesis of the objective object (*objektiven Gegenstandes*). All other phases serve as indices for the normal one.

Thus: to bring an object in general, an objective unity, to intuition means *successively* to intuit, from out of the Ideal union of constituents to whose conceptual synthesis the object owes its unity, those constituents (parts or properties) which provide a completeness that satisfies our interest.

In this sense an object is non-intuitive if its *constituents* are not made intuitive, or cannot be (are unintuitable). Of course the psychologist certainly will distinguish here between actual and presumed intuition. I cannot have an intuition of the route from Berlin to Rome, not even in the form of a sequence of intuitions – that is to say, within one act. But I certainly can do so of individual segments of the road and of the unity of sufficiently small adjacent segments. Thus, in general, only relatively little of the spatial is intuitable, and especially so in our narrower sense. Already in the case of a moderately complicated figure, surveyable in one glance, an instantaneous intuition that is attainable under the most favorable of circumstances does not contain all of the parts and relationships that we thereby presume to intuit – it does not *actually* contain them, or at least not actually *as* they are presumed. The representation – for the most part unclear – which we have when viewing the figure from some given angle of vision is not what we mean by "intuition of the figure." It really only serves us as a basis for the production of the objective unity in which, Ideally, all of the parts and Moments are contained just as we obtain them through selection of the angle of vision most favorable to each and different from one to the other. The *alleged* instantaneous intuition of the object as it actually is therefore turns out to be a sequence of intuitions in which we ascertain the various aspects, parts and relations of the object in their most perfect stages of modification, i.e., those most satisfying to our predominant interest. [275]

Subsection 2. Subjective Conditions of Seeing and Objective Unity of the Thing

It will now be our task to study more closely these intuitional sequences and the constructive operations resting upon them, out of which representations of objective or thinglike unities result.

What lies closest to hand for our analysis is *the thing*. Thus we must set out from it. Of a thing we have, even when we perceive it, no intuition in the strong sense of the word. There is no *instantaneous* act which would totally take in, as actually Present (*präsent*),[2] the manifold parts, properties and unifications of which it objectively consists, simultaneously perceiving and conceptualizing (*auffassen*) them. If a perception takes the form of an instantaneous act, then that multiplicity is mere intention, which is why we say here and in similar cases: We do not have a true intuition of the intended, but rather a mere Representation (in this sense of "mere representation," namely, an *inauthentic* representation). Similarly in the case of a phantasy representation, whose inauthenticity is mediated to a yet higher degree inasmuch as it *directly* refers back to a corresponding perception. But all Representation rests upon intuition. The intuitions in which objective (*objektive*) thing unity is grounded are acquired when we see or touch the thing. Let us first of all consider seeing. If seeing is to provide us with a full intuition of the visible object as such, then there is required a certain intuitional sequence, which we now have to reflect upon. To this end it will be useful to distinguish several things:

I. Those intuitional sequences which are possible without moving the head or the rest of the body, and which depend only on the movement of the eyes, or even only on the movement of the *inner* glance of perceiving or attending.

II. Those intuitional sequences which occur when, in addition, there are movements of the head and of the rest of the body.

In speaking here of the movement of the eyes, as well as of those of the head and the rest of the body, we have drawn into consideration the "conditions" of seeing that are derived from objective knowledge and that do not seem to belong within a purely descriptive psychology of elements. However, let these expressions serve only as convenient and hardly avoidable signs for the subjective lived experiences [276] which always underlie the objective knowledge mentioned, and only through it acquire their interpretation (*Deutung*). It is precisely this interpretation that we disregard. A child that has not yet objectified its body, and in this sense has not yet become acquainted with it, can

[2] Usually "*präsent*" is translated by "Present," with capital "P.

and will nevertheless possess the psychical experiences which very much later it will interpret in judgment as the raising and lowering of the glance, of the head, or as moving around the object, and the like. To each determinate posture and movement of the body, the head and the eyes there corresponds a quite determinate psychical content (*Gehalt*). However complex it may be, or from however many sensate components – well known to be accessible through analysis – it may have originated, this content first confronts analysis as a unified whole or as a constituent of such a whole. Depending upon the circumstances the individual content (e.g., the movement of the eyes) confronts us as a separate unit, or it remains blended into an unanalyzed unity along with simultaneous lived experiences given from other quarters. But this latter unity is perceived as a unity for itself, or else the whole of it remains unperceived – without, however, losing the influences which we shall consider, and to which the individual components make their determinate contribution.[3]

We will designate these concomitant conditions of seeing as the *subjective* conditions.

Let us now begin with I. With the head and other parts of the body unmoved, the gaze is directed upon an object, say the inkwell standing before us. As soon as the seeing begins, there also begins a play of content modifications and of the acts of noticing that specifically grasp them. These acts, occurring within an overarching act of noticing, are nevertheless not arrested by it. The gaze wanders, we say, over the object, and now this, now that part comes into focus, and simultaneously thereby into the mental focal point of the act of perceiving. But this is an unpsychological way of speaking. Subjectively, nothing more is present than a temporalized content-continuum of intuitions, in which particular parts especially stand out, whereas the intervening parts that fill out the rapidly passing [277 transition to them are usually not especially considered. In virtue of the continuity within the variations of the content primarily noticed, we can speak simply of modification of content, and this already expresses the similarity of the particular momentary phases to one another. Thus I first fix, say, upon the one corner of the inkwell, and then my gaze runs hurriedly along the edge and fixes upon the other corner. The total content has been modified. In the modification the

[3] Cf. here selection 22g below, "Conditions of Seeing."

beginning and ending stages stand out, but we pay no special attention to the characteristic content modification *within* the transition. If previously corner *A* was primarily perceived – and in a certain manner brought forth from the unity of the total impression noticed, though not detached from it – and if it previously was distinct in the manner characteristic of what is focussed upon, then all of that no longer holds true – whereas there now holds true of corner *B* what we just asserted of *A*. Now *A* is, to begin with, more dependent, less distinct, and perhaps finally ceases to be an emphatic Moment at all in the new total intuition: one strongly modified as to content, but tied to the earlier one through the phenomenon of transition within the variation appearing (obviously a movement phenomenon). Thus with the wandering of the glance one modification after another occurs. Before the glance moves from one part of the intuition to another, and modifies it in the way described, we notice (if the process unfolds slowly enough) that of the "indirectly seen" parts of the total content one stands out (is perceived); and then we feel a striving, which we will not hesitate to designate as a striving toward clarification. The indirectly seen object appears to us as subject to a certain deficiency, which first appears removed when (where contrary factors are lacking, say the will to not change the gaze) the inevitable redirection of the gaze ensues, and the clarification is given therewith. And thus we can say that there adheres to each of the indirectly seen and analyzable parts of the intuition a certain "intention." This intention is actually *felt* in the analysis and stimulates that movement of the gaze which culminates in the satisfying of the intention. Or, stated in purely psychological terms, in the normal case it has the corresponding clarification as its inevitable consequence. The clarified content is, [278] however, a different content, which is merely similar to the unclarified one; but our interest is more satisfied through the greater fullness of the intention, greater precision of its delimitation, and the like. If we deliberately hold our gaze fixed, the intention of the partial content noticed peripherally often manifests its effect in that a phantasm of the corresponding clarified content flashes, but has no sustained existence. Other intentions which adhere to the partial contents of the intuited, even to the directly seen, and which can find fulfilment only through a modified bodily position, we will leave aside for now.

It will now be necessary to distinguish two components in the modifications of content which we have considered. These also can be separated in reality:

1) The content modifications produced by the change in conditions which we called "shift of the glance."

2) Those which come about through differences in perceiving and attending.

In the normal case the point fixed upon, thus the focal point in the strongest sense of the word, is at the same time what is primarily noticed, and is often the object of an act of attending as well. But not always. Let us consider what kinds of modifications are linked to perceiving and attending. Among them I make the distinction formulated by *Marty* and *Stumpf.* Not every act of perceiving is combined with an act of attending. Attending is a kind of tension toward a content to which there adheres an intention striving for satisfaction. Perceiving is representing in the authentic (*eigentlichen*) sense of the term. It is the simple taking in of a content, the simply-being-turned-to-it. If a content is Present (*präsent*) to us, then we perceive it. Whether or not perceiving is a special type of act is something we will not now debate. But we do have here highly noteworthy distinctions within content: Our total consciousness offers us at any instant a multiplicity in unity. A wholly unanalyzed unity cannot be *perceived.* There is always at least one partial content perceived for itself, set off from the "background" of the remaining contents of consciousness. We can be wholly turned to a content; but at no time is it wholly isolated. The background is attached to it as "fringe." And we can be conscious of the total content of consciousness only by [279]
emphasizing some part or a small number of them. But the "emphasizing" is a modification of content. The total impression is modified depending on whether we emphasize this or that. But frequently we are simultaneously turned to a plurality of things. Even then there is, each time, *one* term distinguished by the fact that we are "primarily" turned to it, while the rest are perceived secondarily, and the background to a certain degree tertiarily. But these are not distinctions of acts – in any case not of them alone. In attending to this pipe I also suddenly perceive the inkwell to the side, and then perhaps in addition the knife lying nearby, and the like; or I simultaneously hear the rumble of a carriage. The distinctions between what I have in the "focal point" of perceiving and what is not there are wholly analogous in kind to those between the seen which we fix upon and that which we do not fix upon – just as between that imaged within

the yellow spot of the eye and that presented through the peripheral regions of the retina. These are once again distinctions of clarity. Each plurality which we perceive all in one instant presupposes a succession within the perceiving, and only in its temporal unification can the perceiving itself be "incidentally" perceived. Moreover, as inner observation teaches, there also exist between things *incidentally* perceived distinctions in degree of clarity and unclarity. We have the capability of transforming, at least for an instant, anything incidentally perceived into something primarily perceived, and thus into something distinct. Again, we find that the modification of content which occurs when what is clear becomes unclear, or conversely, is a continuous one, transforming similar into similar. Here too we find that characteristic feeling of "transition." Distinctness in the sense of the primarily perceived is, in the normal case of seeing, at the same time distinctness in the sense of what is fixed upon. But it may also be that we fix upon one point, but primarily attend to a point off to the side. The bond of habit therefore can be loosened by choice. Of the indistinct in the sense of the peripherally perceived, the same holds true as we previously stated of the indirectly seen. It has a certain intention, a certain deficiency which drives toward elimination, and in virtue of which the incidentally perceived has a Representative function, and thereby a dispositional reference to the primarily perceived corresponding to it. – [280]

After discussing the two components which contribute to the intuitional sequence or the modification of intuition, we must still emphasize a few additional points. *The intuitional sequence is a temporal sequence.* All the same, the temporal aspect plays no role at all in it. It is eliminated. The peculiar relationships involved do not draw attention to it. In that now this part and now that of the total intuition given at the moment brightens up to distinctness, and thereby a corresponding modification of the whole results, it over and over again happens that the old intuition, with its distinctive part, returns. The stages of the intuitional sequence thus flow over into each other in a cyclical manner, but without any fixed cycle, without any determinate order. According to accidental stimulations of contents and the chance turn of interest, the process time and again follows different sequences. Also the will can be determinative in that, from those indirectly seen Moments which are noticed for themselves and which stand in competition, it can select these or those according to preference; and the routes of modification once followed, which easily standardize themselves, can be repeated or altered according to preference. With each recurrence of the same conditions the same intuition occurs, and is acknowledged as such and as a reproduction. And each recurrence of the same routes of modification (or sequences of change in the conditions) induces the same, recognized intuitional sequences, with the recognized distinctive features. This act of recognition does not require mediation through language. That recognition is, however, the presupposition for our appraisal of the object given in the intuitional sequence as something remaining self-identical.

Whichever subjective "standpoint" we may adopt, recognition always sets in; and however we change the standpoint, we find nothing absolutely new, but rather only something clarified – and, indeed, something clarified within a fixed scope. In fact, the total impression, except for the 'clarification' of the part in question, undergoes no basic modification – or none that would introduce something novel. And in any case it undergoes only the kind of modification that was already intended, pre-signified in the earlier impression. Whatever standpoint we adopt, by adhering to the same external conditions of seeing we can pick out from the total impression the various parts [281]
which pre-signify the corresponding clarified parts, or which intend them. Thus, step by step, only a fulfilment of the intention occurs, and therewith identification, so far as the intending representation flows into union with the intended intuition. So step by step identification takes place, for identification in fact signifies nothing other than lived experience of recognition in the case where an intending representation flows over into its intended intuition. Only where a sequence of intuitions repeatedly presents us with new objects do we have, not one intuition, but several – and as many as there are different, separately grasped, new things. But where we always find "the same" over again, there we have, precisely, one thing as our object. It is always the same because we always move within the same *cohering* group of intuitions. Within it a well-trusted passage from one element over to another leads onward, from known to known, in such a way that the total content which we have in each instant already includes in itself the intention upon all further ones, and consequently the passage to any given element of the group presents us with a fulfilment of a part of the intention which encompasses it.

In a certain sense a melody is also an objective unity, but here the temporal succession belongs *essentially* within its content. Only in a determinate temporal succession does the continued recognition and the continued satisfaction of the intention take place. In this case there is also a lack of similarity between each separately noticed part of the intuition and each other one, and, further, a lack of the identity of the intention which permits each part to appear as Representative of the whole. The melody is one, but it is not something known as identical in each of the apprehensions of its parts. Each apprehension of a part apprehends, precisely, a *part* of the melody. Quite otherwise in our present case of the *thing*. Each partial apprehension apprehends the thing, but from a particular "standpoint." And that derives precisely from the fact that each member of the intuitional series – i.e., each static and distinctive modificational state in the chain – includes in itself the intention upon *any arbitrary* other one, and can occasion its revival and identification according to chance or choice. The *thing* is, therefore, given identically the same in each stage, not subjectively but objectively, i.e., according to intention and judgment. The subjective element is taken into account through the judgment concerning standpoint. [282]

Thus if we have the chain $A^\frown B^\frown C^\frown D$..., where the letters signify the clarified Moments of the respective instantaneous intuitions, then when *A* is clearly apprehended we will have *B*, *C*, *D* set into relief either unclearly or not at all, blended into the background of the total intuition. But given appropriate direction of interest, *B* or *C* ... is separately perceived, not as what it is when clarified, but rather as B^1 or C^1 ... perhaps – though endowed with the intention upon *B* or *C* ..., which it respectively signifies, or upon which it is directed. And so the objective thing includes in itself *A*, *B*, *C* ..., each in its full distinctness, and without a determinate order of transition. Does it therefore include them in the manner of a *totality*? In invoking totality (*Inbegriff*) it might seem as if, accordingly, each *thing* were a mere *set*. However the thing is in fact not lacking in relations and combinations, which we also apprehend in the transition. These too belong to the objective unity, and they furnish the relational judgments through which we recognize the members of the totality as absolute Moments reciprocally united. The combinations can confer a certain order upon the contents, but they do not give a predeterminate order to the intuitional sequence, which remains arbitrary in its forms. Taking all of this into consideration, we bring the objective unity of the thing to full consciousness in an arbitrary sequence of judgments. These judgments throw into relief the particular absolute Moments and relations which we find among the contents of the group of instantaneous intuitions and refer to the whole *Represented* (*repräsentierte*) by means of the respective total content, recognizing that whole as appertaining to that content. The objective unity is, thus, a unity in virtue of *judgment*, not mere intuition – though judgment on the basis of intuition.

But now there arises the question: How is it, since certain correlated conditions are also linked with the respective partial contents, that these "conditions" are not blended together with those contents into the unity of the thing, rather than appearing as something "merely subjective" over against the thing as the objective? From the descriptively psychological viewpoint there is at the outset only one difference to be indicated between the two: The "conditions" are indeed conscious phenomena, but they are not objects of the primary perceiving. We directly perceive only the accentuated constituents of the content, and annex to them – unified with them and supporting them – the remaining content of the instantaneous act of perceiving, [283 with a certain accentuation of particular Moments. The conditions are in a certain manner fused into the contents as Moments unattended to – James would say as "*fringes*"[4] – and first stand forth in psychological analysis, as non-real (*unreale*) accompaniments. Thus it is with them just as with the fleeting sensations of transition, which are complex phenomena consisting of the shift in the content, i.e., of the phenomenon of content modification which occurs with the transformation of A into A^1 and of B into B^1; or consisting, further, of the phenomenon of modification in the objective (*objektiven*) conditions (sensations of movement, of convergence, of accomodation, etc.); or consisting, finally, of the phenomenon of modification of the act of perceiving itself. These *fringes* contribute essentially to the consciousness of identity, but they do not belong to the "content"; they do not belong to the *thing*, to which there belongs only the *primarily* perceived and intended.[5]

After these analyses it will not be difficult to understand the psychological origination of those remarkably complicated intentions

---

[4] Cf. W. James, *Principles of Psychology*, Vol. I, New York, 1890, pp. 258ff.

[5] Cf. selection 22h below.

which we call thing-like unities. These are the ones with a content (*Gehalt*) that unfolds in intuitional series, on the basis of which – through an arbitrary sequence of judgments with respect to the identified total content of the respective instantaneous intuitions – the various clarified Moments are linked up as correlated parts and properties.

II. Representation as Repräsentation

Subsection 3. Forms of the Repräsentation[6]

A totally distinct signification belongs to the term "representation" when we take as its correlate the represented in the sense of what is *designated*, or in the sense of what is Represented through *analogy*, or in the sense of what is *defined* through intrinsic or relative properties, or even of what is *surrogated for* by means of analogues, symbols, complexes of signs and elements of a definition and the like. In these [284] cases it will generally be felt unnatural to speak of "a representation content." Rather, we speak of an "object" (*Objekt*), and since this word also is commonly used in the case of representing in the earlier signification, one will therefore have to distinguish between the immanent and the intended (or Represented) object. Representing in this second sense – that of the "Representation," as we perhaps could call it – does not signify the simple state of being-turned-to a Present content, but instead signifies a reference to what is by no means always Present. It is to intend (*meinen*) it, appropriately define it, stand in for it as substitute. Insofar as in this psychical acts, and not mere dispositions, come into play, they are obviously of a wholly different type than that act of being-turned-to. The heterogeneity and multiplicity of psychical acts and dispositions which come to light in this quarter stand in contrast to the homogeneity and unity of being-turned-to. In the latter case the representing is *authentic*, but in the former it is inauthentic – inauthentic in a manifold sense. The concept of the representation in the sense of the Representation is quite certainly not a genuinely simple one. We can distinguish:

1) A Representation in the sense of the sign. By a sign we under-

[6] The following text is not directly continuous with the foregoing in the manuscripts.

stand a content which exercises the specific function of turning our representing primarily to something other than itself: whether to another content, or to a dispositionally unified sequence of contents, or to certain conceptualizations constructed upon the latter. The re-directing toward the designated is to be taken literally when the designated was present to us from the outset – be it even in the manner of something incidentally perceived, so that there exists no psychological identity between what is primarily perceived only as a result of the designation and the factual content.

If the designated is not present to us, then the sign mostly carries out the important function of mentally procuring it, or, at least, of serving as a substitute that at the moment most satisfies our interests.

Psychologically the relationship between the sign and the designated is clear. Disregarding the customary linkage between the two, there still remains the *habit* of allowing the primary interest and perceiving to pass from the sign and exclusively fix upon the designated just as it has been reproduced and perceived. Of course it is also possible for us to arrange by convention for this transition of exclusive interest from an associatively attached content to one intended, and [285] to heighten the relevant disposition through practice. Thus, besides naturally evolved signs, artificial ones become possible.

Repräsentationen in the sense of the sign usually are not contrasted with intuitions because it is in general not customary to speak of the *designated* as something "represented" through signs.

We must set aside the Representation in the sense of the sign as something inauthentic. But it is pertinent to what follows that a sign also can *take the place of* the designated and can represent it Representingly. The equivocation in the expression "represent" caused *Kant* to confuse designation with *analogical* Representation. Along side the conceptual Representation one certainly must distinguish Representation by analogy, although this latter has the tendency to pass over into *conceptual* Representation.

2) *Repräsentation* in the sense of the concept.

First some preliminary observations. Concepts must, I would like to believe, absolutely not be identified with abstract representations. I find it altogether inconvenient to call concepts representations when one takes the word "representation" seriously, as the *actual* representing of a certain (namely a Present) content. To have a content Present – of whatever nature, whether abstract or concrete – does not yet mean to have a concept. We first have a concept where with a certain content (or rather, with a certain representational disposition which that content in fact actualizes) certain further dispositions are linked, in virtue of which it can accomplish all of that which the concept accomplishes for us in knowledge – whether directly, or else indirectly (by means of the inauthentic representations that it allows us to construct). The main emphasis here obviously falls on the domain of dispositions.

If now we confine ourselves to that which the concept-word means when the concept is functioning, and to what we *actually* represent thereby, then it is probably as follows: We have a certain concretum, absolute or relative, which we consider exclusively with respect to its possession of certain partial contents. This involves our attending to these parts primarily, and relating them to the whole, thus also [286] attending primarily to the abstract *relation* of the parts to the whole. In this we position ourselves at the "standpoint" of the whole, so that it becomes the subject term which bears in itself, as a property, the having of the predicate terms: a "having" specifically determined by the type of the parts. It is this property that is of primary interest to us, and the subject term is for us a, so to speak, "empty" or indifferent point of reference. It is no "this-here" for us, but rather is a "something," insofar as we are interested in the concept *alone*. For all concreta which fall under the concept, the "represented" here described is the same – and, indeed, exactly so far as the represented is attended to. But now I think that the same concretum could, in the given case, be present and attended to without a concept being present. Not every peculiarity conspicuous *in concreto* in an object, nor every *abstractum*, is a concept – or, more precisely, belongs to a concept. On the one hand this may be because the dispositions to be linked to it, which would appertain to the relevant concept, are lacking. On the other hand it may be because these dispositions indeed do exist, but are not "stimulated" in such a way that they could attain to actualization at the moment. I think I am not wrong in saying that these dispositions, given that they exist and come into play, also stamp a certain character upon the represented content (*Gehalt*) of the concept. In the first place there is recognition of the property (*Merkmal*), and the "quality of familiarity" with which, in consequence, it appears imbued. (In fact, a determinate recollection of past experiences does not, as a rule, occur where we "recognize" the property.) Further, there is the influence upon the represented that is exerted by the reproduction of the concept word and the excitation of *its* dispositions. My impression is completely as if the word *overlay* the named in the manner of a quality, in accordance with the intended (signified) Moments, and in fusion with them – wholly as a tactile quality appears to suffuse a visual object, in that *it* is, as it were, *fused* with certain visual Moments (glossiness, roughness, and the like).

To that end there occurs the reflection of interest back to its signification which is peculiar to the word as sign. The "complication" (to use *Herbart*'s language)[7] does not indeed consist only in the firmness of the associational union in virtue of which the one draws [287] the other into consciousness, but consists, rather, in a peculiar and easily noticeable union of content that has developed. Enough on this, however. We cannot here pursue these considerations further, and we turn to a distinction valuable for our purposes: the cases where a concept mediates the knowledge of some definite intuitive object (a "this") as *an A*, over against the cases where the concept does not have this function, but rather finds in the possibly co-given concretum only the Representant for some indeterminate, or only specifically determined, conceptual object, for some *A* as such; or where, wholly without such an attendant example or analogue, its application "represents" some *A* as such in a totally inauthentic manner. In a word, we must distinguish the cases where the concept has no Representative function – e.g., when with reference to a given *intuition* we judge, "A lion!" or also, "This is a lion," and so forth – from the cases where it has that function. And only the latter concern us here.

Subsection 4. The Conceptual Repräsentation

Let us consider somewhat more closely these conceptual Representations (which as a rule is what is meant where we speak of conceptual representations). As we indicated above, they can occur in two ways:

1) The concept is given in and with an illustrative concrete Representation. E.g., we represent a black lion (leaping through a red hoop) by representing, through a more or less vivid phantasm, an

[7] Cf. J. F. Herbart, *Lehrbuch zur Psychologie: Sammtliche Werke*, Vol. V, ed. G. Hartenstein, Leipzig, 1850, pp. 21ff.

object in which the respective intrinsic and extrinsic properties are found (recognized). In this case we have an intuition, of not simply each single property, but also of their *unification* within the intuited concretum, which then is itself represented in turn as Representant for any arbitrary or specifically determinate thing that possesses the same properties. But, one could ask, how is this psychologically possible? How is the one thing, alone given, to be thought as Representant for others, often innumerable, and in any case not thought. What, from [288]
the viewpoint of descriptive psychology, constitutes the distinction between the case of the Representation in contrast to that of the plain and simple intuition? If the concretum *is* given to us, we must still find something about it which indicates whether it is the Representing example or not, and, in the former case, whether it stands for an *A* as such, or for a definite *A*. At first one might be inclined to refer to the different ways in which our interest is apportioned in the one and the other case. Where the intuited object as such is of interest to us – and also in the respect, among others, that it is a black lion which is related in such and such ways to other intuited things – there it does *not* Represent. On the other hand, when it does Represent, it has no intrinsic interest at all for us, but *only* with reference to the possession of those properties. But the solution cannot lie here. As a rule a match is of interest to us only as such. But is that sufficient reason to think of it as Representant for an arbitrary match as such, or as Representant for some specific match? In any case, one will have to agree that it is a different psychical posture when we take something as what it is, and when we take it as a mere Representant (as "mere representation") of something else which is not itself given. This holds true for any kind of Representation. If we have a sign which is operative, it induces a certain psychological process: the designated is perhaps reproduced; and if it is, we then have that characteristic transition of interest and perception from the sign to the designated. But if the designated is not there, and we are left with the mere sign, which nonetheless has been acknowledged as a sign, then we feel ourselves simultaneously stretched and constricted; and as a result the intuited sign itself appears to us tinged (*tingiert*) in a certain manner. Diagrams, arabesques, and the like, which have affected us purely aesthetically, immediately appear to us wholly modified when we find occasion to regard them as probable symbols or linguistic signs – even though at first we may think of nothing at

all under them, and even if subsequently we do not become aware of *any* intuitive example whatever of the relationship between the sign and the designated, the symbol and the symbolized. One should therefore perhaps not say: The sign does indeed demand fulfilment, the lacking correlate, and of the latter we have an accompanying, if ever so fleeting, awareness. For since we cannot actually represent the lacking as such, it would have to be thought as Represented [289] through a surrogating correlate. But then the difficulty would only be postponed. In what then do we see the Representing function of the surrogate? Certainly not again in virtue of a conceptualized relationship to a second surrogate, and so on *in infinitum*. Rather, we will have to judge that that peculiar character with which the empty sign is affected enters into associative combinations in virtue of which the thoughts can be reproduced in which the *full* understanding of the state of affairs displays itself. Such associations *must* develop, since it often enough happens that to an empty sign subsequent fulfilment is imparted. Similarly also in the case of our conceptual Representations. It is a different psychical posture when we take a concrete *A* for what it is, even though it may interest us only as the possessor of the properties of *A*, and when we take it as Representant for an *A* as such, and thus for any arbitrary *A*. If several *A*'s are intuitively given to us, then they may perhaps stimulate explicit thoughts that, given the direction of our interest, each is equivalent to any other, that what in the case of one is linked to the possession of the emphasized property is also linked in the case of every other, and the like. But if only *one* concretum is given, then such thoughts find no basis, and *yet* we sense a peculiar mental condition in the face of this concretum, which mediates an understanding, even if inauthentic, of the expression "any arbitrary *A*." We cannot seek the support of those thoughts in the words themselves; for in the case of equivocal expressions, as for example "an *A*," the understanding that results takes on now the one sense, now the other. The difference lies precisely – observation of the inner state of affairs also teaches this clearly – in the way we "view" or "take" (or whatever else one may call it) the concretum. To the intuitive character which it thereby receives the explicating thoughts can then, in turn, associatively attach themselves. Those explicating thoughts are at our disposal whenever we may need them and whenever our dispositions may lean toward them. – Certainly matters are otherwise where a determinate object (thing or type) is to be Represented. What is intuitively Present (*präsent*) in this case not only bears its peculiar character of the Representation, which characterizes it as something referring beyond itself. There is also here an affirmative existential judgment, the matter of which is precisely [290] this Representing content as such. The "represented" can naturally be affirmed only in the form in which we actually represent it. However, what we genuinely represent is only the Representing content. But just as this replaces the unrepresented object, so the existence assertion linked to it replaces the unrealizable existence intuition of the object. If an intuition of the object subsequently occurs, then there results an identification of what was "represented" and affirmed *as* existing with the object intuited. In such a way we obtain intuitive knowledge of existence, we grasp the "correspondence" between "representation" and "object" ("*Sache*"). And conversely: because we have frequently lived through (*erlebt*) such correspondence, we are capable of linking the merely "represented" up with the representation, expectation or objective assertion of such correspondence (the second term of which can then of course also be represented only inauthentically, in the "Representing" sense). And it then follows that merely the affirmation or a Representing representation of that correspondence serves as surrogate. The correspondence is assigned to the Representing content (in which case we have the "mere representation" of a definite, existing object).

2) In the case of the Representing concepts that we have considered up to this point, a concretum was given in which the possession of the properties constituting the conceptual content was intuitively embodied in their unification. We now pass on to consider cases that are constituted quite differently, where such a concretum is lacking, and where indeed each particular property is intuited in a concretum, but not *all* in the *same* one, so that an intuition of the unity and congruity of the properties is impossible. On the other hand, we by no means have a mere collocation in our representation. The synthesis of the properties can come about through their relation to the object of representation, thought in a wholly indeterminate fashion. We have a "thought" which should perhaps be expressed in this way: something which has the property α (intuited in concretum A) likewise has the property β (intuited in B) and so on. For example, something at once round, blue-eyed and square. Or the synthesis is realized in other forms, in that the properties β, γ, ... are adjoined to

an A determined by α. For example, a gnat which is an elephant. [291]

Now how are such syntheses possible? Wherein does the unification of the properties consist, which is after all no truly objective unity of them, and which, with concepts that equally include the examples of contradictory determinations just selected, also never *can* be. How, on the one hand, does the understanding of the concepts become intelligible, and how, on the other hand, are we to take the consciousness of their Representation of *one* object (*wahrer Einheiten*)? How does the consciousness that what we represent is only a Representant of another, of an objectively cohering unity, become possible?

In order to clear this up it will be necessary to consider closely the phenomena here before us. First of all we need to point out the complex and characteristically structured linguistic expressions. The respective expression presses itself upon us, however many words it may involve, as a relative unity. I say "relative" unity, because it can occur as constituent in a higher unity, that of a sentence or string of sentences.

Secondly, I call attention to the accompanying contents: to the actual ones, of course, and not, perhaps, to the intended ones. They do not stand to the words in the relationship of a merely extrinsic, contingent accompaniment. Each one from among the contents comprises a bit of the signification of the whole expression, and therefore must, whether through words or syntax, be represented in it. To the content the expression thus stands in that intimate relationship that generally holds between expression and signification. By contrast, that relationship does not exist between the whole expression and the contents *as* a totality. What the former means, intends, is actually given in the latter only in disconnected fragments.

Third. The disconnectedness of which I have just spoken concerns the *intended* connection, which is precisely what we lack. The contents are nevertheless not without unification for all that. Their relationship to the correlated constituents of the expression bound to them, themselves grasped as a cohesive unit, indirectly confers unity upon the contents as well. The expression possesses a visual or acoustical unification, which is intuited after the fashion of a *Gestalt* quality. The unified expression evokes the unified interest, which supervenes to encompass and hold together all the contents, as we may observe at any time.

Fourth. The expression at the same time evokes – whether through the syntactical form, or through that form plus word content – a certain [292] consciousness of familiarity. *It* is re-cognized. The same often holds true of the whole phenomenon, composed of the expression together with the signification parts. If this latter is the case, then in the rapid flow of thinking, or with laxity of interest, there may be nothing further given at all. The familiarity evokes the semblance of understanding, and it does not allow the fact that we have an inauthentic representation to rise to awareness at all, not even inauthentically. It Represents – so we judge in reflection *after the fact*. But taken by itself there is nothing Representing about it. The consciousness involved lacks a Representing function. Now we want to exclude these cases and restrict ourselves to such as ....[8]

Subsection 5. Repräsentation as a 'Restriction' Phenomenon

The situation certainly is a very different one. Here we have a complex of linguistic expression *and* thought. The former, running its course in quite familiar syntactical forms, would not disturb us so long as there was nothing wrong in the signification contents fused with it. If, as a result of concentrated attention, the full signification intended finally comes forth after all, then we get a clear sense of how the defect is eliminated, how the empty is filled out, how the 'restriction' is dissolved, the dissatisfaction turned to satisfaction. Attention is here turned upon the overall *habitus* of the expression, as well as upon its constituent expressions – but not upon individual phonemes. It is not a matter of an analysis of the words, but only of attaining a representation that is as clear and meaningful as possible. (We also prefer to use only words through which reproduction [of the thought] becomes unnecessary.) We speak slowly, word for word, etc. What I now have in mind will perhaps stand forth in complete clarity by contrast: the specifically determinate *feeling* that the phenomenon carries with respect to the signification. It is a specifically determinate "lack." We sense it, to speak more precisely, in the (characteristic) transition from expression to signification. The interest, habitually [293] passing from expression to signification, does not find its accustomed satisfaction. Something is lacking. So reflection turns back to the

[8] The text is interrupted here.

expression, from there to the signification again, and thus back and forth.

An objection might block our way here, presented in terms of the related phenomenon [the intuitional] previously referred to. Is the feeling of dissatisfaction to stamp that to which it relates with the character of a certain lack, and thus characterize it as something referring beyond itself? But in that related phenomenon the *object* of attention does not appear as deficient. We do not find *it* to be "lacking," let alone that it refers beyond itself. Nonetheless, I hold to my thesis and generalize it still further to maintain that *every* feeling of non-satisfaction causes its object to appear as deficient and as referring beyond itself in a certain manner. Except that under "object" there must be understood nothing other than is here meant and can be meant: namely, the *directly immanent* object of the feeling. But the object to which we attend is not the immanent object of the feeling of attentiveness, as *Stumpf* has demonstrated in his masterful analyses. Intent curiosity is the negative feeling that corresponds to satisfied curiosity as the correlative positive feeling. But this latter is pleasure in the perceiving, *not* pleasure in the perceived, which in fact is often enough disagreeable to us. And thus intent curiosity is eagerness for the perceiving of the new. Its "tension" is dissatisfaction with the fact that nothing new has yet been noticed, not dissatisfaction concerning the object.

Thus this object does not appear as deficient, and accordingly also cannot refer beyond itself. It is something characteristically different whether we attentively consider an object in order to analyze it, and thus to find in it what is new, or whether we consider the object in order to find what customarily is joined thereto according to habit. In the latter the tension is not directed upon the content, but out beyond it instead. It proceeds from the content, with the wish or expectation that something will link up with it, not that a part in it will come to be noticed. The "look beyond" is a characteristic phenomenon quite in the same way as is the "look within": the rising above the object, taking it as a stepping stone, as well as the immersion of oneself into it. [294]

I have now a few more words to add concerning the "referring beyond itself" of the deficient. Each lack is the feeling of something not being as it ought to be. It presupposes either a certain preference or a certain goal as what ought to be. It thus refers beyond itself (of course only implicitly or dispositionally, for the most part). Now whether it is a preference to be represented in generality to which reference is made, or is a correlated specific one (thus, a goal), that depends wholly upon the specific character of the lack, upon the total state of consciousness, or upon the existing and stimulated dispositions.

Now we must move on to study the psychological genesis of the feeling which imprints the phenomenon with the character of an intention or Representation, and also, in particular, to show how that feeling can offer to reflection after the fact the basis for introducing those explicating thoughts which the reflection usually reads, quite uncritically, into the original phenomenon. As a starting point I take a general fact from the psychology of habit. When, in the case of a strongly established habit, the customary object does not show up in the context of intuitions that we usually designate as the correlated habituated "environment," a characteristic feeling of emptiness, of lack, of restriction sets in, even without there being present an (authentic or inauthentic) representation of that which is lacking. If instead of the customary, something else, more or less related to it, turns up, we are at least unsettled (which also is, by the way, a state of restriction); and where the new does not excite an exceptional and enticing interest in virtue of its own peculiar character, the contradiction against the habit of feeling becomes *persistent* – otherwise quickly passing away, directly sensed (i.e., without any intervening representations) as a certain lack or as an incongruity. These phenomena are of interest to us chiefly within the domain of representation. If a familiar melody is broken off in the middle, or is incorrectly played further, if an unusual letter strays into a printed word, if an easily observable detail in the familiar arrangement of objects in our room is changed, and so forth, then we *directly* have the feeling of a certain lack. We feel ourselves to be restricted in a certain way – not restricted in general, but rather restricted by what is intuited. The feeling seems to emanate from it. [295]

Let us now make use of this. Already in infancy we have acquired the habit of linking significational thoughts up with words and with characteristically articulated word combinations, and, upon the hearing of a particular expression, of turning our interest to the thoughts that occur to us and are recognized as pertinent. If the effect is not achieved, in that the expression induces nothing at all of the relevant signification, or does so only in bits and pieces that are intermingled with foreign elements, then here too sets in that feeling of restriction which imprints the character of lack, whether upon the phenomenon as a whole or upon the incomplete significational content (toward which the interest will in fact first spill over).

Perhaps it will be objected that this suits the cases where we hear without full comprehension an expression already understood earlier, but which we cannot at the moment recall. But it does not deal with the other cases, where we are presented with an expression that is unfamiliar, although not in its form or manner of composition. Often such an expression cannot induce the signification exactly as intended in it, simply because that signification is absolutely unrealizable – which holds true, for example, in the case of all absurdities. On these cases, which present the real difficulty for clarification, our fact [from the psychology of habit] which we wished to introduce here can thus cast no light. My reply to this is: Here too we have a case of habituation. The syntactical forms and the individual words are familiar to us, and there is a habit of combining with word combinations thus formed, as their signification, a certain intuitive representation, or a certain sequence of intuitive representations. In this connection it is still to be emphasized that expression and signification correspond to each other univocally and reciprocally in terms of their separately considered parts and forms. Each word and each syntactical form refers to a certain part or Moment of the signification specifically correlated to it. If each word and each element of syntax meets with its proper correlate, then the expression operates in the customary sense. If the proper correlate is missing in some respect, e.g., the syntactical elements do not function in the typical manner, then there is a contradiction to strongly established habits, which must lead to the expected disturbance of feeling mentioned, and – as inner experience moreover clearly verifies – also actually so leads.

If the linguistic expression is accompanied by the signification, then there occurs, as I stated above, a routine deflection of the primary [296
interest from the former to the latter. Since now, in the case where an imperfect or inappropriate significational representation turns up, the interest does not find what it reckons on – since to such representations there attaches nothing of that which usually does, according to habit – then this too must contribute to the occurrence and to the character of the feeling of restriction.

Subsection 6. The Understanding of the Expression in the Transition from the Intention to Fulfilment

It is now to be shown how this specifically determinate feeling character becomes capable of mediating those important epistemic functions which the Representing representations are commonly known to exercise. It all comes down to the fact that this feeling constitutes the characteristic trait from which we recognize the significational content as a "representation" – more clearly, as a Representation. But how does this recognition come about? How does the concept of the Representation arise at all? How *in concreto* do we get the idea that something deficient given to us intends some other thing that is complete?

Clarity is provided by the well-known fact of experience that after the imperfect thought (perhaps even the bare expression) has formed the interest, very often the perfect and satisfactory understanding subsequently shines forth. Each child must, in the course of its development, learn to discern the signification of the words as well as of the syntactical forms. The understanding of a characteristically phrased expression, made possible with the aid of extrinsic factors, establishes in the child the dispositional ability to discern the meaning of the same expression on other occasions, first of all, and then in further consequence of expressions of the same form; and this "discerning" will – in the beginning at least, as a rule – not come about instantaneously. Thus, the situation which we are here considering reoccurs innumerable times. The entrance of understanding is then sensed as the dissolution of the 'restriction,' as removal of the 'lack.' The feeling of restriction gives way to a feeling of familiarity, one which is heightened by means of their contrast. (The temporal contiguity makes the contrast between the two noticeable to the highest degree.) And in the contrast of the phenomena, which are naturally related to each other and distinguished through contrasting feeling characteristics, the one – the satisfactory, the completed – appears as [297]
the goal to which the other – the restricting, the deficient – refers, or which it *intends*. Restriction and resolution, incompleteness and completeness, are quite certainly feelings and feeling characteristics that are necessarily related to one another, the one pressing toward the other, and therefore referring to it, and the negative of course toward the positive. The latter is satisfaction, and thus the goal.

The frequency with which these experiences show up then conditions powerful habits. In virtue of them we become disposed, in the cases where the incomplete phenomenon is at first given alone, to attach explicating thoughts, which, in various levels of authenticity, offer the more or less full understanding of its intention.

At first, therefore, we have accompanied the incomplete phenomenon, the expression, by a deficient (and in any case inauthentic) significational representation, if by any at all. We have an understanding in the inauthentic sense when there is a feeling of familiarity in relation to this complex as a whole, and when the movement of thought progresses unrestrictedly, unaware of the incompleteness. Given very little stimulation of interest in the significational representation, the expression so far predominates in that complex that the familiarity which *it* causes us to feel is able to expand over the whole complex. The feeling of restriction which perhaps is incited, is in any case too weak to become noticeable and to hinder the expansion. For the rest it is to be noticed, by way of supplement, that the attainment of very many goals of thought is untroubled by the deficiency of the significational contents. In certain contexts this justifies habits of abandonment to the surrogates, untroubled by their deficiency, so that in these cases the familiarity character directly has to do with the contents themselves – presupposing, of course, that the same contents always serve as surrogate significations. The restriction feeling, originally perhaps very noticeable, thus becomes diminished to the point of unnoticeability, in virtue of routine inattentiveness and of the familiarity feeling evoked in those contexts.

This actually signifies an advance toward understanding, in that given a more attentive focusing upon the phenomenon all illusory understanding fades away, i.e., a clear consciousness of the deficiency [298] of the significational content at hand emerges. If in consequence of intensely stimulated attention the movement of thought comes to a standstill, then it may happen that owing to a propitious awakening of acquired dispositions, perhaps with the co-operation of the will, a full understanding shines forth. To restrict ourselves to the simplest of cases, there occurs an intuition in which the intended part or property, and indeed in the form intended (just as the expression means it), is encountered. The transition that has been lived through, from restriction to release, from surprise to familiarity, from lack to the corresponding fullness, characterizes the intuited as the intended, as we saw above.

If this result does not occur, then perhaps another does, which at least procures for us an understanding through analogy. The reproduction of a *similar* case, in which full understanding did take place, brings what is authentically intended to a certain clarity for us. In the example, one lives through the occurrence of the intended understanding. One intuits in it at least the typical character of the fulfilled intention, which remains unfulfilled in the case actually at hand. We transfer what the example exhibits over to our case. We think: As it is there, so it must also be here. Transference signifies a kind of coincidence or superimposition of the two phenomena. In it we join to that which the intuition of the one phenomenon offers that which continues to be yet noticeable in the second, though fading out in the process of overlapping: precisely the accentuated Moments that we wish to "relate" to that first phenomenon. But in our case, a conflict sets in between the adjoined fulfilment of intention, taken over from the example, and the restriction actually lived through – a conflict in which the restriction unconditionally triumphs. Thus the restriction appears as what really is, but ought not to be, and the fulfilment as what is not, but ought to be. Consciousness of the *ought* and *ought not* rests here as everywhere upon the relation of a negative feeling to the positive feeling correlative to it, and requires, in order to come to authentic consciousness, the presence of both terms of the relation. – But an understanding through analogy can occur in yet another way. Instead of an example of a full understanding in its *beginnings* there serves us now – and this is the usual case – an example of an expres- [299]
sion merely understood: in particular, an example of an expression of analogous formation. If I want to make completely clear to myself what the expression "round square" signifies, then I intuitively represent to myself a red square. Here I live through the familiarity – which is, once again, heightened by means of the contrast – and I "refer it" to my opposed case of the round square. I transfer it to the round square for myself, and feel it as the non-existent goal.

Subsection 7. 'Realization' of Inauthentic Understanding Through Superimposition or Coincidence

I now add a few supplemental observations concerning the manner in which our inauthentic understanding turns into authentic. In many cases the transition occurs so quickly that one cannot speak more fittingly than of a "lighting up" or "flashing" of the understanding. In other cases, by contrast, it proceeds slowly enough in order to be observed and described ....[9] Of course we have no guarantee at all that what has thus been observed would also have to occur in the same way in all cases, even those that evade observation.

How do we go about obtaining the authentic representation, and in that sense bring the inauthentic representation to realization? To acquire authentic understanding of a conceptual representation means to acquire the intuition of what it intends – inasmuch as we also here leave aside the cases involving many levels of inauthenticities. After all, the representation was already previously accompanied by intuition in the past. If, perhaps, the expression were not acting exclusively as Representant, then the properties and forms given along with it would be abstracted from intuitions now departed, in which they were embedded. Let us now consider an example. In order to represent a *red inkstand*, of which I at the moment have no intuition, there by chance serves me for the abstraction of the color the red writing pad at hand, and, for the abstraction of the external form, the black inkstand nearby. Both Moments are circumscribed by a unified interest, though objectively they are separated at the outset. Then the redness seems to break loose from the pad, to wander over [300]
to the inkstand and suffuse it. And with one stroke an intuitive representation of a red inkstand is there – unsteady and, in contrast to the stable intuition of the black inkstand, hardly graspable, fleeting. However this description is not sufficiently exact. The redness does not actually break loose. Rather, an indeterminately delimited (vacillating, blurry) bit of the surface which bears it does. It also does not continue identically the same, as would at first appear. In the transition it fades toward the point of being unnoticeable, provided it does not disappear entirely, only to revive in the intuition of the form of the inkstand, which during the transition has been shifted into the mental focal point and, at the same time, into the focal point of the perceiving. What is further to be verified here we can designate as the "coincidence" – or, better, as the superimposition or covering – whereby the two Moments that were separately attended to triumph, at least for the instant, in competition with the ones unattended to

[9] There is a discontinuity in the text at this point.

and inconsistent with them; and in this fashion the intuition of a red inkstand arises. The concentration and continuity of the interest thus acquires dispositional effect. It conditions the *revival* of the accentuated Moments, perhaps up to the point of the original fullness and freshness of content – i.e., of the kind that is intuitively grasped with the most authentic and complete attentiveness. The interest does not, of course, accomplish this by itself. It must precisely be an interest upon an articulated unity of the sort which is obviously not present in actual representation. But the contents *also* are in fact not there by themselves. They accompany a linguistic expression, and this total complex is in turn interlinked with certain habits, indicated and genetically explained above, which make the outcome intelligible. The habit of representing precisely thus articulated unities of properties and parts with expressions formed thus and so disposes us to do it also in the cases at hand. If the understanding has not yet set in, and the interest not yet been satisfied (think here of the direct feeling of the intention), then we will naturally allow our mental glance to wander from one bit of the signification to another. The memory images which thus arise and overlay the newly grasped contents will, in an intelligible way, be extraordinarily conducive to the origination of the intended unity, especially since they would be strengthened by the earlier interest concentrated upon their prototypes, and would be more strongly disposed to revival. This holds true not merely for thing-like unities, but rather for many relational unities also. Resemblances, similarities, distinctions and gradations stand forth with special clarity through superimposition, partly in virtue of the resulting full coincidence, partly in virtue of the resulting discrepancy.

In addition I would point out that we are often able so to interrelate parts and abstracta belonging to different intuitions that they unite into a new intuition; or, more precisely stated, that in consequence of this interrelating a new intuition occurs, which possesses the same (i.e., similar, in part perhaps identically[10] the same) parts and abstracta. If

[10] One also must not take the words "same" and "identical" here in the wholly rigorous sense. When we consider a figure and persistently fix upon it, we may have a continuous, and, to that extent, identical, act; but we do not have a self-identical *content* from moment to moment (as we run over the contours, varying it with gaze fixed, etc.). Thus we have in mind here only the, so to speak, "personal" type of identity. Where this is lacking – where, e.g., in the transferral of the property out of the one intuition to become the property of another it fades out with time or disappears, and then suddenly revives united with the latter – we speak of mere

in phantasy I have a red square and a grey circle, then I can form the intuition of a grey square and of a red circle. Here the new seems to be composed of elements dissolved out of the old. The redness on the square, for example, seems to wander over to the circle and become one with it. Still, a retention of the genuinely identical is mere appearance, and so is the dissolution. In the transition from the square to the circle the former loses in fullness and vivacity, in virtue of the "narrowness of consciousness," what the latter gains. It then comes to coincidence (or, if one prefers, to superimposition), whereby both of the accentuated elements triumph in competition with the unaccentuated elements which are incompatible with them. The elements weakened by the transition – if still actually present at all – thereby revive to full freshness. The earlier accentuation thus asserts its claim dispositionally. It conditions not merely self-subsistence, but also revivification up to the original, accentuated fullness and liveliness which was there before the process of coinciding . In that [302 I here spoke of coincidence or superimposition, I used expressions which, for the most part, can be regarded as only figurative. They cannot be avoided, since a special term for the peculiar phenomenon is lacking. In fact the analogy is also palpable enough.

One must not, since in our example unification and opposition on the basis of a superimposition occurs in the literal sense, deny them in the analogical sense. Consider only arbitrary cases of the comparison of non-extensive contents. Where resemblance comes to be known, not directly as a figural moment, but rather through an act of comparison, such a superimposition takes place. E.g., in the case of contents of hearing, in the case of all non-sensuous contents, and so forth. Moreover it is already a characteristic distinction whether the superimposition, and the phenomena linked thereto, concern the attraction and repulsion of only visual extension and quality or those of the various senses. The unification ensues steadily according to types, which are grounded in the nature of the partial contents, and which we have found realized innumerable times in experience. Thus the habit of finding such and similar partial contents *combined* in certain typical forms of unity can prove effective in cases where a unified interest bears upon parts and characteristics *in isolation.* [303

similarity, which moreover will be only a great degree of likeness.

INTENTIONAL OBJECTS[1]

I. Immanent and Real Objects[2]

Section 1. The Paradox of So-Called Objectless Representations

If, after these reflections,[3] we can assume that to each representation there belongs a signification content, it then remains to discuss the incomparably more difficult question of whether each one also refers to objects. Also here opinion is divided, although the dominant tendency obviously is to say that each does. It is held to be obvious that every representation refers, whether in a determinate or in an indeterminate manner, to some sort of object – precisely to that one which the representation is said to represent. But remarkable difficulties are bound up with all of this. *If each representation represents an object*, then surely for each representation *there is* an object, hence: *to each representation there corresponds an object.* But on the other hand there is also the indubitable truth that each representation *does not* correspond to an object. There are, in *Bolzano*'s language, "objectless representations."[4] In fact it is Evident, for example, that to the

---

[1] Part I is from 1894 and 1895, Part II from 1898. First published in *Husserliana XXII*, pp. 303–348. Professor Karl *Schuhmann* points out that this title is correct only for Part I of the text. I am greatly indebted to him for a number of other corrections to the published text. [D.W.]

[2] Compare the contents of this part to the outline of a letter from *Husserl* to *Marty*, of July 7, 1901, Appendix IX below.

[3] Husserl here refers to the first part of the manuscript, which was not preserved. See comments in *Husserliana XXII*, pp. 455 ff.

[4] Cf. B. Bolzano, *Wissenschaftslehre. Versuch einer ausführlichen und grösstentheils neuen Darstellung der Logik mit steter Rücksicht auf deren bisherige Bearbeiter*, Sulzbach, 1837, Volume I, p. 304.

representation “a round square” no object corresponds. We are also certain of this with representations such as “present King of France,” etc. Accordingly it seems that we indeed can ascribe a signification to each representation, but may not ascribe to each a relation to an “object.” And a further consideration of equal weight supports us in this. Meaningfully, and no doubt correctly, we can say that “a [304] round square” represents an object which is simultaneously round and square. But there certainly is no such object. Similarly in other cases as well. We speak of “imaginary” numbers, such as $\sqrt{-1}$, and of the fictional objects of mythology, such as “many-headed lions.” In the representations in question the impossible or fictional objects are represented, but they do not exist. Let it be explicitly noted that such considerations bear not merely upon “representations” in the ordinary and narrower meaning of the word, which *excludes* propositions, but also upon propositions themselves, and thus upon the entire extension of the more inclusive concept of the representation, which we prefer. One can say that every proposition, even the false or perhaps absurd, represents a state of affairs (*Sachverhalt*) as its “object,” and nevertheless there is not for each proposition a corresponding state of affairs. An invalid proposition represents a state of affairs which does not exist, does not subsist.

It is odd that these difficulties, over the resolution of which Scholasticism had already earnestly toiled, are still not settled. At all events we are today far removed from agreement on what we should hold to be a correct resolution.

No scruple or doubt bothers the masses on this point. They easily reach a decision: – Of *course* we can represent an object even when it does not exist at all. For to represent it means to have a mental image corresponding to it. And as, in general, a picture can exist while the pictured does not, so here too. The content of the representation is untouched by the being or non-being of the object. Within the representation there is the phantasy image, and outside it is the object – or else not. In any case, the representation is totally unaffected, whether the object is, will be, was, or not.

The popularity of this resolution could mislead us – if there did not remain in the comparison to the picture the same problem as is supposed to be resolved, and if, in other respects, such decisive objections did not speak against it. We must first raise the objection that here the facts (with regard to a few arbitrarily favored and con-

venient examples) are adjusted to fit the theory, instead of the theory being adapted to the facts in a prejudice-free and comprehensive val- [305]
idation. That each representation relates to its object by means of a "mental image," we hold to be a fiction generated by theory. There are innumerable representations in which a pictorial Representation (*Repräsentation*) in fact does not occur – and innumerable ones in which they absolutely cannot occur. I would like to be introduced to those "mental images" which are supposed to reside in the concepts *Art*, *Literature*, *Science*, and the like. Or are we still talking sense in designating the fleeting phantasm of a golden frame (which happens to come to me) as an image of art, or the memory image of my bookshelves as an image of literature and science? I also would like to meet the mental images of objects thought in absurd representations; and, again, those that come before the mathematician in reading a treatise filled with complicated systems of formulae. Veritable cyclones of phantasms must play themselves out in his consciousness. Think of the enormous web of concepts which, being stored up in just *one* formula of the theory of functions, is often to be explicated only through voluminous enumerations of definitional propositions. *Then* still hold fast to the assumption that to each such partial concept a corresponding (hence, not just any arbitrary) intuition would have to be presented in the form of phantasy images that confer concretion upon the respective abstract Moments. However, this repeats what has already been discussed earlier. It suffices that experience has never verified this quixotic assumption; and if recourse is had to the unconscious or the unnoiced, then one surely makes a use of these ever-ready hypotheses that is rather presumptuous and is untroubled about the rules of sound probability.

But the theory does wholly unnecessary violence to the facts. Through it nothing is gained and nothing is explained. The images are said to be *the* represented objects of which it is truly said: Each representation represents an object. The corresponding *things*, on the other hand, are to be *the* represented objects of which it again is validly claimed: There is not an object corresponding to each representation. But how is it not inherent in the sense of the apparently, or actually, contradictory propositions cited, that in the given case *the same* object which is represented exists or does not exist? The same Berlin which I represent also exists, and the same would no longer exist if judgment [306]
fell upon it as upon Sodom and Gomorrah. The same Centaur,

Cheiron, of which I now speak, and which I accordingly represent, does not exist. And similarly in each case where the representation is univocal – and we also can, for the sake of simplicity, restrict ourselves to these. Accordingly, in no case can the phantasm be what is intended. It in fact always exists, and provided that it intervenes at all, it never holds any interest for us in objective (*objektiven*) thinking – unless as psychologists, and then with representations directed specifically upon phantasma. The difficulty thus remains undiminished. The talk about representations refers to the same object in both cases, and the whole appearance of a resolution derives from the fact that by "representing" we also occasionally understand "picturing something to ourselves." But that is not intended here.

It is also justifiably objected against this view that it proceeds as if the phenomenon which we call "the representation of the object by means of the image" were exhausted by the very *existence* of the image *in* consciousness. In having the image, this subjective content, in the representing act, we still do not mind (*meinen*) the image, but rather the corresponding "external" object, which precisely pulls on us the trick of often not existing. We represent and make judgments about the object itself. To it joy and sorrow, wish and will, relate. It is overlooked that the phantasy content first must become the Representing image *of* something or other, and that this "pointing-beyond-itself" in the image – which first makes it into *an image*, and which distinguishes it from the mere content that in intuition we take just as it is – is a "more" that is essential to consider. And this "pointing-beyond-itself" of course does not mean a certain type of reflexion that fastens upon the content and *conceptually* relates it to objects. Such reflexion would in turn arise from representations which, for their part, could not resolve into Present (*präsente*) imagery-contents, and so an endless regress would be unavoidable. But one also cannot get anywhere with mere dispositions. They are unconscious real possibilities, and their distinctions thus cannot constitute distinctions between experienced (*erlebter*) actualities. They can at most have the value of causal factors upon which the phenomenal distinctions depend. Whether we take a content as what it is, or [307]
whether we "grasp" or "understand" it as a picture, is a distinction within direct lived experience (*Erlebnis*). In the latter case there is, precisely, a peculiar new act-Moment present (which of course is dispositionally conditioned, resting upon the after effect of earlier

lived experiences); and that new Moment confers upon the Present content a new psychical *habitus*, which we (descriptively) express in the formula: The content Present Represents something. That content is not what we mind. By means of it we represent a certain object.

Section 2. Attempt at a Resolution of the Paradox through Differentiation of True from Intentional Existence

Several investigators have adopted the scholastic distinction between intentional and true existence, and conceived of the relationship between them in the following manner: Each representation has an object, even the representation of the absurd. We understand what "round square" means, and so we have a representation. In that we now deny the existence of a round square, we have not denied the content of the representation, for of it we have an immediate knowledge. (This obviously holds for the content in the sense of the subjective ingredients [*Gehalts*] of the representation as well as for the content as objective signification.) The denial thus can only concern the object represented in the representation. It would make no sense at all if the representation were objectless, if no object "appeared as given" in it, to which the rejection could then apply. Thus there *is* for each representation an object (represented by it). But this existence is not "true," but rather is "merely intentional" existence. It consists in "mere being-represented." "True" existence only comes into consideration in the affirmative existential judgment, and already presupposes the "intentional." Recently *Twardowski*, for example, has advanced this view (in close connection with *Brentano*).[5] [308]

[5] K. Twardowski, *Zur Lehre vom Inhalt und Gegenstand der Vorstellungen. Eine psychologische Untersuchung*. Wien, 1894.

To justify the proposition that each representation has its object, Twardowski relies in addition upon an argument that I preferred not to cite above: To what, he asks, do the incompatible characteristics belong in the case of a self-contradictory representation? Certainly not to the representation content, since then *it* would not exist. And therefore to the representation's object. This object is the – indeed non-existent, but nevertheless represented – bearer of those characteristics. Here one may well object: The contradictory properties also do not belong to the represented object, since what does not exist also can have no properties. Those properties are only *represented as* belonging to the object. Some allowance is indeed made for this in the argument, through the inference to, or through the talk of, the "non-existent and merely represented bearer" – but in a manner which deprives the argument of

One can, to begin with, understand this view in such a way that it merely repeats the difficulty, only in different words. Each representation has an object. No doubt this expresses a truth. We call "intentional" existence what this "having" presupposes. And again: There is not an object corresponding to each representation; each one does not found an affirmative existential judgment – since many found a negative one. This too is undeniable; and we call that which the judgment of existence here presupposes "true" existence. But we see no way in which the obvious contradiction between the two truths is to be resolved. Only a terminological distinction has been created, with reference to which it then is questionable what really lies behind it.

But some still believe this provides a resolution. They believe it possible to give this distinction a substance such that to each representation there is assigned an immanent object, but not to each a true object. Therefore I cannot help but see here once again that false duplication which also doomed the image theory: The immanent object can (univocality again presupposed) be none other than the true object in the cases where truth corresponds to the representation. The *same* object which in the representation is merely represented is set forth as truly existing in the respective affirmative judgment. Whether we merely represent Berlin, or judge it to be existing, in *either* case we are dealing with Berlin itself. Whoever merely imagines that gold is yellow has in mind identically the same state of affairs which another acknowledges as existing in the judgment that gold is yellow. And in exactly the same way, of course, the objects [309] of the representations "Cerberus," "$2 \times 2 = 5$," and so forth, are identical with the objects set forth as not existing in the judgments: "Cerberus does not exist," "It is not the case that $2 \times 2 = 5$," and the like.

Perhaps the opposition will agree with the identities here emphasized, and will attempt to escape from the objection of duplication in

---

its overall consistency. One could only say this much with some plausibility: In the inconsistent representation *something* is represented as the unitary subject of inconsistent or conflicting properties, as both round and angular, and the like. This 'something' cannot be the content, which exists, while such a subject cannot exist. On the other hand it surely is *something*, for if it were *nothing*, how could it be represented as bearer of properties. And so on. In this formulation the argument is very closely related with the one used in the text above.

the following way: The objects are actually the same on both sides; only the object which is first merely represented receives, through the affirmative or negative existential judgment, the seal of validity or invalidity, which stamps it as a true or false object. In this sense true existence must be separated from the merely "intentional," the merely "modified" existence. In this way, therefore, one continues to conceive of the represented object as one *literally* residing in the representation. If representing and judging are indeed activities that relate to objects, one tells himself, then in them objects are either merely represented, or are *also* simultaneously acknowledged (or else rejected). How would that be possible if the psychical activities did not contain the objects to which – precisely in the representing, acknowledging or rejecting manner – they could be applied?

Section 3. Critique of This Attempt at a Solution

I hardly need to say that I regard this position as untenable. If the "immanent" (as is flatly stated, instead of saying the "intentional") object is *literally* immanent in the representation, then its existence is just as genuine and respectable as that of the representation itself, and it then becomes totally incomprehensible why we downgrade it and allow it to be regarded as only of a 'modified' existence.[6] It also becomes incomprehensible what the modification here in question really is and means. There *are*, without doubt, cases of genuine immanence. But these are also *eo ipso* cases of genuine existence. If I represent a color on which I am *intuitively* focussed, then this [310]
color is something that exists in as good a right as any entity (*Sache*) in the world. If it is true that there exists no real colored *thing*, then this truth takes nothing away from the existence of the intuited color itself. It is no *thing*, just as little as is the whole representation in which it is a real (*reeller*) constituent. But now, whether thing or non-thing, it exists as this certain part of the representation. Why would anyone think of downgrading its existence? We need not consider whether repudiation of the intuited color is psychologically possible (thus, in the real or causal sense). The *logical* impossibility is Evident. It is Evident that a negative existential judgment in this

[6] Cf. K. Twardowski, *loc. cit.*, p. 25. On the general differentiation between modifying and determining attribution cf. *ibid.*, p. 12f., and F. Brentano, *Psychologie vom empirischen Standpunkt*, I, 2nd ed., edited by O. Kraus, Leipzig, 1924, p. 288.

case would be false – here, as in each case of genuine immanence of the represented; and consequently in every case whatsoever, if the view we are opposing is correct. And the unbridgeable chasm which divides representations of the class just illustrated by means of the example of the color from that other class – for which "a million-angled figure," "a ten-headed snake" and "a square with unequal sides" provide examples – has no bearing here. Some do indeed distinguish intuitive from non-intuitive representations, but vainly struggle to formulate the distinction correctly, since it is overlooked that the case of intuition (that is, of the fuller, stronger type) coincides with that of the immanence of the (or of a) representation-object, whereas in the other case immanence cannot genuinely be spoken of. What nonsense we otherwise would have to champion! If a round square is immanent in the representation in the same sense as the intuited color, then there would *be* a round square in the representation. The truth, "No round square exists," would no longer be valid in rigorous universality, or would be incorrectly formulated, and should state only that there are no round squares outside of representations. That is, as constituents of representations there very well are such things. Thus, there are round squares as often as "they" are represented. Since the representations actually exist, the existence of any and every absurdity would also have to be fully and completely conceded. The domain of the objects and states of affairs immanent to representations would not be subject to the laws of mathematics and logic. But is it not more rational to judge: As [311]
the Evident incompatibilities established in logic and mathematics are *intended* with the most rigorous generality, so they are valid with Evidence (*Evidenz*) in the total scope of their intention? They take no cognizance of the distinction between that which occurs and is valid within the subjective lived experiences of men and that which otherwise occurs and is valid in the world.

Section 4. The Inauthentic Character of Talk of Immanent Objects

In any case, before we decide to accept such potent consequences, it is worth considering whether the talk of "immanent objects" of representations and judgments can be understood as inauthentic (*uneigentliche*). This might be true in such a manner that nothing at all of the object resides in the acts themselves; that in them, in the *authentic* sense, there is nothing of which one could say: *This* is the object that the act represents (or also acknowledges or rejects). It might be that the acts, if in the mode of their engagements they *do* require an existing stuff upon which to bear, could accordingly not find the required stuff in the objects "to which they refer"; and that, instead, the talk of "residing in," and the entire distinction between "real" and "intentional," reduces to certain peculiarities and distinctions in the *logical function* of the representations, i.e., in the forms of possible valid contexts into which the representations, considered exclusively in terms of their objective content (*objektiven Gehalt*), can enter. The talk of the 'modifying' power of the attribute "*represents*" then fits in quite well with such an interpretation. For in every case the modification points to the *inauthentic character* of the mode of expression, which essentially changes the normal signification and function of a grammatical attribute in some respect or other. It is the differentiation of the Ideal from the psychological content (*Gehalt*) of the representational act that impells us from the outset toward such an interpretation. The former content does indeed point to certain contexts of identification, in which we grasp [312] the identity of the intention (perhaps do so with Evidence), while the individual representations nevertheless would have no psychologically identical constituent whatever in common. From the outset[7] we have assigned the objectual (*gegenständliche*) reference of the representations to their *Ideal* content (*Gehalt*). Representations with identically the same signification were yet able to point to objective (*objektive*) variety, and representations with distinct significations yet to identity. Here the talk of representations which represent the same object was grounded on nothing other than synthesis in judgment or knowledge. Closely examined, this grounds the posited unification in the objective content (*objektiven Gehalt*) of the respective representational act, in its essence. But then the question remains as to how the paradox at hand is to be resolved: how, when considered closely, those judgment contexts look that reconcile the apparently contradictory assertions, "Each representation represents an object" and "An object does not correspond to each representation," and that provide the details of the distinction – so easily understandable and convenient to work with, and yet demanding explanation – between

[7] A reference to a first part of this manuscript, which was not retained.

merely intentional and genuine objects.

To begin with we must place ourselves under a convenient restriction – indeed, in a certain respect an indispensable one. According to it, in the following reflections by "representations," without qualification, is to be understood exclusively that distinctive representation class which corresponds to "names" (in the extended sense of J. St. *Mill*).[8] In the debate over the questions dealt with here this restriction is always, though tacitly, presupposed. However, let us at once observe that our investigation is transferable, in its method and results, to representations of *all* types, and thus also to propositional representations (primitive representations of states of affairs) – although with certain modifications tied to the fact that what is understood as "relation to the object" exhibits quite essential distinctions of type which determine, precisely, the types of representation. [313]

Section 5. Quasi-classification of Objects as Classification of Representations

First, in order to clarify the essential point of the inauthentic ways of speaking that come into play here, we make a survey of related cases. The quasi-classification of objects into genuine and intentional is analogous to that of objects into determinate and indeterminate. Representations such as "Charles the Vth" represent determinate objects. By contrast, representations like "a lion" represent indeterminate ones. If we understand the expression "a lion," or judge "A lion exists," then we indeed do represent a lion – but not an individually determinate one – and the judgment also bears upon it. Accordingly one might also want to divide objects into the determinate and the indeterminate. But how are we to have, besides the determinate lions, also indeterminate ones running around in the world? Does the extension of the concept *lion* correspondingly break down into these two species? Within the extension there certainly belongs whatever exists as lion – consequently those indeterminate lions also, of which the above-mentioned judgment speaks. It will naturally be replied: Every object is, in itself, determinate. The classification of lions into determinate and indeterminate is no classification of lions – as, say, that into African and Asiatic – but is, rather, a classification of rep-

[8] Cf. J. St. Mill, *System of Logic*, Book I, Chapter ii.

resentations (and, of course, of objective representations) into those where the *reference to* objects is determinate and those where it is indeterminate (something, an *A*).

Wholly analogous is the classification of objects into the possible and the impossible (of numbers, for example, into real and imaginary). Here we have, not a classification of objects, but one of representations, depending upon the valid judgments of compatibility or incompatibility into which the representations enter. "A number which when squared yields the result –1" cannot be. The representation indicated within quotation marks combines contradictory properties. That it does so, and that therefore an object corresponding to it does not exist, is precisely what the judgment says; and with regard to the fact that it is constituent in such a valid judgment, it itself is said to be an impossible representation, and its object in turn is said to be an impossible one. Obviously matters are no different with the classification or division of objects into existing and non-existing. It is [314] a mere classification *of representations* into representations *A*, which fit into valid existential judgments of the form "A exists," and again into representations *B*, which fit into valid existential judgments of the correlative form "*B* does not exist." We do in fact know, from the principle of contradiction, etc., that every representation *V* must belong either in class *A* or class *B*, and can never belong in both at once. With this the representations themselves always exist, whether they enter into the one or the other judgment context.

Here one must of course guard against misunderstandings that the talk of representations *A*, which fit into judgments of the form "*A* exists," could induce. If I say: "God exists," then the representation "God" – but by no means the representation "the representation God" – obviously is a constituent of the judgment (or of the propositional representation). The sentence does not run: "The representation God exists." In the one case we say God is judged about, in the other case the representation *God*. In the former case the representation "God" enters into the judgment, in the latter case the representation "the representation of God" does. The easily understandable confusion of the two thoughts, "Every judgment combines representations" and "Every judgment judges about representations," was certainly the source of the error, widely diffused in the older logic, that in all cases only representations are judged about, rather than the matters (*Sachen*) represented. And so also in our case, the classifications of representations according to the forms of judgments or of truths into which they fit must not be misunderstood.

With the quasi-classifications of objects here introduced, behind which there in truth lay concealed classifications of representations, the subjective acts of representation were *totally* left out of consideration. Rather, only their possible objective contents – thus, representations in the objective sense, in one objective context or another – were considered. Should we now wish to reduce the classification of objects into the "true" and the "intentional" to distinctions within the realm of representations, and in doing so exclusively take *objective* representations into consideration, then it is clear that we revert to the just discussed classification of objects into the non-existing and the existing – if we understand under "intentional" objects *merely* [315]
intentional ones. A merely intentional object is a non-existent one, a true object an existent one. The concepts certainly have the same extension. The exclusion of existence is expressed in the former case by the little word "merely." If we omit it, then the attribute "intentional" no longer "modifies" the object into a non-existent one. Rather, it is now an object meant in the sense supposed to be proper to every representation, the valid as well as the invalid; or an object in the sense where the relevant existence question is totally ignored. Since the extension of the concept "intentional" then also embraces the true objects, a classification can no longer be spoken of.

Section 6. Preliminary Analysis of the Two Theses in the Paradox of Objectless Representations

Of our two apparently contradictory assertions, the one is already completely clarified through our deliberations up to this point: "An object does not correspond to every representation *V*." That is, not every representation *V* grounds the truth of the proposition "*A* exists"; not every representation is a "valid" representation; not every one has a "true," an "existing" object; or, briefly and *authentically*: not every representation has an object. As a matter of fact the expressions "an object" and "an existing, true, actual, authentic object" are fully equivalent; and if the signification of the word "has" also remains unmodified, it includes the being of what is had. Every true identity refers to objects in the authentic sense. It combines two representations and identifies their object. For being identical (and

likewise: being different) presupposes the being of the identical (or of the different). Similarly for any genuine relational truth; for the being of the relation includes the being of the relation's terms. The reference to objects rests, in all these cases, upon the *objectuality* (*Gegenständlichkeit*) *of the representations* which enter into such judgment combinations. We note here in addition that the distinction between the propositions, "The representation *V* is valid" and "Its object exists," or – in order to bring in the direct expression which [316]
corresponds to the guiding intention – "V exists," consists only in this: that, through an equivalence-preserving shift, for the representation of a representation there is substituted in the judgment the straightforward representation, just as the reflexive judgment "directed upon the representation" is converted into the correlative and direct judgment "directed upon the object."

The second of the repeatedly mentioned, apparently contradictory assertions – namely, "every representation represents an object" – refers to "intentional" objects as the first one does to true objects. In its sense, "a round square" is said to represent an object just as well as "a square," and "Cerberus" just as well as "Bismarck's river Tyras." In order to extract the objective content (*Gehalt*) of this assertion we engage in the following reflection. If, now, it is to be possible in every case to speak of "the" object of the representation, then, above all, it must also be possible to identify these "intentional" objects, or to differentiate them. If it is now asked what it means, in the sense here in question, for two identical representations to represent – disregarding existence and non-existence – the same object, then one immediately sees that here identity only can be intended under a hypothesis or something equivalent thereto: under an assumption, even if unexpressed.[9] Unconditionally posited identity in fact includes the existence of the identified. "Disregarding existence and non-existence" therefore can here only mean that no decision on that point is necessary. With such identification it is in every case a matter of the relationship of the representations, with no prejudice as to their validity or invalidity; and that identification is appropriate where we maintain the objective identity as a hypothesis, whereas it is not when the identity is set forth absolutely. The assertion "*V* is identical with V^1" is a completely determinate one

[9] Cf. On the "Identity of the Represented Object," Appendix X below.

if the representations V and V^1 are, in general, given determinately. It expresses identity of objects, i.e., of true objects; and, while assuming the same representations, to want to substitute for it yet another identity directed upon "intentional" objects, would precisely mean to change its sense. The identity would no longer be identity. Thus if one says that the representations V and V^1 represent the same [317] object, that they have the same intentional object, then nothing other is meant than that – under an assumption to be further detailed, but obvious in the present circumstances – they have the same objects. For example, it is meant (to take the simplest case) that they have the same objects, *if* any at all. "The" object is *in any case* the same; and "it" is, as required, the questionable, the dubious, falsely believed, non-existent, impossible object – pure and simple inauthenticities in expressing attribution and identification. The representations "Zeus" and "the highest of the Olympian gods" have the same intentional object; i.e., Zeus is the highest of the Olympian gods – "according to Greek mythology." As a rule we omit this postscript, and do not miss it where it is lacking. It is obvious that whoever judges about mythological objects places himself upon the grounds of the myth, without actually *claiming* it for himself. Should truth correspond to the myths of Ancient Greece, then the reciprocal representations would also be valid and the identity posited would obtain. Herein lies the sense of the *seemingly* absolute identity assertion, which would not be valid if understood to be absolute. Or the sense lies in an assertion equivalent to this hypothetical assertion: Namely, that the ancient Greeks *believed* that there is a god, Zeus, and that *this same* god is the highest of the Olympian gods *assumed* by them at the time, and so forth. And in like manner, of course, is to be understood the grammatical assertion that the two names *named* the same object. Taken literally and authentically, it is false; but its natural function resides in the inauthenticity with which it is generally understood. A name names something. That is, there belongs to it a representation whose object is precisely that which it names. The talk of the identity of the object meant thus follows precisely, in its authenticity or inauthenticity, that of the identity of *the represented object*; and the inauthentic usage predominates in both cases.

The judgment about the represented object is, as to its objective (*objektiven*) value, a judgment about the representations; and that which is expressed about them is the circumstance that those representations have the same objectual (*gegenständliche*) intention, an identical property of the content, which is called precisely the rep- [318] resenting or meaning of an object. But we do not explicitly judge *this*. We judge about the objects. But our judging is then a "modified" one, a judging which seems to be about the represented objects, insofar as we place ourselves on (phantasizing our way into, etc.) the grounds of the existence of the objects, upon which we in truth do not stand at all. The identification does not rest on actual positings. The foundational positings which the identity judgment presupposes are "modified," are *mere* representations.

Section 7. Transition from Singular to General Representations

Our interpretation is suited not merely to individual representations, to which our last examples belonged, but also to the *general* representations of traditional logic, which were frequently defined as representations that represent several objects. It is certainly not without violence when, in such a manner, the objects belonging to the extension of the general representation are characterized as represented by it. In any case, this talk of several represented objects would be much better suited to representations of totalities and collectives, which, however, are to be distinguished from general representations. Depending on the context, the same general representation, "man" or "a man," refers now to a determinate individual man, such as Socrates ("Socrates is a man"), now to an indeterminate multiplicity of men ("There are men"), now to the sum total of men (precisely as in the expression, "The sum total of men"). And if, now, the extension is defined as the entirety of objects that can be truly subsumed under the general representation, it nonetheless seems unnatural to say that the general representation *represents* the objects in its extension. As if there belongs to it, like its everywhere identical signification content (*Gehalt*), an everywhere identical objectual reference to this entirety of objects. Whatever the case, we find in relation to the extension the same ambiguity or inauthenticity of language, which permits formulation of two assertions: "Every representation (more precisely and especially, every *general* representation) has an extension"/"Not every representation has an extension." Analogously one [319] could speak here of true and of merely intentional *extensions*: the former where the context concerned presupposes the existence of an extension, and where therefore talk of "the" extension is to be understood as authentic; and the latter, by contrast, where the existence of an extension remains undecided, in virtue of an unexpressed assumption that is to be supplied from the intention of the overall context of thought. This inauthentic mode of discourse is obviously the one which almost totally dominates where relationships between extensions of representations (or concepts) are dealt with. The assertion that the extensions of the representations *A* and *B* are identical (that the two representations refer to the same objects, and so forth) is equivalent in signification with the assertion that the two representations have the same extension – *if* any extension at all. Or, in an equivalent rephrasing: If something is *A*, it is *B*, and if something is *B*, it is *A*. In other words, the two representations are equivalent. If there is added the existential proposition, "An *A* exists," then the 'intentional' extensions pass over into true ones.

Nothing essential is modified in these relationships if we consider indeterminately attributive representations to which only *one* object corresponds or can correspond. Thus we have representations of the form "an *A*," where it is true that only one object that is an *A* exists, if any at all. For example, "smallest prime number," "power of the rational numbers," "root of a (namely, a given) linear equation," and so forth. The essential *logical* character of the indeterminately attributive representation remains unchanged if the number of objects to which it is to be predicatively affixed is reduced to one. The concept of extension also retains in that case its normal application. The "general" representations, defined as representations which refer to *several* objects, precisely do not form a natural class.

Things are otherwise with *singular* representations, in the sense which grounds an exclusionary contrast to indeterminately attributive representations. Representations presented through proper names (perhaps in conjunction with attributive determinations) – above all, however, perceptual representations – provide examples of such singular representations. As they are of an essentially different character, talk of an extension, although it is generally usable, no longer finds in them the basis for natural application. Indeterminately attributive [320]
representations have only *indirect* reference to objects. They indicate certain predications, propositions of the form, "*X* is an *A*," "*Y* is an *A*," etc., where the symbols *X*, *Y*, ... indicate representations of the other group, which represent an object in a direct and authentic

manner, without such mediation through attributes and propositional thoughts. In perception, on the other hand, the object itself is represented. In the perception it of course has the attributes that belong to it. But *that* the object is something having them does not belong to the content (*Gehalt*) of the representation. The representation itself does not have the form, "Something which is α, β, .." Rather, it is exhausted in the plain and simple representing of the object "as that which it is." The logico-objective relationships bridge these essential distinctions in certain respects. They interlink direct and indirect representations, and permit concepts which were developed for the one type to carry over by extension to the other. Where in the representation (signification) no attributes show up, there the concept of the extension also loses its hold – as is immediately clear from the definition of that concept. And nevertheless, as already noted, we also speak in the case of direct representations of an "extension" to which the *one* object that they represent is assigned. One in fact also calls them singular representations, in the same manner as those indirect representations with a one-membered extension. In spite of the essentially different signification which is borne in the two cases by the vague expression "reference to the object," one in both cases speaks in the same manner of representations that represent only one object. Along with the concept of extension, the sense of the relationships *between* extensions of course also expands: The formula for the reciprocal hypothetical judgment, "If something is *A*, then it is *B*, and conversely," gives in the simplest way the equivalent for extension identity in the expanded domain. But certainly the so-called copula, "is," then takes on an equivocal sense which, in virtue of certain similarities in the formal relationships, is not objectionable, but rather arises out of them in a natural way – since, in general, it is precisely in these formal equivalences that there lie the chief stimuli for the practically useful transferences here in play. If, say, we replace "*A*" with "Socrates," then the hypothetical antecedent [321]
reads, "If something is Socrates." But we cannot say "Something is Socrates" in the same sense as "Something is red" or "Something is a tree." "Socrates" is not an attribute that could belong to an object precisely as an attribute does. Again, the representation "Socrates" is no subsumptive representation like "a tree," so that it could serve, in the manner of such representations, to relate a "concept" to a "conceptual object." Over against these significations of the copula

(which commonly are themselves not distinguished by logicians) a new one obviously shows up here: "Something is Socrates" can now indeed mean that something is identical with Socrates.[10] Thus, in our hypothetical formula it must be left open whether the little word "*is*" functions in the one sense or the other. Or, what amounts to the same thing: The signification of the copula in our formula embraces, through disjunctive expansion of the concept, the different particular significations. That this expansion, this "remaining open," is admissible, has its basis in the fact that, within the law governed order of hypothetical contexts, the disjunctively expanded concept of the relations corresponding to the "is" functions formally just like the narrower concepts. The law of transitivity, for example, is valid for the "is" of identity, as well as for that of attribution and subsumption. If the propositions, "If something is *A*, it is *B*," and "If something is *B*, it is *C*," are jointly valid, then the proposition, "If something is *A*, it is *C*," is also valid. And not merely this. The law also continues to hold for the disjunctively expanded "is," insofar as the combination of hypothetical premises of the given form, even with different significations of the copula, also yields an appropriate consequent, in which the copula possesses one or the other of these significations. Similarly also for the other logical laws that come into consideration for the propositional forms which turn up here.

Accordingly we therefore can hold fast to the central hypothetical form as criterion for the identity (or non-identity) of extensions for [322
all forms of representation in general, and in similar manner regulate the remaining objectual or, what is the same, the remaining extension relationships between arbitrary representations.

The talk of every representation having an extension, of every one representing objects, would accordingly have to be interpreted as merely inauthentic. Its genuine meaning would be expressed or indirectly indicated in the proposition: Every representation would, granted that for it the corresponding affirmative existential judgment holds true, have an extension, and would represent one or several

[10] That all significations of the copula are based upon a single one – which, fused with the various significations, founds the many-sided signification of the "is" (the temporal among them) in categorical assertions – is here in no way to be denied. Thus the existence concept, the "is" of the existential proposition, is in fact also equivocal, depending upon whether or not it is tied to the "pure positing" of the concept of the real (being as real actuality, or as being *in* it).

objects. The purpose and support for this inauthentic claim could only reside in its practical utility. Thus it resides in the fact that every representation can enter into hypothetical relationships of the forms discussed, that with extensions standing under the assumption of such existential judgments (which of course do not have to be formulated), one can, across a broad and rigorously delimited range, operate in exactly the same way as with actual extensions. More precisely, that equivalence has exactly the formal characteristics of an identity relation, that subordination has the formal characteristics of an inclusion relation, and so forth; so that, provided only the domain of the hypothetical relationships not be exceeded, and all judging concerning existence or non-existence of the respective objects and extensions be omitted, one can operate with the talk of extensions and extension relationships in exactly the same way as if it were authentic and absolute. And also connected with this is the fact that perceptible images of spheres can find application as characteristic symbols and as technical aids for formation of judgment. Obviously there reside in this state of affairs (I do not mean in the knowledge of it *in abstracto*) highly powerful motives for the natural generation and establishment of the inauthentic mode of speech, within logic as well as outside of it.

Section 8. Existential Assertions and "General Assumption" in Mathematical Axiomatics

Moreover, one can relate this mode of speech to deeper-lying and general logical grounds. Considering, within a framework of thought of the type here in question, the existential propositions which – taking them conjunctively united to form a single coherent assumption – would be necessary in order to confer authenticity upon the inauthen- [323]
tic passages of thought and language, we can tie onto a logical law that is valid in complete generality and is of extraordinary importance for the understanding of scientific methodology. Namely this: The formal laws which govern thought under a fixed assumption are identically the same with those which are valid for a, so to speak, "free" thought, i.e., thought restricted by no conditions. Psychologically considered, an assumption which we have fixed "once and for all," or upon which our further thought transitions "obviously" are based, but without express formulation (this indicating certain dispositions

toward deliberations and restrictions at a later point): – such an assumption we often wholly ignore in the course of those transitions, and really can ignore it without damage to the truth. To transform the respective judgments standing under the assumption according to purely logical laws, or to deduce from them conclusions that would be justified *if* those judgments were unconditioned: – every method of doing this is also valid in the present case where the judgments depend upon the hidden or unexpressed assumption. But the derived judgments themselves are, in turn, dependent upon it. Or, in logical terms: The minor premisses to one and the same assumption stand under the same deduction laws which hold for independent premisses. Any conclusion which would be valid from such minor premisses by themselves, is also valid under the assumption, and indeed *as* standing under it. With this is connected the fact that all conditioned talk of truth and falsehood, existence and non-existence, consequence and non-consequence, genuine and fictive objects, and so forth, can be totally *treated as* unconditioned, so long as it is certain that we have not stepped beyond the framework of the governing assumption. This is of especially great consequence for the deduction system of so-called "formal arithmetic." There the hypothetical orderings of basic formal concepts, and of the axioms relevant to them, serve in the constitution of a closed mathematical domain, of an "algebra," whose content forms the infinitely branching system of the formal consequences to be deductively derived from such foundations. Here we not only speak, but rather also *judge*, as though deduced truths, [324]
existences, relationships and incompatibilities were absolutely valid.

Another, more authentic, mode of talk and judgment would be quite impracticable, considering the amount of (useless) thought-labor that would have to be performed. The inviolable rule – living in the well-trained investigator as a complex of active dispositions – to judge nothing which does not follow purely from the axioms, governs the validity of the statements and determines in practice the boundaries of free logical movement. If, then – and this actually happens – an investigator misconstrues the hypothetical character of the basis, and perhaps in all seriousness supposes the definitions (which create "the" domain) to have the power to set up *existences*, to *create* essences, then this has no influence on the internal soundness of the mathematical system. If one misunderstands the character of the basis – if one takes the truths deduced, since they in fact are *judged* absolutely, likewise as absolute in their objective validity standing independently – then the only correction required is mere subordination to the conditioning assumption. And similarly in general, we can in the main rely upon inauthentic language, and upon the inauthentic thinking usually given with it. This is so inasmuch as in it there is at work logical thinking based on the logical law discussed above. And likewise in other cases, where the interconnections of the judgments with one another and with the assumptions are of a merely factual type, because the thinking is based on the certainty of our memory – which by the way also comes into play in the former cases so far as it makes possible the linking of the results to the assumption. These inauthenticities of thinking, this fact – that we make innumerable judgments which we nevertheless do not "really" bring to mind, and that as there is a proxy representing there also is a proxy judging, the one inseparable from the other –: Here is a major aspect of the "economy of thought." Science as an achievement of human thought is not intelligible without it.

We even found in mathematics a support for our view, and yet some will wish to derive opposing arguments precisely from this science.

It could be objected that, according to our view, the distinction between mathematical existence and non-existence (thus, the distinction between the real and the imaginary, in the sense of the older mathematics and logic) is not given its due. The question whether objects do in truth correspond to the Ideal concepts of geometry falls, some will say, wholly outside the domain of geometrical research. And nevertheless geometricians distinguish between valid and invalid concepts. They correspondingly speak of the existence and non-existence of geometrical structures, and carry out the most detailed proofs or constructions to support the one or the other. Thus, with reference to representations to which objects do not necessarily correspond in the real world – and, according to the views of most, also absolutely do not so correspond – there obtains, as it seems, an essential distinction, wholly analogous to the usual distinction between valid and invalid representations: the one type presents geometric objects, the other does not. The identity judgment, "Two representations have the same object," cannot here rest upon the hypothesis, "Given that any objects at all correspond to them." For the judgment of comparison does not refer to real, but rather to math- [325]

ematical objects, and in the case of "real" concepts in mathematics there truly are such objects. Thus, to ground the concept of the represented object on the merely hypothetical relationships between judgments seems to have its problems; or it seems proper only if one wants to speak of objects in the case of all representations whatever, thus in the case of "round squares" as well. But one surely will not set concepts of this latter type on the same level with legitimate geometric concepts, the "reality" of which is demonstrated through construction, say, in the course of which we move not on the ground of reality proper, but rather on that of, precisely, mathematical 'reality.'

If we appealed in the above to *formal* mathematics, our opponents appeal to a mathematics of the real. The first is a type of reasoning which is disengaged from all intuition and which considers the pure forms of mathematical relationships and systems in their highest generality, and investigates their multifarious possible variations. Thus, along with the deepest possible insight into the technique of mathematical relationships, it also establishes a greater technical mastery over mathematical objects and a greater freedom and virtuosity of method. By contrast, the mathematics of the real rests upon intuition. Its concepts are grounded in it and from it they also receive the [326] "reality" peculiar to them.

Thus we have recent attempts along these lines to salvage a type of existence for intentional objects – at least in the domain of mathematics. But here too the deceptive appearance is dispersed through attending to certain multi-tiered inauthenticities. There are no "different modes" of existence and validity. And where we think we find such, there either our language is inauthentic, or the differences concern the range of application, which through customary association *seems* constricted to a dominant sphere of interest, whereby the concept of existence simultaneously undergoes an unexpressed enrichment.

Often the term "existence" is used in the sense of "there-being" (*Dasein*), of existence within real actuality. Here the primitive and more general concept of existence is enriched as to content and its extension narrowed down to objects that are real. Truths, propositions and concepts also are objects (*Gegenstände*). Also in their case we speak of existence (*Existenz*) in the full and authentic sense. But they are nothing which would be encountered in the domain of the actually real. As widely as the expression, "There is an *A*," can lay claim to meaning and truth, so widely the domain of the concept

of existence also extends. Accordingly, the assertion that objects corresponding to mathematical concepts do not exist is ambiguous, depending on whether or not one has in mind the actually real. If we had, as many believe we do, adequate intuition of geometrical structures in phantasy, then truth would correspond to their representations, and yet perhaps no *actuality*, as for example in the case where the projection of an actual world would be an illusive appearance. The situation would be similar to the way it is with the concepts of sensible qualities. Within the phantasizing of a red object, the color red truly exists, even if in external *actuality* a red thing – and consequently also the individuated redness inherent in it – did not exist. The Representing (*repräsentative*) relation of the intuitive image to an actuality not intuited indeed permits us to say: A merely phantasized object is no object. But that only means that the phantasm is not what it represents. But in its own right it exists, and with it all of the abstract Moments which make it up and which we intuitively grasp *in* it. Certainly one can doubt whether intuition of geometrical [327]
objects is in truth possible. Or, rather, it is certain that it is not possible. Intuition and the empirical apprehension of the spatial contains the starting points and the leading motifs for geometric concept formation. But the abstract objects appertaining to those concepts, and the attributes of those objects, are not to be obtained *simply* through "abstraction" from the intuitions (in the now customary sense of an emphatic attentiveness upon particular features). Those objects and their attributes do not lie 'embedded' in these intuitions as the visible form does in the "surface" seen. The triangle as intuited *abstractum* is no geometrical figure. It serves the geometer as a mere symbol, the characteristic type of which has a dispositional connection in his mind with the relevant pure concept and its Ideal object that is merely "thought." The intuited shape (*Gestalt*) has genuine existence, just like the intuition itself. Thus that shape grounds the evident possibility in general of the representation of objects of such a character. The geometric shape, on the contrary, has merely represented existence, existence in virtue of definition and valid deduction from the axiomatic principles. The possibility of *geometric* structures – thus, the consistency of the determinations unified in representations of them – is not guaranteed by means of the intuition of those structures (for we do in fact lack such intuition). Rather it is guaranteed through the consistency of the elemental determinations brought together in

the axiomatic principles, and through the pure deduction that supplies the "existence proof."

The principles of a complete and pure geometric system are nothing but the definition or hypothetical postulation – laid out in a series of separate assertions – of the manifold which must be regarded as geometrical space. These assertions are of two types: existential assertions and general (nomological) assertions concerning the objects thereof. "We *postulate*" (or: "*Let us consider*") a manifold, which we call "space," and its elements the "points," with the following properties: Any two of its points determine a "straight line," "any two straight lines intersect at *one* point," etc. The system of pure consequences (or, in relation to the space considered, the totality of [328 its sequential characteristics) constitutes the content of the geometry. In the space there "exists" each structure whose existence is a pure inference from the principles, thus from the space definition or postulation. All propositions of the geometry, the existential as well as the nomological, stand under a *general assumption*, never expressed because self-evident: – *Granted* that *there is* a space, a manifold of such-and-such a determinate type (exactly defined in the principles), then in it there exist these and those structures, for which these and those propositions hold true, and so forth.

Section 9. The "Universe of Discourse"

Mathematical existence and non-existence is, thus, existence and non-existence under hypothetical postulation of the relevant principles: The existential propositions are, like all mathematical theorems collectively, incomplete. They are mere apodosises in hypothetical propositions with always the same protasis. Only the existences posited in the principles constitute an exception. They are not existences under hypothesis, but rather are *hypothetical*, and thus once more inauthentic, existences. Geometrical truth or existence therefore does not signify possibility or consistency, whether absolute or under presupposition of the principles. A red square is no geometrical existence, and a square red and black (all over) is no geometrical non-existence. Rather the geometrical is a matter of an existence or non-existence conditioned purely deductively (a logical necessity) under hypothesis of the principles, if it is not a matter of the explicitly assumed existential content of the principles themselves. "In the

(defined) space" exists a square, but not a round square, a triangle, but not an equilateral right triangle, and so forth. Similarly, we do indeed say there are nymphs "in Greek mythology," or a Little Red Riding Hood "in the German folktales," and so forth. Only here we are not dealing with scientific hypotheses and pure deductions. Thus we will not consent to the obscure talk of diverse domains of existence, [329]
of different "worlds" (*universes of discourse*) which make different arrangements for the existence and non-existence of the same object. The "world" of the myth, the world of poetry, the world of geometry, the actual world: these are not "worlds" of equal status. There is only *one* truth and *one* world, but manifold representations, religious or mythical convictions, hypotheses, fictions; and the entire distinction comes down to the fact that, perhaps for reasons of practical convenience, we often speak *as if the judgments we make were without condition* – to the fact that we *use* absolute existential assertions – whereas the logically correct expression would require hypothetical propositions (with these existential assertions as apodosises), or also require other propositional forms equivalently reducible to hypothetical propositions of this sort. This takes place naturally where an extensive group of assertions stand under one and the same *assumption* – whether *fictive* (*arbitrary*, *poetic*, *mythological*) or scientific. And indeed this assumption is not only not expressed, but as a rule it is not actually thought. Hence it often remains in a state of readiness for action, and in many cases bestows upon the conditioned judgments a subjective psychical character under which they become recognizable to us as conditioned. But this also often does not take place. The judgments are made *as* unconditioned, while it is only in the context of reflection upon them in conjunction with judgments of other groups that the necessity stands forth of taking them to be only conditionally valid. The geometer does not make conditioned judgments. If at all, only critical reflection on their status leads him to knowledge of the true situation.[11]

Section 10. Explication of the Representation Content

After having discussed talk of "represented objects" as it refers to contexts involving an extension, we still want to look at the cases

[11] Cf. Appendices XI and XII below.

where a series of judgments is articulated as the "explication of the content of a representation" – judgments which concern the representation's possibly non-existent object. We explicate the content (*Inhalt*) of the representation "a lion" by saying: This representation represents an object, and moreover that object *is* an animal of the cat [330] family, and so forth. By "content" is obviously here understood the signification content (*Gehalt*) of the representation. We are explicating what we mean by a lion, what we wish to designate and know with understanding as a lion. In precisely the same way we also explicate "objectless" concepts, such as "the squaring of the circle," "continuously curved polygon," and the like. These also represent objects; and these non-existing objects appear as subjects of judgments whose predicates are the properties that constitute the content. With these attributive representations there is associated a specious argument for the positing of "intentional objects."[12] Namely, one could argue: In the invalid representation something is represented as the unified subject of incompatible, perhaps even *a priori* contradictory, properties, e.g., as being equally curved and yet having angles. This 'something' cannot be the "content" of the representation itself; for that content does indeed exist, while a subject of incompatible properties does not exist. On the other hand, the 'something' certainly is not merely *nothing* – otherwise how could it be represented as the identical bearer of the properties and be judged about in the chain of explicating judgments? It is, therefore, a something that is thought in the representation: a 'something' which in a certain manner – even if not in that of the content – the representation carries in itself, and to which there only corresponds nothing in the actual world.

This argument of course lies back of the repeatedly stressed protestation of the identity of the represented and the true object. To that extent it also is absurd, since the same reason which makes it seem impossible that the content should be the true subject of the contradictory properties in the judgment also would apply to any *other* 'something' immanent in the representation, indifferently of whether it resides in it in the manner of the content or not. In strict truth, the represented object of invalid representations is therefore *nothing*, that is, it is not *something*.

Something and *something existent* are equivalent concepts, and

[12] Cf. Twardowski, *op. cit.*, p. 31.

"not something" is "something which is not." But how do things stand with those judgments explicating the representation content? If they can lay claim to an absolute (i.e., unconditioned) validity, then their subjects would also have to exist; for what is not also can in truth have no characteristics. And conversely, since their subjects [331] do not genuinely exist, we shall conclude that they can and will lay claim to only a conditional validity. This is confirmed by closer consideration of the sense of those analyses of content. Assume for simplicity's sake that the expression "a lion" is a mere abbreviation for a definitional formula having a fixed content. Then it would constitute an analysis or explication if we were to single out the separate parts of this objective representation by analysis. Thus, for example, pointing out the representation "a" (= something), the representation "animal," and so forth. An analysis in this authentic sense obviously is not what we have here. Rather we "dissolve" the unified attribution into a chain of predications, and it is beyond all question that these latter are not true constituents of the representation in question, that the transition from attribution to predication presents us with a modification of signification – although equivalent seems obviously converted into its equivalent. In "explicating the content of the representation" we thus produce a *new* representation which is directly equivalent to the original one and is, for certain logical purposes, more advantageously constituted as to content. How do things stand, then, with the *predications* that have been newly formed here, those constituting the explication? In explicating we judge: A lion is an animal, is of the cat family, and so on. Or: A lion is something of which it is true that it is an animal, of the cat family, and so on.[13] It is clear that these judgments have no existential character, as if they were co-affirming the existence of lions. They indeed mean only to say: By a lion we understand, we *have in mind*, something of such and such a character. Or: The representation (that is to say, the objective representation = signification) is the representation of something which, etc. But in bringing up this thought we actualize the "*representing postulation*" (assumption) of a something in general, or of an animal; and the associated predications (the something is of the cat family, and the like) stand under that postulate. They are conditioned by it. The terms "understand," "represent to oneself" and

[13] Cf. E. Schröder, *Vorlesungen über die Algebra der Logik (exakte Logik)*, Volume I, Leipzig, 1890, p. 231.

"think" are modifying, and they lower the judgment to one of merely conditional validity. Therefore many will not hesitate to locate the [332] true sense of these explications in the hypothetical assertions: A lion as such is an animal, etc. – i.e., if anything is a lion, then it necessarily has the respective properties. Nevertheless, this interpretation would be no simple statement of the sense, with only a change in the mode of expression. Nor is an equivalent representation given therewith. For not all of the properties necessarily belonging to *A* belong to the *sense* (*Sinn*) of *A*, but rather those alone which are partial representations contained analytically in the representation of *A* and represent parts and characteristics of *A*, or are reference points for the framework of relations that are represented in the representation *A*.

Of course it not infrequently occurs that someone wishing to explicate a signification will incidentally state the individual explicative predicates as being quite absolute, although the one judging well "knows" that they are not to be claimed as valid without qualification. Whoever has, for example, intuitively represented a lion, and then associates with the phantasy image such unconditioned judgments, does not lapse into error; for he is disposed in such a way that the judgments reviving in the broader contexts of thought indeed are only to be taken as conditioned. And he will claim them to be such because he carries out the founding postulations for the "unconditioned" judgment in *modified* form, through which any synthetic judgment built upon those postulations quite certainly assumes a changed character – precisely that of a judgment 'under assumption.' If our logical interests are attuned to the actual existent and its properties, then the standard assertion about an individual being has, accordingly, the standard sense, which implies the existence of that being. Thus the asserting itself requires *unmodified* founding postulations of actualities. And thus the expression of a judgment *under assumption* further requires that the modified founding postulate be expressed as such, thus that the judgment be appropriately changed – formulated as a hypothetical judgment, or as a judgment *about* mere nominal or propositional *representations*, or *about judgments, statements*, etc.

From these reflections we learn that also here the talk of represented objects does not draw us out beyond the signification content (*Gehalt*) of the representation, and that such talk is rendered determinate through certain valid networks of judgments into which the [333] signification integrates itself. What the representation itself contains is at best the *representation* "object" (*Gegenstand*) – it is the "a" or "something" in the attributive representation. And if the "something" is lacking as an explicit part in the representation, then the representation includes it implicitly, namely, in the logical sense. Of any and every thing we can say that it is something. A lion is something, a proposition is something, Socrates is something, etc., assertions which of course are to be interpreted not in the existential, but rather in the hypothetical (or some therewith equivalent) sense. With this in mind, the following propositions are equivalent: – "A representational lived experience has a content," and: "It represents an object." Or, objectively formulated, "It is an (objective) representation of something," and: "It represents an object." For every representation (= content, signification) validly adapts itself to the formula, "If something is *A*, then it is something." We here have to do with one of those primitive axioms in which the essence of the objective representation (the essence of the "content") is defined.

Finally, it hardly need be noted that all cases where one speaks of represented objects without considering their existence or non-existence lead to relationships that are marked by logical equivalence only. Whether we say: Each representation represents an object, each has an extension, any and everything is something, and so on, it "all comes out to the same" – although the various forms of assertion do not merely have their different psychological origins, but rather also are *not* identical as to signification content.

Section 11. The Differentiation of True from Intentional Object in the Case of Representations in the Subjective Sense

We still must add a supplementary comment. We have related the distinction between true and intentional objects exclusively to the objective content (*objektiven Gehalt*) of representations. Even if it were psychological *motives* that determined the modes of expression which are so easily misleading, their *signification* would still contain nothing of those motives or of the psychological peculiarities of representations in general. – "A representation has a true object; that is, an object corresponds to it." "A representation has an intentional object; that is, it represents an object." "A representation has a merely [334]
intentional object; that is, it has no true object, it merely represents."

Normally in such cases we are speaking of objective representations. Whoever says that the representation "a tree" has true objects, but the representation "a dragon" merely intentional ones, certainly is not speaking of his subjective and momentary representations.

But the distinction can also be applied to *subjective* representations; and, indeed, this in such a way that its signification undergoes an essential modification. "A representation 'merely' represents an object" is now to mean: the object "itself" or "in itself" is not given with the representation, but rather only the object thought, or the object *as* thought, is there. "The representation has a (not merely intentional, but rather also) true object," means that it carries within it its object "itself"; that it does not merely represent it, but rather the object itself is there in it.

This distinction presents itself chiefly in the case of representations to which, in fact, objects correspond. But it has also been extended to invalid representations, since with them we likewise speak of objects which they represent. If we focus upon its enduring core, the distinction separates subjective representations into (i) those which are related to adequate intuition, and thus are posited in union with their objects, and (ii) those lacking adequate intuition, possibly because the objects in fact do not exist. Anything which goes beyond this is false and misleading.

Talk of merely intentional objects, of "represented objects as such," is reprehensible in the sense and context here in question. It sounds as though *in* the non-intuiting representation (as a subjective lived experience) the represented object were Present (*präsent*) as something in its own right, although only in the form of a ghost-like shadow image to which one would be justified in denying "true" existence. But the represented object and the true object are identically the same; and, however "shadowy" one takes the intentional object to be, one only separates it all the more from the true object. If *it* is an object in the representation, then it no longer is identical with the object outside.

The subjective representation represents an object – this assertion [335] is to be interpreted precisely by the standard of the corresponding assertion referring to the objective representation. Moreover it is a distortion to turn back to the subjective representation with the statement: In thinking we are always directed upon objective relationships. This statement concerns once again the mere function of representations. It does *not* say that object and objective context lead some peculiar sort of "mental existence" in representation and judgment. Talk of the psychical activity (e.g., of representing) *bearing upon* the intentional object, of the representation *being directed upon* its object, and the like, is *always* improper where the object is not intuitively present, or else it has only an inauthentic or figurative signification. What has most contributed to the confusion, and to the form of the expressions themselves, is the circumstance that the most primary and the most common representations, those of sense-perceptible objects, usually are accompanied by Representing image-representations, wherewith the cognitive activity frequently is directed upon the images in a manner similar to its direction upon the facts (*Sachen*) in the case of authentic intuition. It appears that these mental images, which are present to us as stand-ins for the "facts," have been regarded as the represented objects, as in other contexts they have in turn been confused with the "content" of the representations, with their signification content (*Bedeutungsgehalt*).

Section 12. Synopsis of Results

After these detailed reflections we can regard as clarified the *inauthenticity* concealed within the distinction between intentional and true objects. Likewise we consider as resolved those apparent contradictions in which the scintillating talk of represented objects entangled us. Since such talk is practically unavoidable, even in cases where objects in truth do not exist, or where one must leave undecided their being or non-being, then the best one can do is to differentiate the mode of expression in such a way that the talk of "corresponding" objects is established as authentic, and that of "represented" objects as inauthentic. The proposition, "To each representation an object *corresponds*," is then false; the proposition, "Each representation represents an object," true. The proper contradictory of the former [336]
states: The relevant existential judgment "*V* exists" is not valid for each representation "*V*." By contrast, the latter proposition states: Each representation can function under an assumption as if it were an unconditioned representation with respect to its object. It can become the subject representation in numerous valid (namely, hypothetically valid) categorical propositions, and to its (hypothetically assumed) objects belong these or those characteristics, and others again not,

quite the same as if the representation were an unconditionally valid one – with the reservation that all of these predications themselves stand forth as only conditioned, or will be downgraded to the rank of the merely conditioned in subsequent critical reflection. Judging about objects whose existence is left undecided is nothing other than a conditioned judging based on representations whose validity remains undecided, thus on an hypothesis. What influences our mode of expression in this context to a special degree is, as was shown, the circumstance that, following the natural thrust of thought economy, we are used to lapsing from conditioned into unconditioned judgments – the corrective for which lies in the disposition to carry out the required curtailments wherever necessary. And, in other contexts of thought, for which the respective assumption no longer forms the obvious basis, we are used to claiming the judgments obtained as only conditioned. And what we in this way naturally do also proves to be practically useful; and it is exactly this circumstance that also causes the common manner of speaking to seem permissible.

As one main result of the reflections of the last section we can designate the knowledge that in fact, just as we had surmised, it is the signification alone that is the inner and essential determination of the representation, whereas the *objectual relationship points to certain frameworks of truths or judgments* which the signification fits into. Of course our line of reflection is restricted to only a certain class of representations, favored from the outset, and externally characterized through the linguistic form of expression which we call "names." But whoever once gains his footing in our way of looking at things will see, without further ado, that our result and also its basis is *mutatis* [337]
mutandis transferable to all of the signification forms which one can designate, in ever so broad a sense, as "representations" of objects of any kind – even allowing their function in the context of knowledge or truth to be quite other than that proper to these nominal representations. If, for example, we impute an object to the *proposition*, as what is represented by means of its signification content – and indeed its *whole* signification content (thus we have in mind not the mere object for which the subject of the proposition stands, the characteristic corresponding to the predicate, and the like) – then by that we pick out the "state of affairs," which subsists if the proposition holds true, and does not subsist if it does not hold true. If the question about the distinction between true and intentional objects in the

case of nominal representations has led us to existential assertions in which those representations function as subject representations, and which, depending on the circumstances, were advanced absolutely or were understood as only conditioned, then all of that carries over analogically to the case now at hand, if only we replace the assertions of *existence* with assertions of *being* (or: "*A* subsists"). But these assertions, too, can be meant, at one time absolutely, and at another time under hypothesis. The circumstance that with reference to each proposition an equivalent existential proposition can be found, which, however its signification content may be modified, represents the same state of affairs as the proposition originally given, in a way reduces the present case back to the earlier one, comprising merely nominal representations. And so the talk of intentional and true objects agrees in the two cases. For example, to the proposition, "*A* is similar to *B*," there pertains the nominal representation, "a similarity between *A* and *B*," or also the other one, "the validity of the proposition '*A* is similar to *B*'," and the objectualness (*Gegenständlichkeit*) of the latter can directly define that of the former.[14]

These terminological discussions call our attention to an important *equivocation* in the assertion, "A representation refers to an object (or represents it)." We had at first looked upon representations as subjective lived experiences, and attributed to them the relation to objects. But when we say, "The representation *lion* represents an object," then not the subjective, but rather the "objective" representation figures as bearer of the relation; and this also harmonizes with the undoubtedly justified saying that the relation to the object is, with [338]
each (subjective) representation, mediated through its "content," i.e., its signification. That in this a definite relationship is expressed, one already notices from the fact that the converse of the saying, through the interchange of "signification" and "object," is absurd. And so we are led to suggest that the objective content (*objektive Gehalt*) of a representational lived experience is not perhaps an intrinsic unity into which signification and objectual relation enter as two "sides" or components; but rather that the objective content primarily, in and for itself, is nothing other than the signification, while the objectual content (*gegenständliche Gehalt*) of the representation indicates certain characteristics of its signification, certain objectively obtaining

[14] End of the 1894 MS. The following paragraph is from 1895.

structures into which the respective signification is integrated. If the question about the objective "what" which we say the representation represents directs attention most immediately and primarily to the objects, and only in the second place to the signification – in order then to find in the one and the other two sides of equal status in the unified objective content of the representation – then it now turns out that the governing role belongs to the signification. And indeed yet more: Already it alone constitutes the objective content of the representation, if under that we choose to think an intrinsic (Ideal) Moment of the representation considered in and for itself; whereas the extension to represented objects seems to turn out to be a secondary and possibly external matter, grounded in truth structures having a signification content (*Bedeutungsgehalt*) to which the signification in question contributes in a determinate manner. This will be substantively confirmed in full detail, and more precisely explained, in further reflections.

First we have to settle a few pressing questions. With suitably chosen examples we have shown that in our representations . . . .[15]

II. The Idea of Truth

Section 13. Ideality of the Objectual Relation. Unity of the Object

All that the object is for our knowledge reduces to certain valid ("correct") judgments which we make concerning it. They contain our entire knowledge of the object. The corresponding judgment contents are objective truths. But not all truths which refer to the object come to knowledge and are thus contents of our actual judgments. If we imagine the closed totality of the true propositions which bear upon the object, it contains within itself all that is valid or not valid of the object. Each property which intrinsically belongs to the object, each relative determination which assigns to it its position in the realm of what there is, finds in that totality its objective imprintation.[16] We can thus say that the objective being of the object displays itself or

[15] Here the brief passage from 1895 breaks off. The following text stems from the year 1898, and has been subsequently added to the original manuscript by Husserl.

[16] In this sense it is obvious that the absolutely complete knowledge of any thing includes the knowledge of the entire world, or that each thing mirrors from its standpoint the entire world.

expresses itself in the system of Ideal truths appertaining to it. The object as the identical intention of this manifold of significations is not thereby something external or contingent in relation to it. In that the manifold intentionally refers to it, and expresses what it is or is not, the two are inseparably *one*. This already holds true of each particular valid signification in relation to its object. Signification and object have not perhaps fallen in with each other by accident. Their unity is not the contingent one within the subjective act of knowledge. Rather, they necessarily go together, so that one can in a certain sense hold the object to be nothing "outside" the truth appertaining to it, and the truth nothing "outside" the object appertaining to it. On the other hand one must not conceive of this necessary and Ideal correlation as an *actual* being-in one another; and one must especially resist the temptation to allow the object to dissolve into that objectively closed (although for all actual knowledge inexhaustible) manifold of interrelated truths which refer to this one and the same object, which spell out its essence in the manner indicated above, and so constitute [340]
what it is for "the" knowledge of it. It would be an obvious error if one, failing to recognize the nature of this situation, wished to deny the "being in itself" of the object and perhaps to say: There is only the "framework of thought," the "unity of knowledge," wherein the *one* object is thought, the object "itself" being nothing over against the unity of the knowledge. But the error would occur in its crudest form if this were then taken in a wholly subjectivistic manner, and so came out in the end as: There exist only these and those judgments or possibilities of judgment, and the so-called objects about which they judge are mere peculiarities of this judgment context, while in and for themselves they are nothing – thus, utterly nothing, in case the relevant judgments and judgment possibilities do not obtain! Judgments are themselves objects, and even real objects, which can only have existence in judging beings. And possibilities of judgment are real tendencies to judge, thus again determinations of real judging beings. Are these beings themselves only determinations of judgments which they make? Surely it is not necessary to enter into criticism of such absurdities, which can hold no enticements for those who distinguish the subjective from the objective aspects of knowledge. It is easily made clear that in judging (as well as in representations) act, content (= signification) and object are completely different things, of which the first two, act and content, under no circumstances can be identi-

cal, and can be (totally or partially) identical with the remaining one only very exceptionally. If we disregard the subjectivistic distortion, then it is no less clear that the object is not in the literal sense something *in* the truths which are valid of it, that object and truth also are distinct entities. One has only to consider the fact that the truth is something Ideal, supra-temporal, whereas the object quite well can be something real.

If every object were a mere determination of the significations which represent it (more specifically: of the truths which are valid of it), then these significations, which indeed are in turn themselves objects, would likewise be only determinations of certain new significations, and so on *in infinitum.* If also, on our view, that which one calls *relation* of the representation to an object (more specifically: of the proposition to a state of affairs, and in a somewhat different sense: of the proposition to the objects about which it makes assertions, [341 e.g., the objects for which the subject of the proposition stands) is nothing other than an objective characteristic of the representation as signification, we do not intend to say thereby that the object itself is only a determination of the signification. Rather we intend only to say this: that object and signification stand in a relationship under objective (i.e., here: *Ideal*) law, that they are *Ideally* related to one another, and inseparable from each other. The object would not be if the significations which refer to it, more specifically the *assertion* significations, had no objective validity. As soon as we say that the object *is*, then we have already given expression to one such truth. The "being of the object" and "the truth that it is" are conceptually identical expressions. They signify that the representation concerned (in this case only indirectly indicated) is an object-bearing one or is valid. Existence is only *apparently* a predicate of the subject which presents itself in terms of the grammatical form. What is expressed is, rather, that the predicate *validity* belongs to the subject *representation* of the grammatically existential proposition. In spite of – indeed, rather precisely because of – that, the existential "is" must not be grammatically interchanged with the predicate "is valid." This latter, as a non-modifying predicate, requires the subject normally accompanying it. But the former, as a modifying predicate, requires the anomalous subject. In other words, it requires a subject expression whose signification must precisely be, not, as in the normal case, the subject signification, but rather the object for which the subject signification stands – and, indeed, the object for the unexpressed predicate "obtains" (*gilt*).

The unity of the object is to be characterized as an *Ideal* unity, and in a twofold respect:

1) In the subjective respect. As the identical unity of the signification stands over against the plurality of possible representing subjects and acts, so the same also holds true for the identical unity of the object. In relation to the subject, just as is the signification, so the object also is an "in itself." To this, even the lived experiences of the subject constitute no exception. Each "subjective" experience is also something objective. It is "in itself," over against the manifold actual and possible acts of judgment which cognitively refer to that experience. Each object (specifically here, each subjective event) is an intentional pole of unity for an unlimited manifold of possible [342] acts. That unity is something Ideal, even if perhaps the object is itself real.

2) From the objective (*objektiver*) point of view, the object is an Ideal unity with regard to the infinity of "objective" representations or significations which represent it: specifically, with regard to the infinity of truths which are valid of it. The unity from the subjective point of view is co-determined by this unity from the objective point of view. The manifold of the acts of representation is co-determined by the manifold of significations. The former is not exclusively determined by the latter, insofar as to each particular signification (e.g., "the" representation "Socrates") an unlimited number of acts of representation correspond. But the unlimited number of *truths* points to the unlimited number of *intuitive* contexts in which the object can be positioned or of the intuitive representations in which it can be represented for itself in terms of its various parts and aspects or in relation and conjunction with other objects.

Section 14. Truth

The discussion of the relationship between representation and object inherently leads to discussion of the concept of truth. In fact our analysis implies that in the *authentic* sense we can speak of a represented *object* only in the case of valid representations. A representation is "valid," "correct" or "true" if there *is* an object to which it refers in the manner of representations. And conversely, so often as a representation has this latter character, the predicates mentioned (and synonymous predicates from ordinary language) apply to it. The equivalence is obviously a completely immediate one. Accordingly, truth seems to be only another word for "characteristic of a representation as having an object," whereby the latter expression is *not* to be taken in that inauthentic sense which permits us to speak also of representations with no objects as if they represented objects.

This interpretation harmonizes very well with the traditional definition of truth as agreement between representation and fact [*Sache*]. For that the representation agrees with the object (the "fact") can surely be little more than an expression for it being precisely a rep- [343] resentation which represents the fact. The entire difference between the two descriptions would lie only in this: that in the latter not the *character* the representation has of participating in this relationship, but rather the *relationship* itself, seems designated as truth – a case of that most ordinary kind of equivocation, through transference, which is hardly avoidable, and in any case is harmless. Certainly it could be objected that representations such as, for example, "the renowned Socrates, a Spartan philosopher ...," indeed refer to a (genuine) object, but yet do not agree with it, inasmuch as they only partially correspond to the object. For the renowned Socrates existed as a philosopher, but not as a philosopher of Sparta. However, it is not the whole representation that represents the renowned Socrates, but rather only one of its parts. Inasmuch as with this part yet other partial representations are linked, which have no representing relation to this object, the *whole* representation is therefore in truth not a representation of Socrates.

Now what sort of a relation is this representing relationship or "agreement"? The latter term, or especially the Latin expressions of the Schoolmen (such as *conformitas*, *adaequatio*), seems to indicate some type of *similarity* relation. In fact, that was the dominant view of those who have upheld the traditional definition of truth. Ueberweg, for example, expressly says: "To agree means to be similar in certain respects."[17] It was, precisely, believed possible to explain the relationship between representation and object on the model of that between image and thing. One does not really need to look deeply in order to recognize that this old, and still influential, theory

[17] F. Ueberweg, *System der Logik und Geschichte der logischen Lehren*, 5th edition, Bonn, 1882, S 3, p. 5.

is fundamentally false, and is possible only from a standpoint for which the distinction between image-Representant and signification (i.e., the representation in the objective, logical sense) is not clear. Take any representations which characterize their objects through remote relative determinations, e.g., "the real root of the equation, $x^3 - 5x^2 + 4x - 20 = 0$," and inquire into the similarity of *this* representation with its object, the number 5. But such blatant examples are [344] not needed. Any arbitrary indeterminate attributive representation, such as "a horse," can serve as a refutation. Notice that the little word "a" also contributes to the signification, and then inquire what in the horse itself is imaged through *this* signification Moment. Consider representations that are equivalent and are distinguished by mere form – as, for example, "something circular that is red/something red that is circular" – and inquire what these distinctions in form concern that could be imaged in the object itself. One will see in general that nothing of what belongs to the "form" of the representation fits in with the convenient model of the image Representation, and that consequently the representation itself cannot be regarded as a mere reduplication of the object. The image theory, as commonly understood, does not even work for image representations.

Similarity cannot constitute the essence of that "agreement" which relates the representation to the object and thus marks out the representation as the representation of *this* object, and the object as the object of *this* representation. If we restrict ourselves to psychical contents as objects, then the representation, taken concretely, is, once again, a psychical content. If mere similarity is the decisive factor, then it would be an arbitrary matter whether we designated α as representation of $\acute{\alpha}$, or $\acute{\alpha}$ as representation of α.[18] Even the imaging relation thus requires something more than similarity. The image "represents the object" precisely in virtue of the fact that the similar serves as the representing Representant of the similar. The "more" here is the psychical bearing of the Representation, which first confers signification upon it, and therewith its character *as* an image.

How then are we ourselves to conceive of the peculiar character

[18] This point may also give special pause for thought to those who think they find the peculiarity of the representation reflected through the identification of representation and phantasm (re-produced content). How then are we to distinguish the representation from the representation of the representation?

of the representing relation, of the "agreement"? To this end it is of course required to look at cases where we can intuitively encounter the agreement between representation and object: hence, where the "object itself" is given, and simultaneously the representation which represents it – both given in *one* act in which the agreement is a lived experience (*Erlebnis*). In other words, we turn back to the cases of Evidence, in which we grasp the truth itself, and thus do [345] not merely represent or hold-to-be-true, but rather live through and behold (*erleben und erschauen*) truth.

In this really consists the "corresponding," in this also the "direction" upon the object, which is imputed to the conceptual act, as to the intending act in general – and, indeed, *could* be imputed to it – only because such acts frequently are fulfilled, and in that fulfilment the relationship between act of meaning and the being-given of the meant is intuited in actuality. In reflexion upon this relational unification there arises the Idea of the possible reciprocal relationship between signification in general and object Ideally belonging to it. For the abstractive consciousness in fact informs us that the relationship is grounded purely in the Idea of the significational: thus in the Ideal signification, on the one hand, and purely in the Ideal essence of the intuition on the other. But in the *ultimately* fulfilled and ultimately fulfilling intuition, the latter coincides with the object itself. Thus, in order to make clear what it means to say that a signification signifies this or that thing, an assertion refers to a state of affairs, a name names an object – and, more generally, a representation represents an object – we can do nothing else than form for ourselves an intuition of a fulfilment. We bring an Evident assertion before us and live through its Evidence; or do so for an Evident nominal fulfilment, as when we say "red," for example, and intuit red *in union with* our utterance.

The signification, in the Ideal sense of the identically one signification, is constituted in the significational essence of the *conceptualizing* act. It is the Idea appertaining to this essence. Just so, the identity of the objectual bearing is constituted in the intentional essence of the *intuitive* act. Thus the two types of acts coincide as to their intentional essence. If we make this relationship the basis of an abstraction, then we discover the lawlike general coincidence or solidarity of the Ideal unities. We have insight into the law that the signification *in specie* and the fulfilling sense *in specie* "agree." But this only says, in different words, that it is absurd to hold there could be a judgment (a statement, assertion) transparent to one person as Evidently true, and to another as Evidently false.

Here offence will immediately be taken. Must we in order to [346] establish the special features of the domain of representations have recourse to Evidence, which after all, it is generally believed, is merely a peculiarity of *judging*? Do we not therewith carry off a fallacious μετάβασις? The entire subsequent reflection will make clear that this is not so. Let it here be observed that, however representation and judgment are to differ, a representation surely underlies each judgment, and in such a way that its signification content is the matter (*Materie*) of the judgment, its identical "what." Further, in absence of actual relation to the "corresponding intuition" we can have, even though not judgments without corresponding representations, yet representations without corresponding judgments (thereby we of course take as "corresponding" a representation and judgment of identical signification content). But as soon as we have produced a representation *with* actual relation to the correlative intuition, it is no longer mere representation, but rather it is lived through (*erlebt*) simultaneously and in unification with the judgment corresponding to it. However, this law-bound connection need not trouble us. If we attend *merely* to the character of the relationship which obtains between representation and corresponding object, and which in fact only in these cases of intuitive agreement we live through as that which it is, then we shall easily abstract from the conjoined judgment Moment. Evidence is not a character of the judgment as such, namely, of the judgment-Moment. Rather it is a complex lived experience, in which a representational act of such and such a signification intention, a judgment-Moment, and a corresponding intuition, are given in a characteristically intermingled style of unification. The analysis of the situation here before us will offset the much more considerable offence that might be taken at our restitution of the ancient concept of truth, with its tie to representations. "Truth" is an ambiguous term. But its equivocations rest completely on transference, and therefore presuppose a first and radical truth concept, from which all the others derive. It is this radical concept of truth, based on representations – ultimately on objective representations in the sense of significations – which we here wish to establish.

We take our starting point from those most simple and transparent intuitive judgments, which *Sigwart* has dealt with under the misleading title of "denominative judgments," but for which he has not [347 provided a satisfactory analysis. There are chiefly two questionable theoretical persuasions – in relation to the representation, the image theory, and in relation to the judgment, the predication theory (according to which there is no judgment without predication, without subject and predicate) – which have here as elsewhere clouded his presentations. Moreover, the denominative term is, on my view, not anything essential for the judgment, and so it can fall away, as in questions, without disturbing the judgment character.[19] Here we are dealing simply with perceptual judgments, and indeed with those more primitive perceptual judgments, which are so little predicative that they are, rather, presupposed in every predication carried out on the basis of intuition – thus, in the genetic respect, in every predication whatsoever.[20] Perceptual judgments thus divide into those which involve naming and those which do not, and again into predicative and non-predicative perceptual judgments.

If my gaze wanders over the surface of this writing table, fleetingly lingering now on this and now on that object, often denominative language does emerge: "my clock," "inkwell," "pen," and the like. The objects are familiar to me, I recognize them, but I also can recognize them *without* an accompanying denomination, which as a rule is absent anyway. If I survey a living room unfamiliar to me, and individually consider the objects – that is, perceptually organize and interpret my total field of intuition in terms of objects – then I do "recognize" these objects, but it is not a recognition of individuals, as in the previous case. Also here denominative expressions occasionally

[19] As Sigwart also portrays it in the *Logik*, Volume I, 2nd edition, Freiburg i. Br., 1889, subsection 9, 1.

[20] With this is connected the multi-tiered nature of any predicative judgment. It consists in this, that the judgment, without harm to its unity, extends itself through all of the indirect partial significations – living in them all, as it were – from which there analytically follows those existential judgments realizable on the basis of any predicative judgment. The *Brentano/Marty* theory of the double judgment, as laudable as its tendency is, does not exhaustively cover the matter. To the single-term subject-judgment is certainly to be added, in my opinion, also a similar predicate-judgment. Moreover, I absolutely cannot agree when the single-term judgments are interpreted as existential judgments. These latter first result through an equivalence transformation into predicative form. For existential judgments are nothing other than predications of objectuality (*Gegenständlichkeit*), but in anomalous grammatical form; and *any* single-term judgment has its equivalent in an existential judgment, only therein the *relation* function is dropped. This is especially to be noted in the case of those predicative judgments to be characterized as non-independent.

occur: “a mirror,” “an odd picture,” and such phrases, are present – [348] but also often not. Such perceptions are not merely “linguistically abridged” predications, as *Sigwart* teaches. They are really no predications at all. To be sure, we can in each case construct an appropriate predication: “This is my clock,” “This is a pen,” and so forth. But now the judgment content is a modified one. Any predication presupposes an articulation within the representation – correspondingly, within the signification – in terms of which a subject signification and a predicate signification are distinguished. Even the empty indexical sign that functions as subject expression has a signification. It is the bearer of a Representative function. It is not merely *there*. It does not merely channel attention off to the respective perceptual object, but rather it is also in this function understood as a referring sign, and is there to ‘express’ that function. In each predication there thus occurs not merely *one* representing (a Representative) apprehension, but rather at least one in the subject and one in the predicate. If therefore the act of representation or of judgment is carried out on the basis of intuition, then intuition is utilized in a twofold manner (at the least). But just as well as there are such multiple perceptions, there are also simple ones.

CRITICAL DISCUSSION OF K. TWARDOWSKI, *ZUR LEHRE VOM INHALT UND GEGENSTAND DER VORSTELLUNGEN. EINE PSYCHOLOGISCHE UNTERSUCHUNG* (Vienna 1894)[1]

It was a good plan, on the author's part, to devote a special investigation to harmful equivocations associated with the terms "representation" and "represented," "content" and "object," and to establish the importance of the corresponding conceptual distinctions, which have been harmfully confused even by distinguished investigators. The act of representing is commonly distinguished from that which is represented in it, which is called now "content" and now "object" – and indeed promiscuously, as if the concept of *the represented* were the same in each case. But there is a more essential distinction yet to be considered here. If I represent a tree, then the psychical "image," the immanent content of my representing, obviously is not the tree itself, the object, of which I say, in the common meaning of the word, that I represent it (Subsection 1). As a corroboration of this distinction between act, content and object for each representation, the author appeals to the three roles which, according to *Marty*, every name plays: it announces that the speaker represents something (the act), it arouses in the hearer a definite psychical content, which constitutes its "signification," and finally, by means of this signification, it names an object (Subsection 3). [2] [350]

[1] From late 1896. First published in *Husserliana XXII*, pp. 349–356.

[2] It is accordingly obvious that the author identifies the signification of a name with the content of the corresponding representation – a fundamental error, which unfortunately considerably diminishes the value of his splendid treatise. That we really are dealing with an error here is proven by the following arguments: 1) The signification of two names can be the same while the present content, bearing the Representation (*Repräsentation*) type of consciousness (thus we mean the *immanent* content of the "representation" – in the sense, here in question, of the Representation of an object), is different, and conversely. If we are speaking of a tree, then for one person the phantasm of a fir, for another that of a linden, and for a third the mere

verbal or written symbols – of many different possible kinds (e.g., italicized, printed type, etc.) – serves as Representing content. That each significant expression must in its normal function carry along with it, in addition to those changing, certain such contents that are constant in all cases, without which no understanding would be possible, and which therefore could be understood as *the* "content" of the respective representation in a more pregnant sense – all of this I hold to be psychological fiction. 2) At best, the Representing contents are more or less similar; but the significations are *identical*. The former are constantly fluctuating; but the latter, on the other hand, are so little so that it is altogether *absurd* to speak of constant variation in the case of significations. 3) The Representing content can stand in various sorts of relations to the Represented object. It can be inherently foreign to it (e.g., the word sign), it can be one part or "side" of the object (as in the case of external perception in the usual sense), or even wholly similar to it or to some aspect of it (external phantasy). It can also stand in other relationships to the object, relationships which association, the main psychological source of our Representations, could seize upon. But in every case the Representing content is, precisely, content, i.e., something perceptible. By contrast, the signification is never something perceptible. The content is, as such, an individual psychical datum, a being which is here-and-now. But the signification is nothing individual, nothing real, never at any time a psychical datum. For it is identically the same "in" an unlimited manifold of individual acts separated in reality. That there is and can be something like minding what is *identically* the same "in" distinct acts – this is the primal fact, certified through Evidence, underlying all knowledge and first giving it any sense whatsoever. – The content resides in the representation as a real constituent, but the signification does so only functionally. It would be absurd to conceive of it as a real fragment or as a *part* of the representation. The function displays itself in discursive thinking, in the interrelationships of judgments; e.g., in identity judgments such as, "The representations *A* and *B* mean the same thing." They mean the same; i.e., here, not that they represent the same object, but rather that they have identical signification. – Also the residing of the object in the representation is no real one, but rather is only a functional one, coming to discursive intelligibility through other suitably framed identity judgments. Accordingly, in the case of a representation, and indeed for every one, we must distinguish: [350n]

I. The *real, psychological content* (*Gehalt*); and indeed a) the Representing act, b) the Representing content.

II. The *Ideal, logical content* (*Gehalt*); and indeed a) the signification, b) the object.

The logical content (*Gehalt*) articulates itself psychologically into two different series of possible lived experiences of judgment: e.g., corresponding to b) "My present representations 'Wilhelm II' and 'the current King of Germany' represent the same object," and, corresponding to a) "They have a different signification, they represent that object in conceptually distinct ways." Psychologically the two representations come before us as detached individual representations. They have no parts in common. They do not coalesce. But they always ground a unitary act of identification or differentiation which points in two "directions," allows two logical "sides" of the representations to stand out from each other: the object-bearing and the conceptual. – Signification, concept in the broadest sense, and representation in the authentically logical sense, are synonymous. – Analogous distinctions exist, and are equally fundamental, with regard to judgments. I will provide more detailed expositions in a larger publication.

The content as well as the object are spoken of as represented. The author illustrates this double sense through an analogous one in which we say, now of a picture and now of a landscape, that it is painted. Thus by a painted landscape we understand now the picture, and now the landscape itself. Just so, with regard to the activity of representation, which moves in a twofold direction, we understand by a *represented* landscape now the representation, i.e., the content, and now again the object, thus the landscape itself. The designation "rep- [351] resented" acts, as previously the designation "painted," to *modify* the one sense, but not the other. Accordingly, the talk of "represented objects" will also have two senses: now it means that an object, besides other relationships, also stands in a determinate relation to a representing being. In this sense the represented object is a genuine object. But now again it means the mere representational content, and thus the opposite of the genuine object. It is the represented object in the first sense which in the corresponding existential judgment is judged as being or not being. Following *Zimmermann*, the author proposes to describe the object as being represented *through* the representation, or also through the content, whereas the content is described as represented *in* the representation, since the content confers the objective reference upon the corresponding expression. The object upon which the representing activity is directed is the primary object, and the content – the means, as it were, through which the representing activity represents the object – is the secondary object of that activity (Subsection 4).[3]

For every representation, a content as well as an object is to be distinguished. But this is contradicted by the widely adopted theory of the existence of "objectless" representations, a theory supported

[3] I cannot concede as valid what this section offers as psychological descriptions. The representational activity does not move in a twofold direction. Only our reflection after the fact does so. Likewise, the "direction upon the object" is no psychological ingredient of the act, but rather a logical function of such an act, coming to light, once again, through reflection after the fact. I also hope that the author's terminological proposal does not catch on. If one speaks of the object as represented, then it is in general inadmissible to speak also of the content as represented. Nothing identical in the two cases is invoked by the same term. Representing as act "exerts" itself upon the content, but it has a logical, functional relation to objects, upon which, by the way, it in no way "exerts itself." But then it is absolutely necessary [with reference to content] to relinquish the term "representation," in order to avoid confusion with noticing (*Bemerken*) it. I have, accordingly, proposed the term "Representation" (*Repräsentation*).

on such examples as *nothing*, *round square*, and *golden mountain*. The author seeks (Subsection 5) to refute this theory. In an excellent discussion he first exhibits the *de facto* sense of "infinitalization," through which non-*A* arises out of *A*: always based on a tacitly present, higher-order concept, partitioned into a dichotomy. (E.g., non-Greeks are *men* who are not Greeks.) But from this the author concludes [352] that, since in the case of "nothing" the higher-order concept is lacking – and thus that expression, understood as a name, is meaningless – it has only a syncategorematic function, similarly as with the little word "no" (*kein*). This line of argument is obviously fallacious. The scholastic interpretation of the non-*A* as something which, in general, is not *A* does not, as it suggests, agree with the way we actually speak. But the interpretation is not for that reason senseless. Accordingly, the word "nothing" also gives, under the interpretation "something which is not," a quite intelligible sense, which certainly does not agree with the normal, syncategorematic function of that word. The author then discusses self-contradictory expressions, like "round square." We understand their signification, and hence *it* exists. So if we say, "A round square does not exist," then this judgment cannot refer to the signification, i.e., the content, but rather only to the object of the representation. So the self-contradictory representations also do not lack an object. More dubious is another line of argument, which begins from the insidious question as to what, then, has the contradictory characteristics. They do not belong to the content, since it, after all, truly does exist. So they belong to the object, as the non-existing but represented bearer of those characteristics (p. 24). – The mistake committed by the advocates of objectless representations is seen by the author to lie in their holding the non-existence of an object to be the same as its not being represented. The difficulty that one might find in the fact that an object that does not exist can be represented all the same, whereby at least existence "in the representation" is given, is resolved by the author by means of the distinction between genuine and intentional existence – which latter consists solely and alone in being represented, and so little has true existence for its presupposition or consequence as, say, being-named does the existence of the named (p. 24f.). [4] [353]

[4] I cannot concede that the difficulties of principle treated in Subsection 5 are fully comprehended and really resolved. According to the author *there is* for every representation an object as well as a content. The object thus exists, and this

The question of whether the distinction between content and object is a "real or merely logical one," i.e., whether the two are in truth *one* entity which is only considered under two points of view, is dealt with in Subsection 6, and of course it is decided in the negative. For it is the content, not the object of each representation, that exists. Contents have other characteristics than the corresponding objects. And equivalent representations have a different content, but refer to the same objects.

With Subsection 7 there begins the "description of the object of the representation," which surely is a *psychological* description of

existence is true existence, inasmuch as we indeed are not just fabricating that every representation has an object. Then it is said: This existence of the object nevertheless is not true existence, but rather is merely intentional existence. What genuinely exists in this case is the represented object as such, not the object itself. Accordingly it seems better to say, not that there is a distinction between one kind of existence and another, but rather between that which exists in the one and in the other case. By this means also the apparent double sense of the "truly" is dissolved. It is precisely the object simply (*schlechthin*) and the object "thought" that are two. But now we are in more difficulty than ever. According to what was just discussed, in the proposition, "In every representation there *is* an object," only the represented object as such can be meant. It always exists, whereas the object itself exists only when the corresponding affirmative existential judgment is correct. But how, since he distinguishes content and object in every representation, could the author not hold that it is the latter which the corresponding existential judgment judges about? Accordingly it could again be only the object itself which is represented, and not the represented object as such, which indeed even 'exists' in the case of self-contradictory representations, and which we therefore certainly do not have in mind when we, for example, truly deny existence to it. Further, how could there be a distinction between represented *A* and *A* itself, since, *when* the proposition "*A* exists" is true, *the same A* which is represented simultaneously exists? *The same* Berlin which I represent also exists. Such difficulties do not trouble those who do not confuse the represented object as such and the content. It seems as if the author also confuses these in Subsection 4; but that certainly cannot be his view, since the represented object that belongs to any representation is, according to him, *never* to be identical with the content, which also is never lacking in any representation. (cf. Subsection 7) – The solution for this and similar difficulties is only then possible when the locutions, "Every representation refers to an object, represents an object," "For every representation there is an intentional object," "The object has intentional existence," and the like, are understood as totally *inauthentic*. And this in such a way that, in the matter of an object, no existence judgment is made at all; but rather only a certain function of all representations is marked out – that is, how they, after a fashion which typically remains the same whether or not the respective existential judgment holds true, enter into judgment contexts. Probably this was precisely the view of *Bolzano*, against whom the author's polemic is mainly directed, and whom he also in other respects has completely misunderstood, in that he interprets his "representation in itself" as *content* (p. 17).

the object in no other sense than the formulation of the elementary propositions of arithmetic is a psychological description of the concept of number. What cannot be psychologically encountered also cannot really be psychologically described. Or are we perhaps, going along with the author, to count the divisions of objects into possible and impossible, existing and not existing, real and not real – as well as the proposition that the concept "something" is the *summum genus* [354] of all objects – as falling within the "description of the object of the representation." It is overlooked, with the first two of these divisions, that in the literal sense they are *not* divisions of objects, but rather of representations, with respect to certain pairs of existential and possibility judgments, one member of which necessarily is always true. In this way is altogether dissolved the delusive appearance that characteristics could be established for objects in that allegedly general sense which would leave open whether or not the "objects" exist in the given case. If one treats inauthentic ways of speaking as authentic, then everything lapses into confusion. Characteristics of objects are *eo ipso* characteristics of existing objects, since the concepts "object" and "existing object" are equivalent. Hence all of the author's ingenious discussions about objects have to do throughout with *existing* objects and with existing parts and relations thereof – but, precisely, discussed in general terms.

In Subsections 9 and 10 the author treats of the general characteristics of objects which result from their being composite. He distinguishes matter and form of the whole, and accordingly material and formal constituents, whereby under the latter are understood the relationships that unite the parts with one another and with the whole, whose manifold combinations he tirelessly elaborates.

In Subsection 11 the author turns to consideration of the constituents of the representation content. Following *Brentano* and *Stumpf*, he differentiates the material parts in it in terms of mutual separability (binding and pages in the representation of the book), mutual inseparability (color and extension), and finally one-way separability (color and red), all depending on whether the parts can or cannot be *represented* without each other – reciprocally, or on one side only. It must now be asked: In what sense "represented"? The author naturally replies: In the sense of the content. "Red cannot be represented *in* a representation (as content) unless *in* the same representation extension (as content) is represented." But it would follow

from this that, since content and signification are supposed to be identical, the concept of extension belongs to the *signification* of the word "red"; and *vice versa*, that "red" and "extension" are thus words [355]
of the same meaning – a difficulty which the author unfortunately has not considered.[5]

Subsection 12 raises the question whether outside of "being represented by means of the content," which is a primitive relationship, yet other relationships between the content and object of a representation are to be detected. Only for the representations of composite objects *as* composite is the answer in the affirmative. For the *way* in which the object parts are united into the unified object simultaneously determines the way in which the object parts are represented by means of corresponding content parts. – The confusion between content (Representing image) and signification (concept) impairs in great measure the subtle discussions of these paragraphs.

In Subsection 13 the author tells us that not all parts of an object find an equivalent in the content of the respective representation. There are no adequate representations. For the designation of the part of an object which comes to separate representation through the content he proposes the term "property" (*Merkmal*), whose ambiguities he earlier (Subsection 8) discussed thoroughly. The question whether unity and plurality, identity and difference, are in this pregnant sense properties of each object is raised, and it is decided that this is to be accepted only with respect to unity.

Subsection 14. It turned out that not all object parts have corresponding content parts. Now, is the converse also true? *Bolzano* has denied it. According to him, "Not every part of a representation is the representation of a part contained in its object." The essential line of argumentation in *Bolzano* – as the pertinent examples: a simple [356]
being, the eye of a man, a land without mountains, may here serve to suggest – rests upon indirect representations, and loses its force,

[5] In truth the above distinctions fall within a totally different domain. They relate to necessary relationships between characteristics (thus, concepts), and it is only inasmuch as contents also have characteristics that laws are thereby decided which find application to contents. In this it is indifferent whether we are dealing with contents of Representing representations or with contents of other sorts. It is certain that the author does not draw the latter distinction; and likewise he identifies (Subsection 10) the characteristics with the corresponding *parts* of the object to which they belong. For example, the characteristic *red* is, for him, a "metaphysical part" of the red object, although identically the same red can belong to many objects.

the author thinks, if with such representations one only takes care not to confuse their content with certain auxiliary representations bound up with the "inner form of speech." Thus for example, according to the author the representation "man" does not belong to the content of the representation (or to the signification of the expression) "eye of a man." It only serves as a sign to evoke the genuinely intended representation, that of an eye with a certain character. – It is regrettable that the author did not, right at the outset of the treatise, investigate the distinction between "auxiliary representations" and significations with the conscientiousness otherwise characteristic of him. Certainly then *his* concept of content would have been different, and he never would have come to confuse, in such an inadmissible fashion, constituents of the objective *concept* with subjective "auxiliary representations."

In conclusion, Subsection 15 treats general representations as representations of general objects, which the author accepts along with *Erdmann.* But the interpretation he gives to the concept "general triangle" astonishes me. Namely, that it is *a triangle.* He is only consistent, then, when he designates it as neither right-angled, nor obtuse, nor acute, and sets it on a par with absurd representations, such as that of a white black-horse and the like – among which, if Geometry is to retain its validity, it also would unquestionably belong. It is also consistent of him, then, to allow *general* representations only the status of *indirect* representations.

In spite of the relative thoroughness with which I have gone into this work – also distinctive for its excellent style of presentation – I could only characterize the main questions which it deals with, and what, roughly, is the position that the author takes on them. The fulness of stimulating and ingenious details must be left unmentioned.

CRITICAL DISCUSSION OF H. CORNELIUS, *VERSUCH EINER THEORIE DER EXISTENTIALURTEILE* (München 1894)[1]

According to an indication given in the opening words (p. 5), we must designate as the *ultimate* goal of these investigations the solution to the general question about the essence of the activity of judging and about the *principium divisionis* of judgments. Setting out from the more specific question about the nature of the existential judgment is justified by the way the problem of the judgment has developed historically, and especially by the circumstance that from this formulation of the question the line of argument takes shape with considerably greater ease. Since the problems here, which the author treats with obvious talent and conscientiousness, currently stand at the focal point of logical interests, a more exact analysis of the contents of these investigations cannot be unwelcome. With regard to the critical doubts which arise in opposition to the theoretical position they take, I do not think I should restrain myself just because the clarity and conciseness of their line of thought makes it easy to detect the fundamental weaknesses – which, it seems to me, also harmfully influence the other theories of judgment currently dominant.

The author takes for a starting point the provisional definition: *Existential judgments* are those acts of knowledge and assertion which in any way concern the being of a conceivable content (6). The work divides into six chapters, the first of which is to consider the activity in virtue of which we directly know the being of the present contents of our consciousness, as well as the forms in which we bring such knowledge to expression. The question of how [358]
existence can in any sense be known and asserted of *other* sorts of

[1] From late 1896. First published in *Husserliana XXII*, pp. 357–380.

contents is dealt with in chapters II–V, while chapter VI draws the consequences for the general theory of judgment.

I. Perception and Perceptual Judgment

Perception is obviously nothing other than the coming upon, the awareness or noticing (*Bemerken*) of a phenomenon factually occurring in our consciousness. Inner perception is the perception of psychical, outer that of physical phenomena, in the sense in which *Brentano* distinguishes them. By the physical phenomenon is to be understood, accordingly, not the extra-psychical object in the sense of physics, or any other thing not itself given; but rather, for example, the sensed tone as such – and thus the very content that is given, which is no less something psychical than the "psychical" phenomena so-called. To the distinction between the two sorts of perception there thus corresponds no distinction of significance for knowledge. In external as in inner perception we have, according to this author, foundations and sources of knowledge that are absolutely on the same level and liable to no deception. In fact, in the light of the difficulties of inner observation, external perception is frankly to be counted as the most certain source of knowledge (10).

Perception as *noticing* of a content is the same as the analysis of our current content of consciousness; and since the analysis of simultaneous partial contents reduces to the analysis of successive contents, i.e., to the *differentiation* of the currently present content from the previous one, it therefore follows that the perception of a content consists in the *differentiation* of it from the previously perceived total content. This also is automatically clear where we are dealing with awareness of a content newly emerging or not perceived in isolation. For with the awareness of this content there is indeed always given a differentiating of the new perceptual content from the previous one (13). Since, thus, perception turns out to be differentiation of successive contents, the author investigates the conditions of the realization of such differentiation. We agree with the author's [359]
claim that differentiation has to do not only with the temporal locus, but also with the qualitative similarity or dissimilarity of the contents, and that it becomes possible through the holding over of a more or less mirror-like image of the past content to which the temporal rela-

tion attaches (14). For clarification of the author's view, which is not totally intelligible at this point, we refer to a passage on page 88. There it is stated that the differentiation upon which any perception of a content as such rests includes in itself the *knowledge* of the temporal relation of the content earlier perceived to the one now present.

In precisely this relationship of the perceived content to the foregoing total content – and thereby indirectly to earlier contents in general – there is obviously also contained all of that which can serve in the characterization of the content now present. Any denomination of and any knowledge about the present content ("knowledge about" in *James*' sense)[2] can only be based on its relation to earlier contents, can consist only in a knowledge of its connection with our data preserved in memory (15–17).

The mere perception of a content does not require the analysis of all these relations. Only the least bit of determinate knowledge, that of the content's difference from the foregoing one, need be present in order for perception to be spoken of (17). This most simple act of taking note of the being of a content is called "the elemental perceptual judgment" by the author – since with it a knowing (*Wissen*), and consequently a judgment in the broader sense of the word, is attained. As a judgment about a being, it is an existential judgment. With the perception of physical phenomena things turn out quite the same as with the perception of psychical phenomena (18). Precisely this inseparable union of the knowledge of the existence of a content with the content's occurrence or "appearance" in consciousness conditions the infallibility of the perceptual judgment. But that "appearing" is identical with the process which psychology designates as the "being represented" of a content in the wider sense of the phrase (19). The author distinguishes three significations of the term "representing": [360
this appearing of a content, remembering and imagining in opposition to sensing, and finally the psychical activity which is required for the voluntary formation of a phantasy representation signified by means of symbols. He adopts the first meaning. In conformity with it, being represented is identical with being perceived (20). If no phenomenon is to be object of the representing activity without at the same time becoming object of knowledge, then the distinction of the representing act from the perceptual judgment is a redundancy.

---

2 Cf. W. James, *The Principles of Psychology*, Vol. I, New York, 1890, p. 221.

The expressions "representation" and "judgment" are here synonymous. If there are single-term judgments at all, then it is certain that elemental perceptual judgments – since they only assert the being of a content, without a union of subject and predicate representations – would have to belong among them (21). But if we attempt to reproduce such a judgment *linguistically* in single-term form, we do not succeed without predicative determination. At the point at which we, for example, call the content a tone, we have therewith carried out a *comparison* of that content with the memory images of contents (tones) sensed earlier, the result of which is a predication (22). If we say, "There is a tone," then that comparison relation is already co-asserted. Only the designation of the content as a current one, as conveyed by a mere "this," can be conceded. Every other formulation of the perceptual judgment is – even if not grammatically, then logically – to be understood as a two-termed judgment. But if we *explicitly* reproduce what the demonstrative expresses, then we must say: "This is other than the previous one." The elemental perceptual judgment therefore is also not single-termed. Also in it we find a *relation* between two representation contents as the object. It, like the ordinary two-termed judgments, is a relational judgment (23 & 31).

The more fully determinate perceptual judgments, in which the content appears as named in some way, are accordingly complexes of judgments. For example, the perceptual judgment "This is red" is composed of the following steps: 1) the elemental perceptual judgment (the "this"); 2) the elemental judgment of comparison, i.e., the noticing of the psychical phenomenon which forms the transition (the "is") from the perceived content to the memory images of similar contents – to which attention is associatively directed by means of the former content; 3) the occurrence of the designation for the perceived [361] content (the "red"). The genuine substrate (*Materie*) of the assertion about the perception is obviously this entire process – possibly a multiply executed repetition, forward and backwards, of the process (24). If the judgment of comparison also comes to expression, then this occurs through the copula or the verb form combined with the predicate. The mere exclamation, e.g., "Fire!", gives expression to the last step only, in the course of which the interjective tone intimates the first step, the perceptual judgment. All of this is less clearly expressed in the corresponding impersonal locutions. Formulations such as "Red is" do violence not only to language, but rather to the logical signification of the respective judgments also. A new complication enters when the subject appears named as well, for with that name an additional judgment of comparison is obviously co-asserted (25). Another modification is that the relative determination appearing in the predicate also comes forward as such *linguistically*. Instead of saying, "This is tone *a*," I also can say that it is the fifth from *d*, and the like. The distinction between absolute and relative attributes can accordingly not pass for an objective one (cf. also 32). Even the apparently absolute attributes arise from comparison. Occasionally we hear that it makes *no sense* to proceed to assign the attribute "different from blue" explicitly to a red object, since that attribute is in any case given along *with* red. The same holds for all predicates that analytically follow from those already known (26). The choice of linguistic expression often may depend upon whether we take note of the individual steps in the act of judgment (27).

As in perceptual judgments, so also in explanatory judgments, the relation asserted between subject and predicate is nothing other than the result of the comparative act, on the basis of which we recognize the more or less exact agreement between the contents designated by means of the subject and predicate terms. If I judge "Blood is red," then the phantasm of blood hovers before me, whose similarity with phantasma of other color perceptions known to be "red" I verify.

It accordingly is *not* to be conceded that in the question and in the negative judgment the same union of representations is present as in the affirmative judgment. Certainly the same *words* are linked together in the question and in the answer (28). The question clearly has only the sense of pointing the hearer to the representations named, [362]
and of causing him to carry out the elemental judgment of comparison between them. That this elemental judgment is not *eo ipso* realized upon hearing the question lies in the fact that in the question, as in any linguistic expression, only symbols are given. And it is therefore that the question can simulate an arbitrary relation. Only when the hearer in actuality perceives and compares the representations merely indicated to him, i.e., takes notice of the relation actually obtaining between them, does he judge. The judgment can obtain the sense of the affirmation or denial only with reference to a question posed (28). If the relation perceived is not similarity, but rather one of considerable difference, then – provided that no question about

sameness is raised – the judgment will assume the positive form of an assertion of the difference of the two contents. Only if that question is expressly or tacitly raised can there appear, in place of the positive judgment "*A* is something different from *B*," the negative "*A* is not the same as *B*," which expressly states that the relation of the two contents is other than the sameness relation (merely simulated in the question). But in order to verify the contrast between the relation encountered and the one simulated, the latter would have to be represented.

In the case of *comparative* perceptual judgments, the possibility of error is given in virtue of the fact that, with inadequate analysis of the relations of a perceived content, the memory image of it remains permeated with inexactitude, in consequence of which the relations of the current content to the ones earlier perceived are more or less inadequately known (32).

II. Sensation and Phantasy Representation

The analysis of the existential judgments which assert the being of a content *not* directly known as present requires the preliminary discussion of the relation between sensation content and correlated phantasm. That the two are *toto genere* different from each other needs no proof, and no one can be in doubt about it – as the author begins on page 33. But there is the question, he thinks, as to wherein lies the matching up of the two types of contents to form pairs. However, even the most fleeting glance at the views of *Berkeley*, [363]
Hume and *Meinong*, among others criticized in this chapter, shows, to the contrary, unanimous *denial* that the two types of contents are "*toto genere* different," and only leaves no doubt, at least in most cases, about the factual possibility of distinguishing them. With that phrase, therefore, it is only the theoretical position of the author that is expressed. It will suffice for now to emphasize a few points: The relation between sensation and corresponding imagination content is a *relatio sui generis*. Nothing bars us from designating it as a type of similarity, but then it is of a type which is not reducible to the other known types (41). Distinctions of intensity and quality, in the sense in which we speak of them in the domain of sense impressions, cannot serve in the explanation of the distinction here in question.

It must be a distinction of a type which is altogether missing from every single one of these domains. But then the difficulty arises of gaining insight into *how* the depicting of the sensation content could still come about by means of an imagination content *toto genere* different from it. The author finds the answer in the general experiential fact that no sensation is perceived without corresponding phantasm. In the hearing of the tone there is present to us, as it continues, not only the tone as sensation content, but rather the memory image of the part of the sensation already passed as well. The knowledge of the intimate connection between sensation and corresponding phantasm is consequently one of the primary facts of the psychical life, which already underlies our most elementary acts of knowledge, and therefore appears to be neither capable nor in need of any further explanation.

Language makes no distinction between the object which I presently perceive by means of my senses and the phantasm of that object, which appears in the remembering of it. This equivocation is to blame for the fact that to this very day no agreement could be attained on the question about the nature of the activity of judgment. [364]

III. Symbolic Existential Judgments

If we are dealing with the existence of an object not presently perceived, then the present content is a more or less determinate phantasm of that object. But the existential judgment does not signify the being of that *phantasm*, although it of course implicitly co-asserts it. What, therefore, is the meaning of the existence concept in the case of such judgments? If they refer primarily to the existence of sensation contents presently not perceived, then such judgments obviously can only have the sense that it is possible for me or others to perceive such sensations under certain conditions (45). They are thus, in essence, anticipatory judgments made on the basis of earlier experience. That they do not linguistically assume the form of assertions about the future derives from the fact that those tacitly presupposed conditions usually involve no temporal determination and from the fact that the one judging gathers up in the sentence uttered mostly perceptions already had (of the kind concerned), along with those to be anticipated. The author subsequently stresses the formulation according to which the existential judgments discussed are, in their sense and logical signification, nothing other than abbreviated modes of expression for experiences had; and he seeks to prove this proposition by means of two series of reflections.[3] The *one* (Subsection 19) aims to do this through answering the question of *right*, the question about the justification of such judgments, which certainly does become confused with the question about their origination. The *second* series of reflections has recourse to the function of the symbols (Subsection 20). We will pursue it.

We can judge concerning objects not present only if they are Represented through contents which are present. The existential judgments here considered are therefore to be characterized as symbolic (53). If the object is a sense content, then the corresponding phantasm serves as Representant. The principle in virtue of which this phantasm appears as symbol for that content is the specific relation obtaining [365] between them – which, like any comparison relation, permits the indirect determination of one of its terms by means of the other. In this case the judgment asserts that a sense content corresponding to the present phantasm is to be met with under certain conditions. Another content, linked to the absent one through a comparability relation, can replace the phantasm. However, also in these cases there is ultimately required mediation by a phantasm of the content (as a symbol). A painting reproduces a phantasm of the painted as Representation of the painted (54). In the spoken judgment the word serves as symbol of what is not present. Symbols [such as words] that work by association are to be subordinate to those which work by comparability. In the former, attention is directed upon a content present in consciousness: upon a phantasm which itself functions as a symbol through comparison. If every such phantasm is lacking, then the word is completely meaningless. This mechanism of the symbolism immediately shows that that interpretation formulated above for symbolic existential judgments, which bear upon sensory and then upon perceivable objects in general, is correct (55). But it also proves correct for cases where existential judgments assert the being of objects that are by nature not perceptible. For the subject term of such a proposition only then has a specifiable sense when cer-

[3] I struggle in vain to understand the author's line of thought. Is it not the same one which at the outset is set forth as self-evident ("obvious") that is later proven in such detail (and unfortunately so unclearly)?

tain phantasma of perceptible contents show up in association with it, and when, thus, its signification is reducible to comparison symbols consisting of sensation contents. (With somewhat more precision it is explained on page 49 that each property through which we determine an object is either perceptible or is defined in some way. Of the definitive properties the same holds true, and so forth. But we ultimately must come upon the directly perceivable.) When one trained in the natural sciences maintains that there exists a magnetic "force" of the planet earth, he can give a sense to his claim only if he represents to himself some sort of example of the *effects* of that force, which are summarily expressed by means of the word referred to. The existential proposition has for him the sense that the represented effects can be factually observed in nature.

If we thus make clear to ourselves what we mean by the existence of something not perceived, and what on the basis of our experience we *can* only mean, then the assertion of the existence of an external [366 world loses the mark of the mysterious. We require no metaphysical hypothesis to understand it.

Similar results carry over to the remaining classes of existential judgments, the itemization of which we omit. The discussion of those which deal with the meaning of symbols is characteristic. Thus, on the author's views, the question is always whether a content actually is associated with the respective symbol, constituting its signification. In case of a simple symbol, it is not an existential proposition that corresponds to the answer, but rather the definition or the denial of one (57ff.). It is otherwise for composite symbols, i.e., those whose individual parts are known to be meaningful. For with them a content is determinately "indicated," and the question posed as to whether such a content is factually to be met with in our consciousness. This question is equivalent in meaning with the demand to form a phantasy representation corresponding to the given combination of representations. If we succeed in encountering or newly forming such a phantasy representation, then the positive existential judgment is made, and otherwise the negative. The former has the character of a perceptual judgment, but the latter not, since it as a rule is intended as a judgment of incompatibility.

The author then goes on to explain that the totality of simple existential judgments divides into three classes: perceptual judgments, memory judgments, and anticipatory judgments. All other existential judgments are composed of these (59, 61). Also, symbolic judgments are of course in every case relational judgments. They assert the subsistence of a comparability relation between the present phantasm and a content not present (61).

IV. Polemic

The critical interchange with *Hume*, *James* and *Brentano* pursues the goal of refuting the *belief* theory of the judgment. Let us cite a few of the author's assertions by way of example. Against a well-known argument in *Hume*, he objects that the expression "to represent a thing as existing" has no sense at all. I can have the same representation whether it exists or not, but in the first case just as [367] little as in the second can I represent it in my phantasy as existing, but rather always precisely as represented – as phantasm (64).

The assumption of a new type of relationship of consciousness to its content is utterly superfluous. The persuasion [in belief] indeed does not relate to the existence of the present content, but rather to another one which is not present. But this persuasion obviously cannot be obtained through any sort of modified grasp of the present phantasm (65). The assumption of that new type of relationship even involves a contradiction: If the existence of each content is given with the representation of it, then in fact there is no need for an additional new activity in order to acknowledge that existence; but if such an activity is *needed*, then existence cannot be simultaneously given from the outset with the content. (66)

Also, judgments which involve a relationship or combination of representations do not first come to be through the attachment of a *belief* to the representational compound. If *Hume* objects that whoever hears a proposition without accepting it, or asks the corresponding question, certainly forms the same combination of representations as when he really judges, then the author replies that in the majority of judgments (e.g., This fruit is a cherry) there is no combination of representations, as, say, in the case of "green tree." In the question, to be sure, the properties of this fruit and of the recollected cherry are externally united by means of the corresponding words in the interrogative sentence. But the traditional theory of judgment has never held *this* unification to be what constitutes the judgment. The question merely calls us to compare the two groups of representations mentioned, to look into their relations, but not to form a complex whole out of them. If the comparison is carried out, then the actual judgment consists in the knowledge of this specific psychical act. Here is the *result* of the comparison, the completed synthesis of the concepts; there in the question is the *call for* comparison (67).

Against *Brentano* the author objects that, if in his theory "is" means the same as "is true," and the latter predicate is only admissible for judgments, then in existential propositions there would be lacking the judgment to which this predicate would be assigned (83). Therefore the reduction of all other propositional forms to the existential "*A* exists" would also be a mere tautology. Under *A* only the original judgment itself could be meant (84). [368]

The author traces the errors of all the theories discussed in this chapter back to the confusion of symbol and symbolized content, on the one hand, and that between combination of representations and comparison, on the other.

V. Memory Judgments

Here the fundamental problem lies in the question about the distinction between those phantasma which are products of our imagination and those which are memory images from earlier lived experiences (*Erlebnisse*) (86). The distinction cannot lie in the contents themselves. The memory image of a tone is, if we set aside the simultaneous partial contents in consciousness, absolutely identical with the corresponding imagination representation. For the assessment of a representation *as* a memory representation, it is also necessary for its background, or at least a number of the partial contents of consciousness associated with it – simultaneously present, but *not* in general examined individually – to come in for essential consideration (87), which permits the inference to the temporal relation of the remembered lived experience to the present. It is, to begin with, a primitive fact that we recognize the relation of a content *just previously* perceived to its presently perceived memory image, and do so directly, from the way and manner in which the latter influences the current state of consciousness. For that relation is already included in the most simple fact of the perception of a content as such. The capa-

bility for differentiation of such a memory image from every other phantasm is therefore just as primordial as the capability for differentiation of different sense contents from each other (88). Memory images which differ in temporal relation to the same object (the same tone *a*, heard a short while ago or heard yesterday) also show, considered in and for themselves, no qualitative distinctions. But the total state of consciousness is influenced differently through the one and the other: The *Gestalt* quality resulting from the past and from the present perceptual content is different in the two cases, as, say, the quality of two colored parts of a surface is different, depending on whether they are closer or more distant. We *designate* this difference [369]
as that of the temporal relation of the contents concerned. The phantasm which turns up as after-effect of a past lived experience obtains a specific "coloration" ("*Färbung*"), which is lacking in the pure product of phantasy (89).

Every phantasm of a simple sensation is a reproduction from an earlier perception. That it nevertheless is not always characterized *as* memory is explained by the lack of differentiation of the various nuances of the background, which nuances correspond to the repeated appearance of the memory image in various environs and are characteristic of it (90). Just so, the mistaken assessment of a product of *imagination* as a memory image is explained by the fact that the distinctions of background, which generally mark out phantasy and memory, are not noticed (90). A more specific ground of error is that the frequent repetition of similar impressions leaves behind the same number of memory images, which indeed are at first distinguishable, but often are not considered with an attention sufficient that they could still be distinguished later on. The memory judgment thus becomes at first indeterminate, and finally completely dubious. On the other hand, the judgment is made with greater certainty the more carefully we have analyzed the distinctions of the impression concerned from other similar ones (91).

Since in the memory judgment we are dealing with the knowledge of the relation of a present content to one not present, that judgment cannot consist in a specific relationship of consciousness to the partial content which we call a memory image (91). *James*, who with his theory of the *relation-fringes* was first to discern the character of memory images, errs in that he still takes *belief* to be necessary to the memory judgment (95). The object of consciousness is different when I recall an event and when I merely make it up. The difference lies in the "fringes." For it is impossible arbitrarily to think these "fringes" onto the representation of a merely pretended event. The "*belief* theory" can specify no distinction between true and false memory judgments. *Belief* would also be present in the false ones. That theory fails to answer the question about the basis for these judgments and also is here guilty of dodging the difficulty to be faced (96). [370]

Memory judgments can be located in *Meinong*'s "*Evidenz* class" of judgments of difference. This is because a memory judgment essentially rests upon *differentiation* of the memory image from a corresponding product of phantasy. The judgment is made with full certainty, thus confidently, in case those "fringes" are noticed at all (99). The question whether that which, on the basis of the 'fringes', we judge as after-effect of past lived experience also really corresponds to past lived experiences, probably amounts to a mere verbal dispute. (100)

VI. Concluding Reflection

Existential judgments in the broader sense which this work has taken as its basis are not to be singled out as a separate class of judgments. All "relational" judgments, commonly so-called, fall under this concept, inasmuch as in them the psychical phenomenon in which the relation affirmed comes to consciousness is the object of consciousness in a similar manner as is the perceived content in the perceptual judgment (100). Given this, it holds true in general that a contrast of principle is *not* to be drawn between representational activity and judgment activity (101).

If by "judgmental activity" is understood the psychical function that is subject to the question of truth, then obviously perceptions are not to be considered judgments. For it would make no sense to ask about the truth of a perception. The perception is simply given to us, and we can just as little put it in question as we can abolish it (103). The concepts *true* and *false* first find their place where a content is used as a symbol of another one.

\+ + + +

After having reported the essential ideas in this stimulating and substantial publication, I must confess that I cannot agree with the author on any point of substance.

1) Already the method evokes doubt. What is today designated as "*the* problem of the judgment" is a whole bundle of problems, which, although interconnected in various ways, yet from methodological [371] and substantive interests are to be carefully held in distinction. Here we have in the first place the general *descriptive* questions concerning which descriptive-psychological characteristics actually belong to every psychical experience (*Erlebnis*) that we call a judgment, which peculiarities distinguish it from other psychical experiences, whether with it a novel primitive fact of inner experience (*Erfahrung*) is to be recorded, or whether it is to be understood as a mere complication of other such facts, and the like. Essentially distinct from these are the genetic problems, which concern the origination of judgments and the system of psychological laws for them. Of yet another character again are the *logical* questions in the narrowest (i.e., non-methodological) sense of the word, which do not have to do with judgments as individual events, but rather with their objective content (*Gehalt*), which can be identically the same in innumerable individual cases, and is, *as* this self-identical One, the object of logical analysis. Whoever judges that $2 \times 2 = 4$ has made "the same judgment." Objective science has to do with *the* judgment $2 \times 2 = 4$, as logic does indirectly: equivocally designating the fleeting acts as "the judgment" in its psychological and methodological segments, and in its pure logic segments so designating their identical objective content. Establishing the intrinsic and, indeed, the *general* characteristics of judgments in the second sense (which are the significations of the corresponding assertions), and formulating the objective laws grounded in those characteristics: – these are investigations of a wholly different type, which are to be carefully separated from the psychological. On the other hand, these and *all* of the other inquiries just mentioned are interrelated with those that have as object the provision of norms for judgments as well as the classification of the "sources of knowledge" in terms of their importance for knowledge.

Certainly these ranges of investigations are not separate at all points in such a manner that an occasional shift of levels (μετάβασις) would not be necessary. But one must once for all

get it clear that there are here distinct goals of investigation, which, if confused with each other, must entangle the process of the investigations and obscure their results. Now it is unfortunate that we do not see this clarity in the author. The refutation of the *belief* theory of the judgment and its replacement by a better theory is the interest that motivates him. He thus seems to have his sights set on the *descriptive* problems. But, as frequently occurs, they seem in [372]
him to be disturbingly mingled in with the problems from the other groups. Connected with this is also the fact that he, again and again, directly identifies psychological preconditions of the act of judgment – or psychical Moments more or less essentially interwoven with the act, or even logical grounds, which alone would justify it – with that which the one judging means, asserts, or denies. For example: Perceiving or distinguishing presupposes the relation of the present content to the preceding one. Thus, the primitive perceptual judgment *means*: "This is something other than before." Any designation and any "knowledge" concerning a present content can only rest upon its relations to earlier contents. Thus, any such must *consist in* the knowledge of these relations. The representing of something unperceivable comes about directly or indirectly by means of unalloyed "perceivable" properties. Thus the corresponding existential judgment can only *mean* that we have had and will have certain perceptions. And so forth.

2) A perilous reef for descriptive analysis is our natural tendency, in the description of acts which were actually given, to mix in various others that first occur in reflection after-the-fact upon the earlier psychical situation. This confusion holds true of the work before us to an exceptional degree, favored by an equivocation with which we shall forthwith become acquainted. –

The main foundation of the theory before us is the identification of noticing or awareness (*Bemerken*) with differentiation. This is not, perhaps, just to fall into an assignment of terminology, but rather therewith an important bit of knowledge is supposed to be obtained. To be aware of a content is to differentiate it from the previous total content. Unfortunately this identification consists in a confusion of two fundamentally different significations of the phrase "to differentiate": namely, the differentiation involved in *relating* and that involved in *analyzing*. The former already presupposes, as does any relating, that the relational terms are noticed in their own right, and thus are differentiated in another sense. Wherever a content is noticed in its own right, there of course it is correct to judge that it is different from the "previously perceived total content." But it does not follow that the noticing consists in this relational judgment (which reflection after-the-fact first generates). If one sees in the "setting off" of a content from a background a peculiar sort of relation, which even [373] admits of gradation, then I have no objection to that. But the noticing of the content is not the noticing of *that* relationship. Rather, the objective subsistence of the relationship (i.e *not* the carrying out of the relating or noticing activity through which *we* grasp it on some occasion) is only the precondition for the noticeability of that content, a precondition knowable by means of psychological reflection and induction. How prepared the author is to introduce his reflections into the original situation is also shown by his making even the knowledge of the temporal relation to what has passed a constituent of plain and simple noticing.

It would be superfluous to further point out the absurd *regressus in infinitum* involved in the differentiation theory: Where a content is not already noticed, none other can be differentiated from it by means of a relation. The current content A leads back to the just past A', this latter to A'', and so on. There is implicated in this a second and no less absurd regress: For the plain and simple awareness of a content the author obviously substitutes the awareness of its relation to the preceding one. To be consistent, we must then substitute for this latter itself the awareness of the relation of this relation to a past relation, and so forth.

Therewith the whole analysis of the perceptual judgment collapses, and all based upon it.

It is further to be pointed out that the same error of the differentiation theory reappears later on, since the absurd insertion of memory judgments within the "*Evidenz* class" of judgments of difference rests upon it (99).

3) The author reduces all judging to mere noticing. In general, noticing, differentiating, perceiving, representing and judging are to be synonymous terms. He not only denies the distinction of two psychical functions within perception, but rather believes it possible to establish this also in every other case where to judging we oppose a mere representing. Even the distinction between affirmation and denial refers only to distinctions within the noticed, not to differences

in a presumably peculiar act of judgment. The author succeeds in appearing to carry through with this view by means of a chain of [374] confusions and mental leaps of which here (so far as they are not already indicated under 1) and 2)) only a few may be emphasized. Above all we stress the confusion between content and object, and correspondingly between representing in the sense of noticing and the utterly heterogeneous representing in the sense of Representing (*Repräsentierens*). This especially stands out throughout Chapter I. To sensibly perceive an object is, for the author, to notice a sense content. But is the tree which I now grasp, as my view wanders over it, identical with the manifold and diverse contents which I notice? Does not a single – and in different cases a different – [content] suffice for the grasp of "the" tree? In the same way the author identifies the encountering of a sensible phantasm with the representation of the corresponding object in phantasy. In consequence of this [he identifies] the phantasm accompanying a name with that which the name names. If one yet further considers that the signification of a name is confused with that which it means, and the designation of something perceived is interpreted as the comparison of it with the "signification" (i.e., with the phantasm accompanying the name), and knowledge interpreted as the agreement of the two – then the theory of the linguistically formulated perceptual judgment is complete. Compare the analysis of the judgment "This is red," on page 24 of the book.

Of course I do not mean to claim that the author has paid no attention at all to the distinction between the noticing of a content which itself serves as a symbol and the Representation of an object by means of that symbol. He in fact has discussed it many times and in detail, and has established the class of symbolic existential judgments upon it. But this changes nothing in the fact that, particularly in Chapter I, the distinction has not been properly accounted for. The results of the analysis would in fact have to run in a totally different direction if it had. The psychological content (*Gehalt*) of the "more determinate" perceptual judgments would stand forth as incomparably more complicated. But if we wish to regard these analyses as merely preliminary, we look in vain in the further course of the essay for the pertinent supplements. If the author had just once recognized that in the linguistic formulation [of the judgment] the "symbolic function" collaborates step by step, he could not have allowed the discussion of that function to come *after* the analysis of linguistically [375] formulated judgments, and even in a very much later portion of the work. In fact, a greater part of the relevant expositions become as good as unintelligible because the function in question now seems to be in action and now not, so that one also never really knows in what manner, to which degree, and through which psychological content, it is supposed to be playing a role.

But even where the author treats the distinction in question *ex professo*, he does not, it would seem to me, do it justice. Of course for him representing in the sense of the Representation reduces again to mere noticing, and the means which is supposed to make this reduction possible is the "indirect" representation. Every relation permits the indirect determination of one of its terms through the other. The relation which comes into consideration here is that between phantasm and corresponding perceptual content. We live through it in each moment of our psychical being, already in virtue of the fact that to every perception a corresponding phantasm is attached ("in primitive association," to use *Brentano*'s language). But what is accomplished with all of this? Is representing then really reduced to a noticing? But let us reflect on the essence of the matter: Wherever we apprehend a relation – or, more generally, a predicative relation "*A* is *B*" – there we can form the corresponding attributive representation: "a *B*," or "something which is *B*." Have we then arrived at something *noticeable* (*Bemerkbaren*)? One merely has to take in thoroughly the sense of the little word "something," with attention to the function which belongs to it, and he will have to acknowledge the abyss which here separates us from "noticing." Only representing, understanding, can give what this "something" means, not intuiting, as of a content (*Inhalt*). The same holds true of the *B*. The predicate is no content, the noticeable, but rather is a *concept*, which is representable (*vorstellbar*) only. If I say, "Something which is red," then what "red" here means cannot be intuited (*anschauen*), but rather only intuitionalized (*veranschaulichen*). But what I 'notice' in the latter case is an individual, an individual *abstractum* within the individual *concretum*, a non-independent Moment of the intuition, enduring so long as the intuition exists, gone when it is gone. If the whole is individual, here and now, so also is each part. No power in the world, no concentration of attentiveness, no ignoring of conjoint [376] Moments, transforms the *part* into a *concept*, the individual into a universal. I may ever so narrowly restrict my interest upon the nose of Mr. *X*. It remains the nose of Mr. *X*, and will itself never become what the expression "nose in general," "the concept *nose*," or the like, means. And what holds true of the part in the sense of a fragment (*Stückes*), holds true also of the part in the sense of an abstract Moment. All that the intuitionalization accomplishes is to present to us something noticeable which (at least in the most favorable case) we can apprehend with Evidence as "object of the concept." I must here be satisfied to give these suggestions, and can only repeatedly express my conviction that one will in all seriousness have to get used to this idea: The Representation is, in contrast to mere noticing, a fundamentally different "mode of being conscious of a content," and indeed precisely that mode which stamps the content as a Representant of an object, which object itself is no part, no fragment, and also no non-independent Moment, of the "representing" act, but rather resides in it only "intentionally"; i.e., it is that mode which only refers to a *function* of the content, a function which spells itself out within certain possible contexts of judgments. (Consider, for example, the identifying judgments: The author of *Faust* is the same as the author of *Götz* – the two "representations" represent the same object.)

In any case, representations in the sense of Representations are the ones which essentially come into consideration in the theory of the judgment, and it was they alone that were always used as an aid to bring to Evidence the peculiar character of the judgment. It was an obvious blunder on the author's part, looking more at the language and at particular confusions than at the overall tendency of the discussions, to constantly impute to a *Hume*, a *James* and a *Brentano* his own concept of representing as *mere encounter* with a content. The greatest part of his critical discussions with these investigators thereby loses its support. Representations in his sense form the basis for the judgment, but the noticing (considered pure of all Representation, e.g., of classificatory ordering) is no judgment, and thus is no perceiving, no knowing. The author himself seems to have sensed the difficulty in simultaneously bringing noticing (*das Bemerken*) and symbolic knowledge under the concept of judgment. It is probably because of that that we have the curious contradiction [377] between Chapter I, where he interprets noticing as infallible knowing (*Wissen*), and Chapter II, where it is supposed to have *no sense* to inquire into the truth of perceptions – on account of which they are not to be reckoned among judgments in the sense of the psychical *function* toward which the question about truth *can* be directed. That in this he also has been led into error by an equivocation in the popular usage of the expression "to have no sense" is clear, as is the fact that this phrase cries out for discussion.

Now the same representation which founds a judgment can also frequently show up without affirmation and denial being referred to its object – whether it shows up in its own right, or as the basis of another act, e.g., in the corresponding question. What is added to it in the judgment cannot be a new representation, which would be linked up with the original one in some way. For to each judgment can be correlated a representation that unites all representations (explicitly) occurring in it, within the same arrangement and unification, but *without* an acknowledging or rejecting of its object. This old and enduring line of argument makes no impression upon the author. In it he misinterprets the "unification of representations" as the formation of a complex phantasm, and so reproaches the investigators named for a confusion which was utterly remote from them. He further seeks to prove the distinction to lie merely in the fact that on the one side a comparison is only intimated or its performance required, but on the other is actually carried out. In the latter case the respective relation is noticed, in the former not. Against this it suffices to object that only where Present (*präsent*) contents are compared could it be considered whether we can speak of a *noticing* of their relation, but not in any other case. Certainly not where, instead of the contents, mere "concepts" of the comparing activity are given. Even where the concepts are accompanied by intuitionalizing phantasma, the situation is no different. One must only take care not to confuse the comparison of phantasma with that of the objects Represented by means of them. And how about when the mathematically untrained person hears a complicated proposition of algebra, understands it, asks about its truth and accepts it on authority – without the least capacity for any actual carrying out of the comparison. Does he then [378]
not judge? Does even the mathematician carry out the *comparison* in every case where he brings up the proposition and makes use of it?

Also in the remaining arguments against the *belief* theory, the author is luckless. "The conviction of the existence of the object not present obviously cannot be attained through a modified grasp of the present phantasm." As if anyone maintained this! As if there ever had been someone who thought we could *arbitrarily* exercise affirmation and denial, and as if the question of how conviction is attained were of any concern to us in the *description* of its peculiar character!

The author even finds a contradiction in the two-fold way in which consciousness relates to a content. If the existence is already given with the representation of the content, then no additional new activity is required in order to apprehend it. As one, with *Brentano* and *Hume*, conceives of noticing as a perceiving whose object is the noticed content itself, it is then only consistent if one permits the two functions of representing and judging – which otherwise do not coincide – to coincide in this case. Now there certainly lies no contradiction in the constant union of two activities within a determinately delimited range. But of course it has not occurred to *Hume* to mean what the author attributes to him: that existence is apprehended not merely with, but rather in virtue of, the representation. He would say: If it subsisted alone, then existence would precisely never be apprehended.

Again, the author invalidly seeks to convict *Brentano* of contradiction and tautology, proceeding from the claim that, according to that investigator, "is" is synonymous with "is true." But *Brentano* never has identified the two. The judgment is true, the object acknowledged in it exists. That is his opinion.

4) Objections can also be raised against the author's remaining theories. In many cases, certainly, I struggle in vain to obtain a clear conception of his view. How, for example, sensation and phantasm are supposed to be *toto genere* different and yet similar, and how he is thinking about the knowledge of their "correlation" – in fact, what he understands by that correlation, and what consequently is the problem it poses to him – this all refuses to come into clear focus for me. He lays great weight upon the equivocation in virtue of which the sensibly perceived object and the phantasm thereof are not [379] differentiated in linguistic expression. But it is wholly unknown to me that, for example, Bismarck himself and some phantasm which someone has of him would be designated the same. When I wish to speak of my phantasm I say, as just now, "my phantasm."

I cannot agree with the author also in other respects where he deals with the function of linguistic symbols. That each word in living discourse is accompanied by phantasma (as symbols of comparison), that where this is not the case the word is meaningless – this is simultaneously a psychological fiction and a logical error.

5) The distinction between memory images and other phantasma is explained by the author, in dependence upon *James*, by means of the *fringes*, which are there in the first case, lacking in the other. But when he interprets the elemental fact here present in such a way that we directly recognize the relation of a just previously perceived content to its present memory image from the way in which the latter influences the present state of consciousness – or, as he says, from a *Gestalt* quality resulting from the past *and* present content of perception – then I do not understand how such things are possible, since the one term of the relation is, precisely, not present. Is direct knowing really a noticing? The author reproaches the *belief* theory in this context with not being able to do justice to the distinction between true and false memory judgments, since also in the latter case the *belief* would be there. A curious objection! Should the *conviction* be lacking when we are convinced? For the author, on the other hand, there truly does lie here a reef upon which *his* theory also founders. If judging = noticing, and noticing is an infallible knowing, through what is the true and the false judgment distinguished? In the case of the memory judgments the author answers: In the one case precisely the *fringes* are noticed, in the other not. But if they are not noticed, then the memory judgment would no doubt be omitted – just *not* made, not made *falsely*! Do we not often enough judge with the most perfect conviction, and yet erroneously, about something past?

6) Concerning negation, the author deals with it only incidentally, in connection with explanatory judgments. It also is to rest only upon a distinction in the content noticed. Instead of similarity a [380] considerable degree of difference is then noticed, and depending on the direction of the questioning (without which the judgment cannot obtain the sense either of affirmation *or* of denial) it is asserted: "*A* is something other than *B*" or "*A* is not the same as *B*," i.e., the relation of the two is something other than the sameness relation. It suffices here to object that considerable difference continuously passes over into similarity – thus denial into affirmation! Beyond this the two assertions are analytically equivalent, and since, according to the author's claims, it is supposed to have no sense to draw a distinction where one side is in every case given with the other, e.g., "*A* is red" and "*A* is other than blue," then we would for the same reason also have to surrender the distinction between affirmation and denial.

B. APPENDICES

[TO *HUSSERLIANA, XXII*]

APPENDIX I

FROM HUSSERL'S SKETCHES FOR HIS REVIEW OF SCHRÖDER[1]

The expression "valid *thinking*" certainly seems hardly suitable, since we cannot very well be referring here to direct cognitions of a correct deduction. Let us nonetheless wait and see how the author defines this concept of *validity*, by means of which the domain of deductive logic is to be delimited. First – under his heading ε), and in a discussion far too wide-ranging – a refutation is sought for the older characterization of the validity of a line of thought as its agreement with itself. Surely it would have been possible to show the utter uselessness of this definition with a few brief comments.

By that "agreement" is understood (and can only be understood) either freedom from contradiction, or else harmony of thinking with the norms of thought. (*Ueberweg*, as is well known, inclines to the latter view, *Drobisch* to the former.) But how then is the cognition of the freedom from contradiction, or of the harmony with the general norms, to establish *itself* as knowledge? And surely it must do that, in either case, in order to be able to serve as a criterion. If it is itself "valid," and thus genuine knowledge, then the consistent answer must read: in turn by means of freedom from self-contradiction or through harmony with those norms. The circularity here is obvious. Mr. *Schröder* is unfamiliar with this approach to the matter. In spite of the wide-ranging resources which he brings to bear – and precisely *because* he goes about it in such a way – he does not succeed with an unconditional and indubitable refutation, as he himself concedes. What, after all, *could* a line of argument prove, that appeals to the possibility of constructing calculi which, while internally consistent,

[1] From *Husserliana XXII*, pp. 381–399. This Appendix contains notes relative to Husserl's review of Schröder, item #5 in this volume. [D.W.]

are nevertheless distinct from the logical calculus (from the "exact" expression of valid thought as such, on the author's view), and which accordingly could be interpreted as the expression of a thinking that is self-consistent but *not* "valid"? (p. 7) Surely arithmetic itself is [382] such a calculus, but is not the thinking in it "valid"? Mr. *Schröder* even maintains the unserviceability of the definition merely within a restricted domain, namely, in that of the "purely deductive processes of thought, whereby reference to new experiences would be excluded from the outset. If, on the other hand, those contradictions are also to be considered which can occur between our thought processes and the testimony of the senses . . . , then the question might present itself differently."

The author holds it to be quite feasible still to posit, "as the criterion of truth for the totality of our beliefs, the universal and contradiction-free agreement with itself of all thinking grounded upon that totality, provided that the agreement also continually holds good for all further expansion of experience . . . " (8) The view implicit in this, that correct judging as such could be confirmed or refuted through experience, seems to me very dubious, above all for a logician. Consider – to leave entirely unmentioned the directly Evident and syllogisms concluded with Evidence – just the case of a legitimate induction. Can it be shown legitimate by means of ever so many experiences, of whatever type, that harmonize with it? Certainly not! The induction could very well be falsely advanced, but the proposition obtained by means of it nevertheless in harmony with experience. Here as elsewhere, judgments derived from false data can be true. And, from the other side, do contradicting experiences suffice in order to show an induction erroneous? That also is not so. Only a contradiction between the experiences upon which the induction was based as its data and its result shows it to be a false one. If the induction was a logical one, then its result could only be a probability judgment; and if the assignment of the probability was correctly derived on the basis of the given data of experience, then the induction could be refuted through no new experiences of any kind whatsoever. New experiences may also, as new data, make possible a new induction – perhaps even one with a contrary probability content. But the old induction remains what it was previously: legitimate. It is an even more severe and more momentous error to allow this *extended* concept of freedom from contradiction to serve as the criterion of truth.

Mr. *Schröder* then decides under point ζ), on page 8, to adopt *Sigwart*'s definition of validity, in which he unfortunately overemphasizes what are precisely the weakest points. He speaks, almost in a remorseful tone, of a "circumstance from which it is absolutely impossible to escape," namely, that "the concept of valid thinking . . . is based upon a dogma which, so to speak, constitutes the faith of the logician." "We may (with *Sigwart*) call thinking valid or logical if, [383] for the critical understanding, it is conjoined with the consciousness . . . of Evidence; if a necessity of thought forces us to carry through that thinking with the conviction of absolute certainty." The "purely personal character," the "subjective moment," in this explanation is then annulled "through the faith that there is a thought-necessity binding upon all intelligences – because it is objectively grounded." (9) These definitions fall under misgivings no less severe than those that earlier served to reject the older logicians. What then, I ask, makes a bit of knowledge into knowledge, according to them? Its Evidence alone is obviously not what does it. Evidence is, in fact, merely something subjective. This subjectivity we are supposed to evade by means of the dogma mentioned. But is this itself a bit of knowledge? One cannot answer with a "Yes"; for in the best case this would lead to an infinite regress when the knowledge criterion is applied. But if one answers "No" – to which modes of expression such as, "faith," "dogma," etc., already point – then one eliminates the merely subjective moment, which is supposed to constitute the alleged lack of Evidence, by reintroducing it again in a stronger form. The subjective Evidence receives its objective grounding through a faith which then itself, as one admitted to be totally groundless, is subjective in the worst sense.

The above definition speaks, in an obscure manner, of an "objectively grounded necessity of thought." Now the author certainly takes adequate pains to bring the corresponding concept to clarity, through its contrast with that of the subjective necessity of thought (11ff.), and not exactly without success! The contrast is identified with the familiar one between psychological and logical necessity. But in the explanation of this latter, recourse is once again had to the concept of knowledge, in the definition of which that of the objective necessity of thought was, however, presupposed. Moreover, we are "certainly never to step outside of the jurisdiction of our subjective necessity of thought" – as if the complex of Moments causally conditioning a judgment, which constitutes its subjective necessity in thought, fell within our consciousness, and as if consequently it were proper to speak of its "jurisdiction."

. . . .

Every pure inference is a hypothetical one. One concept among others is that of non-existence. Under it falls any and everything, provided it is the object of a correct denial. The extension of the concept *non-existence* – and that is precisely the class 0 – is then in no wise itself something non-existing, an "empty" class. Under it falls the celebrated round square and the like. It would be a [384] sophism to conclude: To the concept of the non-existing it belongs, precisely, not to exist; and thus there exists nothing which falls under this concept. All of that which is represented, but to which in truth nothing corresponds, is object of this concept. "Round square," for example, refers symbolically to something which possesses the properties of roundness and squareness. But to this, which is intended as existing, there corresponds nothing in truth; and for that reason it is object to the concept mentioned. And indeed it holds true quite generally that subsumption under a concept does not presuppose the existence of the subsumed: – that there are no triangles in the sense of geometry does not at all prevent their being subsumed under the concept of that figure.

In this way we immediately obtain the correct "interpretation" of the formula $0 = a$, which states that if something falls under the concept of the non-existent, then it also falls under any other concept. And that is correct: If there is no Pegasus, then there also is no Pegasus which would be blue or not blue, or made of gold or not made of gold. The extension of the concept non-existence really does fall, in the sense of the formal theory of inference, within the extension of each single concept (thus there are no "disjunctive" conceptual extensions!). And that is precisely what the above formula states.

On the other hand, it is also false to translate the formula $a = 1$ as "a is everything," whereas its only correct translation reads: "a is something." *Schröder*, to be sure, also gives this interpretation, but without observing its essential distinction from the other.

Our language is, therefore, fully capable of expressing the judgment contents corresponding to those class formulae, and it would also be odd if it were not.

. . . .

After *Schröder* sets forth the distinction of word from name, of single-termed from many-termed name, and of syncategorematic from categorematic expressions (42–46), he proceeds to develop the logical requirements which an adequate symbolism must satisfy. The first requirement to be fulfilled is that of an appropriate abbreviation (44; 0_1). As the second, the symbolism must be rationally composed out of simpler and simplest signs (47; $_1$) This vague specification is only illustrated by means of a reference to descriptions and definitions, without, however, it being explained wherein consists their "rational" mode of composition. In any case, *Schröder* already has that rational mode of composition in mind – that one which the algorithmic sign complexes exhibit. For they are suited, in the most perfect manner, to mirror the thought content which comes into consideration precisely in the application of the algorithm – to mirror it unambiguously, and in faithfulness to its structure (that is, exactly so [385] far as that structure is of interest in the application of the algorithm for the making of judgments). *Schröder* of course makes reference to the "innumerable imperfections of the language of words," because of which it is in many domains frankly an art "to say with certainty of what one really intends to speak." Out of this comes his aim of "working for the perfection of the elementary system of designations for our world of ideas . . . ." I probably do not need to say that also here, once again, there is a huge chasm between what *Schröder* claims to strive toward as a goal, and that which he actually strives for. In his calculus we have to do with, not the construction of a new language, but rather the construction of a logical algorithm. And this latter so little has the task of eliminating the innumerable imperfections of language, that it, instead, already *presupposes prior to its application* a rigorous analysis of the linguistic expressions. Only this much is correct, that so far as its symbols have the character of names, the calculus, to the extent that it is artificially devised, is of course held free from the outset from those imperfections which have been supposed of names and other expressions of language in the course of its natural growth.

The third requirement concerns "the consistency" and "discipline, with which the symbolism is operated" (47; σ_1). The name is supposed to be univocal, "of one sense." *Schröder* surely labors wholly in vain to set in clear concepts the distinction between univocal and equivocal names. He does not get beyond long-winded obscurities, since all of his wide-ranging reflections concerning names, and the associated ones concerning concepts, are permeated with innumerable errors and misunderstandings.

. . . .

In the course of the further discussions of the Second Section of the "Introduction," there first comes the treatment of the proper name (56–62), simultaneously with the distinction of concrete from abstract names, and the distinction of parts from properties. Under the heading "proper names," two types of names are run together, whose separation is not entirely a matter of indifference: namely, singular terms and proper names in the common sense of the word. The explanation which *Schröder* first gives for the phrase "proper name" actually concerns proper names of individual sense perceptible things. But later on (notice, if you will, the examples of abstract proper names explained on p. 58; as, say, in "the color of this corrola") proper names are spoken of in a sense which comprises the [386] totality of singular terms. One cannot help but find this unsuitable. The phrase "singular term" serves for the whole class. The singular term is a name which in truth can be applied only to what is an object, in the sense cogently defined by *Mill.* (p. 10)[2] And whether it is so applied by indirectly designating the particular thing under mediation of reciprocally delimiting general representations and names, or whether it is applied directly under the simple mediation of the (allegedly) full concept of the object itself, is indifferent here. But nevertheless, through the circumstance just mentioned a not unimportant division is effected, in which singular terms break down into two classes: into the proper name, i.e., the non-connotative singular term, and into the singular term with a connotation. Certainly this is a division which rests on the further one between connotative and non-connotative names – a division which *Schröder* (perhaps influenced by *Jevons*' book, *Pure Logic*, p. 6) calls an utterly insignificant

[2] See J. St. Mill, *System der deduktiven und induktiven Logik*, in the Gomperz translation: *Gesammelte Werke*, Vol. II, 2nd edition, Leipzig, 1884, p. 26.

one, since it coincides with that between common names and proper names. "When I carefully consider it, there remains to me nothing of the contrast but the – certainly very important – distinction between the proper name and the . . . common name; the rest dissolves into a haze." It at first seems quite puzzling how *Schröder* – of whom we certainly must assume that he really has read the relevant parts of Mill's *Logic* – could come to claim such a thing, since Mill explicitly gives examples of singular terms that are connotative (I, 17); and to this class there even belong all of *Schröder*'s examples of "abstract proper names" (58). I said here, certainly correctly, "singular terms," and not, like *Schröder*, "proper names." Proper names really are all non-connotative, and generic names all connotative. So *Schröder* may seem to be justified. However, this classification is in fact not exhaustive. It still lacks the singular terms which are not proper names; and precisely these, which *Schröder* collects under his broader term "proper name," are connotative. One does not deviate with impunity from a common terminology. Obviously, as he writes down his claim, *Schröder* lapses back into the old and generally common signification of the phrase "proper name," and hence his attacks upon Mill are highly unsuccessful.

What *Schröder* in other respects alleges in refutation of Mill's division is more than a little flimsy. He imputes to Mill the opinion that a connotative name is in itself meaningless, does not designate, because [387] he falsely translates the word "non-connotative," and in general does not attend to the explanations which Mill provides with all desirable clarity. A proper name is, according to Mill, a name without connotation, and, in this "literal sense," without signification. Whoever reads Mill's explanations cannot possibly find them objectionable. It is not meant that the name designates nothing, that it is only a apparent name, but rather that it connotes no attributes, that the name therefore is not conferred upon the object through the mediation of certain properties, and consequently that the naming of the object by the name can express no knowledge of any value. It is therefore a crude misunderstanding, when *Schröder* objects: "The proper name is so little without signification that it rather, quite to the contrary, is to be described as one which expresses as much as is possible. This is because it designates a completely determinate thing, with all its properties, known as well as unknown . . . . In *Mill*'s sense, all of that, though it belongs to the *notation*, yet does not belong to the *conno-*

tation." What consequences would not follow for *Schröder* himself, if the two were not held in distinction. Not merely of proper names, but rather of all concrete names – thus also of the concretely general – it would hold true that they designate a determinate thing with all properties, known as well as unknown (cf. the explanation of the concept of the concrete name on p. 57); and if that designation were also its signification, then all of the infinitely many names which we can attach to the same thing, would also really be synonymous. The distinction "over which *Mill* makes such a fuss" surely is also *worth* the fuss. Hence *Schröder* also will have to discover, with reference to it, that science correspondingly maintains *Mill*'s "scholarly sounding" and "pompous" denominations as a requirement, since they do serve essential distinctions.

Having considered *Schröder*'s characteristic statements about common names, and the related statements about concepts, we must also subject the distinction between concreta and abstracta, interwoven with the treatment of proper names, to a brief discussion. "A thing is said to be concrete if it, on the one hand, is thinkable in complete isolation, and, on the other hand, is intended or is to be taken with all of its properties (parts, attributes and relationships)." The name of such a thing is a concrete name. I am unable to find this definition to be of value. If we are to take an object with all its properties, then we must either be in possession of those properties, or must have the capability of reaching them. But neither the one nor the other is correct. It is not correct that when we speak of the terrestrial ball we *mean* all of "what is in or around it." It is first through the activity of knowing that we acquire a – moreover, vanishingly small – part of those valid predicates which gain our interest, while innumerable others remain forever hidden to us. Accordingly, we also cannot consent to the way *Schröder* introduces the concept of the property and its specifications "part" and "abstractum." [388] "Out of the representation of a concrete thing we are able also to single out certain elements and to isolate them more or less completely in our minds .... Such partial representations, in the broadest of senses, or else the conceptualized reality underlying them, we designate as the *properties* of the object." But this view of the conceptual property marks every categorical judgment as an analytical judgment in *Kant*'s narrowest sense. Wherever we analytically extract a property from an objective concept which we ourselves possess, there we may speak of an "isolating." But wherever to an objective concept there is linked a relational determination *not* contained within it, or, more precisely, where such is acknowledged of it (that requiring, under certain circumstances, the most highly complicated of inferences) – there we do not arrive at a property of the object by means of analysis of it. More appropriate is another definition, which *Schröder* himself occasionally mentions (336). According to it, all that can be truly asserted of a thing is to be designated as a property of it. If that "isolation" of an element from the concrete thing fully succeeds, then the property is a part. If it does not fully succeed, it is an abstractum.

That the differentiation of the abstractum from the concretum is not practicable in every case is something which *Schröder* himself senses; and he finds himself forced to speak – with regard to examples such as concrete geometrical figures, and the geometrical abstracta position, form, distance, etc. – of abstract and concrete to the first and to the "second power." All difficulty is eluded, it seems to me, by conceiving of "concrete" and "abstract" as correlatives, in such a way that not the concretum, but rather the abstractum is taken as starting point for the interpretation.

. . . .

The aversion of many logicians to the *Euler* diagrams indeed originated from very respectable motives. The obscure utilization of them, half as proof, half as intuitionalization (*Versinnlichung*), which for so long was customary, must seem flatly repugnant to any friend of pure method, to any logician intent upon a deeply penetrating scientific understanding of the mind's deductive processes. Justified objections might also be raised against them from the pedagogical point of view. The acuteness and freedom of logical thinking can only be learned by means of a fullest possible exercise with concepts themselves, by means of genuine inferences; and it must be characterized as a hazardous, indeed ruinous, practice, if the symbolic, blind, technical schemata are offered to the student right from the outset as a crib, and thus mental laziness is put to rest in the lap of idleness. However, these artificial tools are, certainly in the domain [389]
of the common "simple" syllogisms, not wholly without value as reliable methods of proof for the valid moods. (Of course one must also understand what it is about the diagrams that provides proof.) It

still remains a matter of some interest that technical methods of proof are possible in this domain. But they are of no practical importance worthy of note. Just as little as all of the specific rules of inference in general; since for the thinking brain subsumption under the rule or the construction of appropriate schemata will take up much more time than the direct deduction.

....

I just now said: As a rule, an algorithmic procedure is not to be accompanied by insight into its justification. However paradoxical this claim is, and however obviously correct the directly opposed precept may seem at first glance, I believe it yet possible easily to demonstrate and verify that the whole advantage of the algorithm rests exactly upon the dispensibility of the knowledge of its logical force during its application. In fact, whoever has at any time made the attempt to think through the exceedingly complicated ideas upon which the logical conclusiveness of an algorithm rests, will immediately agree that the algorithmic procedure never could become a mechanical one, and therefore never could fulfil its purposes, if that complex of substantiating ideas had to accompany us in each step. It suffices that we have once familiarized ourselves with the theory back of the practice, or otherwise secured for ourselves the verified conviction of its soundness, i.e., thus of the fitness of the algorithm to be substituted for those domains of relations which it is supposed to serve. If this has taken place, then our sole aim is the practical mastery of the algorithm and the mechanization of the deductions to which it corresponds. However, this is only incidental. Let us again turn back to the question about the relationship of language and algorithm. *Schröder* speaks as if, with the introduction of a symbolic language in place of the natural language of words, the algorithmic procedure were already given. It is this same confusion which we already find in *Leibniz* – and in fact even, after a fashion, in *Hobbes*. Thinking is a type of calculating, the eccentric philosopher last-named supposes. Expressions in terms of symbols are essential to thinking. Judging is realized with and in linguistic signs. But language utilizes signs in many ways, as the special language of arithmetic utilizes the number signs. And in both cases the combination of the signs corresponds to the combination of thoughts – or rather the two are one, as, in the particular thought, external sign and interior correlate are one. Thus, for him, speaking and calculating coincide. *Leibniz* has run the two [390] together in a somewhat different way. There are two ideas which move in confusion through his celebrated attempts at a *characteristica universalis*. The various names which he uses for it already point up this fact. Besides the one mentioned there also occur the names *speciosa generalis*, a name imitating the *arithmetica speciosa* of *Vieta*. Further, *mathesis generalis*, *calculus ratiocinator* and so forth, *lingua adamica*.

On the one hand, the *characteristica universalis*, through analysis of all concepts into their ultimate, irreducible elementary concepts, and through the introduction of signs corresponding to them (simple signs for the simple concepts, and, for the composite characteristics, signs constructed from the simple signs), is supposed to become a language of characteristics. In such a language the sign (insofar as it is not a sign of something simple) is a faithful image of the fact. (Taken strictly, of course, only of our concepts; but of course there is operative here the thought that thereby could be disclosed the inner essence of the thing as well.) Through simple combinations of the simple signs we gain the further possibility of being able to acquire all conceivable concepts which are accessible through our mental faculties.

From this Leibniz in turn constructs the Idea that all conceivable necessary judgments about things (as conceptual objects) would have to be discoverable through a mere comparison of the signs. In this he certainly has been guided by the prejudice that all necessary judgments are analytical, whereas he overlooks the fact that there are innumerable necessary judgments which are not to be derived from the composition of the signs, even under the presupposition of the realized ideal of a perfect notation, ... The analysis of the concept of the straight line into its ultimate constitutive properties would, in fact, never at any point ground the proposition, which nevertheless is a necessary one, that two straight lines can have only one or no point in common. The analysis and character designation of concepts like red and green would never at any time yield the proposition that a red point can never be simultaneously red and green, and the like. But that is the Idea which the names *characteristica universalis*, etc., express.

However, quite another thought is implicit in the expressions *math-*

esis universalis, etc. *Leibniz* holds it to be obvious that with the completed analysis and character notation of concepts – and then also with a *calculus ratiocinator* based thereon – a deductive algorithm would have to be given. But what justifies him in this? In a biased fashion he lets himself be guided by the analogy with mathematics, and in addition to that by an imperfect and inappropriate analogy. It is highly regrettable that *Leibniz*, such a great mathematician and keen thinker, neglected to bestow upon the logic of that science a little of his astonishing capacity of work. For him as for *Descartes*, *Hobbes*, *Spinoza* (in fact, for this whole time-period), the model of mathematics is the one also methodologically decisive in philosophy. But no one attempted to ground the logic of that discipline more deeply, [391] and so all are guided only by vague, superficial and – when closely examined – false analogies. *Leibniz*'s greatest discoveries depended upon the introduction of algorithmic methods into analysis. Not so very long before him, general arithmetic had been founded by the introduction of general signs into arithmetic, and geometry had been guided into new pathways by *Descartes* – once again through algorithmic methods, through the introduction of general arithmetic as a tool of geometrical analysis. In every case the essentials of the new methods appear right off to lie in the proper analysis of the basic concepts of the discipline concerned, and then in adequate symbolization of them. This is most clear in the case of ordinary arithmetic. The numerical symbol system was a logical notation for the number concepts, on the basis of a logical analysis of them. In all of these cases there is immediately founded in the characterized concepts numerous operations of combination and transformation, and therewith rules of mechanical calculation. Algorithms result.

However, if one looks more closely, it turns out that that analysis which in the mathematical disciplines led to a *calculus ratiocinator* is removed as wide as the heavens from the analysis which *Leibniz* had in mind with his character notation for concepts. Taking the case of the numbers, it is not analysis into ultimate properties which underlies the notation, but rather an artificial, systematic analysis into natural parts, into partial numbers, which is not pre-designated from the outset with the concept of number, but is led into determinate pathways by peculiar logical interests. We lack authentic number concepts for the most part, and have to operate with symbolic representations. But as soon as we do so, as soon as we indirectly describe the concepts which we intend, it immediately becomes clear that a whole series of equivalent formations or symbolizations are possible – as, for example, 8 + 3, 7 + 4, 6 + 5, and so forth. The decadel number formations are, then, a matter of a selection according to a uniform, systematic principle. From all of the modes of formation for a number (the diversity of possibilities indeed increases more and more with the magnitude of the number) a single one is selected, according to a simple, clear and systematic principle; and thus the series of symbolic number concepts is simultaneously formed and ordered in terms of a classification. But here are quite different interests than that of an analysis of the concepts into characteristics. And if, on the basis of the modes of formation thus ordered, a calculus arises, that of the *arithmetica numerosa*, then the inner grounds for that lie in their systematic classification within a progressive series in conformity with a synoptical principle of classification, that of the decadel number series; and something similar can be identified everywhere that calculi show up. Not every analysis, with the symbolization of concepts reflecting it, yields a calculus, but rather only certain ones, conditional upon their inner systematic structure. To the systematic structure of the concepts there corresponds the systematic structure of the signs; and upon this latter, in turn, there rests the possibility of uniform paths of deduction, which are reflected in uniform sign combinations, and thereby permit a blind calculative procedure. So matters stood at least initially. I do not mean to claim that the parallelism between the inner and the outer systematic structure really must be complete. But it would lead too far afield to discuss the meaning and importance of this observation. If, now, such points are attended to, one will then find it understandable when I consider the emergence of a *calculus ratiocinator* from the *characteristica universalis* as so little obvious that, to the contrary, I believe that, if it emerges, this would be something quite odd, most noteworthy – resting upon relationships of the ultimate conceptual elements which, at least in the beginning, are by no means to be overlooked. It is truly astonishing that *Leibniz* himself missed this, since he surely was already on the path most favorable for the discovery of the logical calculus, the elementary propositions and basic operations of which he almost completely established – notwithstanding that the *characteristica universalis* as a characteristic still did not even exist. In the bits and pieces relevant to the characteristic he in fact operates with algorithmic, but not characteristic, symbols; and hence it surely was obvious to observe that a *characteristica universalis* and a *calculus ratiocinator* rest upon different principles.

If we find that in *Leibniz* the Idea of a universal language (in him, specifically in the form of a *characteristica universalis*) is not sharply distinguished from a deductive calculus, this all the more holds true of the period following. In the conversion of given arithmetical problems into the formulae of the calculus, mathematicians like to speak of a translation or rewording of them in the "language" of the calculus. *Newton* already directly speaks of a *traductio ex lingua latina* into the *lingua algebraica* – a mode of expression which also is so widespread in our time that one can scarcely escape it. But the blending of language and calculus (similarly as already in *Hobbes* earlier) was driven to the limit by *Condillac*, who was in a certain manner a kindred mind to *Newton*. In saying this I am not perhaps denying that from a certain viewpoint algorithmic symbols possess the character of linguistic signs, whether syncategorematic or categorematic. Indeed, the strange thing is that they even possess it in several senses. Consider general arithmetic, for example. The symbols *a*, *b*, *c* all have the signification of symbolically represented objects of the concept number. *a* designates the same as "a number," and likewise *b*, *c*, etc. Under the number concept there fall all of the general concepts contained in the series of positive numbers, and it can be any one of these which is meant – and indeed not perhaps as an abstraction, but rather as a general concept. That is, *a* can, as "a number," signify either "a three," "a four," etc. Likewise for *b*, *c*, so that, of course, *a* and *b* also are possibly symbols for the same conceptual objects. Add to this syncategorematic symbols, [393]
$\pm$, :, $\times$, $\gtreqless$, which, when set between the number symbols, indicate summative or subductive combinations and so forth, or the relations of equality, etc. But if the symbols mentioned were really linguistic signs, then they would always and everywhere have the function of the mere expression of the intended objects or relations. But this is absolutely not the case. In the course of the calculation the symbols do not serve at all in the expression of concepts, and whoever considered them as doing so, and always burdened them with the concepts, would precisely not calculate. But I do not wish with this to claim that, in the process of calculation, the symbols are *totally* blind and arbitrary. To the contrary, they are determinate symbols, symbols of

signification. But the signification has now become a wholly different one. The symbols of the calculus, or the arithmetical signs as algorithmic symbols, have their signification (connotation in *Mill's* sense) exclusively in the rules of combination, dissociation, transformation – in short, in the rules of operation – which, in their totality, alone make the algorithm into an algorithm. The capacity to be subject to these combinations under these, and to other combinations again under other rules – the permission to replace combinations of a determinate type with combinations of some other determinate type, and the prohibition against making combinations and transformations which contradict the rules: – therein resides the general signification which gives the symbols their general signification. Now add to that, further, their special signification, which arises out of given restrictions to which given signs are subjected in consequence of given formulae – as, for example, with the signs of an equation expressing the algorithmic content of a problem. From all of this is composed that signification of the symbols which we can term "algorithmic." Here, therefore, the signification is constituted exclusively of wholly external formulae (for the combination, dissociation, re-arrangement, equivalence substitution) that are collectively based upon the external conventions for the symbols. The dissociation from the conceptual signification of the first type is a complete one. So much so that the same algorithm and the same formulae often can serve totally distinct conceptual domains, can solve totally distinct problems of knowledge, so far as the objective relationships of the one admit of an algorithmic treatment – and consequently possess an algorithmic content – formally identical with that of the other conceptual domain.

Finally, let us consider the application of the algorithm to determinate problems about numbers. We coordinate with the determinate number signs indeterminate general symbols, carry out the algebraic calculation, substitute the specific number signs back into the result, and the problem is solved. Through this coordination the general number symbols acquire a new, third signification, consisting in the coordinated determinate numbers – a signification, but one which, again, does not give the symbols the sense of linguistic signs, since it is permitted to fall away and, during the calculation, is replaced [394] by the second signification, in order to be resumed again only at the conclusion.

I cannot here go further into the theory of the algorithm. But one will understand if I now state: Linguistic signs have the function of sensory means of expression constantly accompanying thoughts. Algorithmic signs have the function of sensory symbols replacing and sparing thoughts, at least temporarily – that is, for the purpose of a symbolic process of deduction. In linguistic thinking the representing and the judging activity derives from the thing itself (*an der Sache selbst*), from the concepts and judgments. To the transformation in genuine thinking the transformation of linguistic expressions merely runs parallel, possibly enriching it, offering it a support. In algorithmic thinking the logical activity derives entirely from the sensory symbols. Or to the contrary, it does not occur at all; but instead of it another one does. It is replaced, in terms of an explicit translation, by a sensory activity following rigorous rules for symbols. Following the mechanism of the symbols, we obtain a result which, transformed into thoughts ("interpreted"), leads to a judgment that states the truth sought. An algorithm is, therefore, a systematic method of symbolic inference; a language, a systematic method of expression of thoughts and psychical phenomena in general by means of signs. How it is possible that a blind mechanism of sensible symbols can replace and spare us logical thinking? That is the great problem for the logic of signs, for semiotics. Upon its correct answer depends the understanding of arithmetic, of analytic geometry, of the theory of magnitudes (*Ausdehnungslehre*), of the logical calculus, and in short, of all disciplines utilizing algorithmic methods. A deductive logic without a semiotic appears to me not to deserve its name at all. The reproach which this involves falls upon the whole of logic, of the present and the past; but in a heavier degree upon symbolic logic, since it itself regards the development of symbolic methods as its chief aim.

. . . .

Schröder has in several passages correctly characterized the mission of the symbols 0 and 1. They alone make possible a serviceable calculus that permits of unrestricted blind calculation following fixed rules. Unrestricted calculation requires, precisely, operations which can be carried out without restriction. But the product of two disjunctive manifolds is meaningless, and thus multiplication is a restricted operation. In the application of the rules of multiplication we would accordingly be disposed in every case first to check whether multi-

plication can be carried out, or we would always have to drag along [395]
the validity condition, which would constantly interrupt the course of pure calculating. The 0 makes multiplication into an operation which can always be carried out, at least symbolically. Even if the factors are disjunct, the product nevertheless has a value, the value 0. It is quite the same as in arithmetic, where the quasi-number 0 and the so-called negative numbers have the function of subtraction, as the "imaginary numbers" have the function of root extraction for negative numbers, and so forth, to form operations which can be unconditionally carried out. This mission is of the highest importance; but it still does not ground the validity of the introduction of these symbols, either in arithmetic or in the calculus.

. . . .

In the fifth lecture, now, the algorithmic consequences of the definitions and propositions advanced are drawn, and there is derived a whole series of laws for multiplication and addition – and, to be sure, partly laws that control *any* case of these operations ("pure laws"), and partly those which are valid where the two operations are mixed together ("mixed laws," p. 270). The exposition here is of the highest quality and is complete; but there is also here from time to time something of that zeal for *proving*, which will not be content with the directly Evident and believes it possible to impart to it a higher degree of certainty by means of a derivation within a symbolic calculus. Of such a kind are, for example, the evident laws: $a \cdot b = b \cdot a$, $a + b = b + a$, $a \cdot a = a$, and the like. The development finally culminates in the "provable subsumption of the laws of distribution": $ab + ac = (= a\ (b + c)$ (I omit to write out the corresponding dual); while the sixth lecture (282–298) is specifically devoted precisely to the question of the non-provability of the second subsumption: $a(b + c) = (= ab + ac$. The two taken together are identical with this major law of the logical calculus itself: namely, $a(b + c) = ab + ac$. In the attempt to prove this law in the purely formal way, exclusively on the basis of the two "principles" and the "definitions" of equality, of 0 and 1, of $a \cdot b$ and $a + b$, the author attains to the insight that the proof is impossible, and can be carried out only for the one part of the law. Thus it is not a matter of non-provability absolutely, but rather only of the demonstration of non-provability understood

in these terms. The author also really is completely successful in this demonstration; since he displays an algorithm for which all of the propositions mentioned have validity, regarded as mere formulae, and not as propositions of the calculus of sets, while the second part of the law of distribution certainly is not valid. This latter is therefore impossible as a mere formal consequence of the principles presupposed in the algorithm. That conceded, I still cannot find anything surprising in this situation. It certainly is a very praiseworthy [396]
aspiration that *Schröder*, with *Peirce*, wishes to support himself on a smallest possible number of basic formulae. But that we could limit ourselves from the outset to only the ones mentioned above, was surely not to be expected. The correct method for the construction of a deductive calculus is to set out at the first from the greatest number of immediately Evident propositions possible, in order thereby to secure a broadest possible basis. But in the systematic development there remains, as I have already indicated, a great interest in making use only of the unconditionally necessary propositions, those irreducible with respect to the others, and therewith so to fashion the course of the derivation that even the consequences of the individual "principles," and of their different combinations, are held distinct from each other so far as possible. Not as if this had the least significance for the usefulness of the calculus – or for the domain itself, which is to be governed by it. The interest in this procedure is, instead, one external to that domain, but yet an eminently fruitful one. Innumerable times in the algorithmic disciplines we have had the experience that the same algorithm governs different domains, or that at least parts of the algorithm comes into play for this or that domain – parts, which are deductive consequences exclusively of sub-sets of the "principles." Through that procedure is thus made possible an extraordinary ease of transfer of an algorithm to domains which are determined by analogous relationships. If one discerns the sameness of the few principles, then with one stroke the entire calculus is carried over; and if one discerns the sameness of a subset of the principles, then at least the part of the calculus dependent upon that sub-set (in case the exposition of the calculus places that pure division clearly in view) is immediately transferable. So also in our calculus, the advantage of *Peirce*'s procedure, expounded by *Schröder*, of limiting ourselves at the outset to those first principles mentioned, is obvious. Only in this way does *Schröder* succeed

with his transition to the "calculus of algorithms." In order, then, to continue with the development of the calculus, *Schröder* adopts the procedure of putting forth the second, non-provable part of the distribution law under a trivial restriction, (bc = 0), as a new axiom (Principle III, p. 293). I find this absolutely inappropriate. The proposition surely is no immediate axiom, and there must be simpler Evidences which it refers back to. Whether this will succeed if we leave completely aside "argumentation from individuals" (the class members), and work exclusively with relationships of classes, as *Schröder* would prefer it, surely is most dubious.

An important thesis which *Schröder* ties in here is the following: Through this proof of non-provability (proof of the partial subsumption in the law of distribution mentioned), "it is made clear that instead of the one, there really exists two calculi, of such a kind that in the one both parts, and in the other only the one part, of the law of distribution unconditionally holds good." For the first calculus [397] *Schröder* retains the name "identity calculus," and for the second he introduces that of the "logical calculus."

Also here, as in almost all fundamental formulations, I must diverge from *Schröder*. If the identity calculus really is a calculus of manifolds, as which it certainly was defined, then there is only one calculus which has the full distributive law. The class calculus, which *Schröder* brings forth as an example of the first type, is indeed a calculus which covers manifolds, but is not one of manifolds *as such*. *Schröder* has confused these two things. So far as formal systems are manifolds, there obtains between them only relationships like those of factual inclusion, exclusion, etc. But the relation of logical deduction of one formal system out of another is no manifold relationship as such, and therefore the class calculus is no calculus of manifolds.

....

The eighth lecture, which deals with the use and interpretation of negation for conceptual extensions, once again gives occasion for a few observations. Since the 1 enters into the definition of negation, the signification of negation acquires the restriction which *Schröder* held to be indispensible for the concept of the 1. If the 1 must not mean the entire totality of all that is conceivable – thus, the extension of the concept *something* – then, in pursuance of the formula, $a + a_1 = 1a_1$, plainly the "negation" of a cannot be the extension of the concept non-(a), whereby we with (a) designate the content of the concept which has a as its extension. For it is certain that the extensions of the concepts (a) and non-(a) add up to the entire totality of all that is conceivable. Accordingly the formulae $aa_1 = 0$ and $a + a_1 = 1$ also are not the class equivalents of the principle of contradiction and of excluded third; whereas *Schröder*, in opposition to this (and in that erroneous way of speaking which we have repeatedly censured), even says as touching the former, for example, that it appears as "the most concise expression for the principle of contradiction in the old logic first . . . upon being restricted to classes and extensions of concepts" (346), and similarly with respect to the principle of excluded third. In fact, *Schröder* even supposes that the curtailment of the concept of the 1 is essential to the latter principle: that it would under certain circumstances lead to falsehood if it were rigorously maintained that any and every thing must be either A or non-A. It may be true that each sheep is either black or not black. "On the other hand it still does not become true that every species of sheep is either black or not black, since such a species may also contain, besides white sheep, black ones also." I am unable to convince myself of the soundness [398 of this logical discovery. What does it mean to say, for example, that a species of sheep is black? If it means that it contains black sheep, then the principle states: Every species of sheep contains black sheep or it does not contain black sheep. If it means that it contains black sheep exclusively, then again it claims: Every species of sheep is either black or not black. So I fear that the logic of the Schröderian discovery – that the principle of excluded third has validity only in the "original" manifold, and not in the "manifold derived" from it – will be of no use.

For the present, therefore, the old principle will retain its validity – also still over the old range. For the logical calculus the restriction on the 1 is, as we have shown above, not necessary. Consequently the definition of negation can also be deemed unrestricted. But this definition then essentially consists in the equivalence, for classes, of the full and genuine principle of contradiction and excluded third.

For the rest, from here on the discussions of matters of principle become more and more infrequent, and therewith the reading of the book more and more gratifying. The almost unbearable prolixity and the torturous progression of the development, interrupted by continual digressions, takes a more lively and more interesting turn. The technical or, as I could more aptly put it, mathematical exposition is exemplary, and, in contrast to the expositions up to this point, rich in fruitful simplifications and elegant developments. For those familiar with the Boolean calculus, especially in *Venn*'s fine exposition, I call attention to the justification, found on p. 368 of the ninth lecture, of the concept of "identical" addition, introduced by *Schröder* in opposition to *Boole*'s "exclusive" addition. *Venn*'s objections are here refuted in a completely convincing fashion. The restriction of the addition concept to classes which are mutually exclusive is, for *Boole*, an essential element of the system. It alone made possible, as I may dare to claim, that ingeneous transference of the arithmetical algorithm over to the domain of classes, through which the class calculus stood forth at one stroke, almost as founded on a miracle. Hence, for those familiar with the arithmetical calculus, the Boolean calculus can be acquired with the greatest rapidity; and if its methods were also most concise, then I would give preference to it in spite of the many reproaches which that transference has brought forth against it. I have not the slightest doubt that the Boolean method admits of a full logical justification. That the transference of the arithmetical algorithm, restricted to 0 and 1, to another domain signifies a confusion of the arithmetical with the non-arithmetical (specifically, the logical), is, as can be rigorously proven, a false conclusion. Also, that the Boolean method so frequently utilizes senseless symbols does not yet in itself serve as the basis for a logical objection. We can only object, rather, that that method does not adequately *justify* the use [399] of such symbols. But in this regard general arithmetic is subject to exactly the same objection. In any case, a construction of the calculus which at every step is grounded upon interpretable, and thus conceptually founded, symbols, circumvents all obstacles. And since, besides this, *Schröder*'s construction retains the nucleus of the elegant and mechanical methods of *Boole* – in fact even finds ways of considerably simplifying them – his construction at present deserves the decisive preference. But, with that, also preference for the inclusive addition, so much better adjusted to natural thinking, over the Boolean exclusive addition.

APPENDIX II

MEMORANDUM OF A VERBAL COMMUNICATION FROM ZERMELO TO HUSSERL[1]

Zermelo comments (April 16, 1902) with regard to page 84 of my review of Schröder:

In fact, Schröder is incorrect in his reasoning. That is:

A set M which contains each of its sub-sets m, m', ... *as elements* is an inconsistent set; i.e., one such that, if it is really treated as a set, leads to contradictions.

Proof:

We consider those subsets m which do not contain themselves as elements.

[M contains as *elements* each of its sub-sets. Thus sub-sets of M will also contain certain sub-sets, but not themselves, as elements; and now consider precisely such sub-sets m, which perhaps contain other sub-sets, but not themselves, as elements.]

These form in their totality a set M_0 [*viz.* the set of all sub-sets of M which do not contain themselves as elements]. And now I prove of M_0: 1) that it does not contain itself as element, 2) that it *does* contain itself as element.

ad 1): M_0 is, as sub-set of M, itself element of M, but *not element of* M_0. For otherwise M_0 would contain as element a sub-set of M (namely M_0 itself) which contains itself as element, and that would contradict the concept of M_0.

ad 2): Thus M_0 itself is a sub-set of M which does not contain itself as element. It would thus have to be element of M_0.

A set of definition M is of course the set of all sets. This also offers an illustration for the fact that a set can contain itself as element: the set of all sets is indeed a set.

[1] From *Husserliana XXII*, p. 399.

LOGICAL CALCULUS AND THE LOGIC OF CONTENTS[1]

The entire calculus of extensional logic (the calculus of classes) can be simply carried over to the "logic of content." Over against the calculus of classes I place a *calculus of conceptual objects*.

I first observe that concepts as complexes of properties (property as partial concept, whether one separately and specifically noticed, whether to be brought out by analysis – but in any case actually contained within the whole) fall, in terms of their reciprocal relationships, wholly within the framework of the calculus of classes. (*Schröder* calls it the calculus of domains.) The name "calculus of classes" is in fact not at all appropriate. It should better be called "calculus of manifolds." And those complexes are also manifolds. They represent totalities of properties whose interrelationships are of no further significance. Rather, all that matters is the relationship of the totalities as such. We find then once again the five well-known content relationships, which of course are also to be made intuitive by pairs of circles.

Judgments about content relationships – as, in general, about the concepts themselves – are uncommon, and likewise for judgments about classes. The latter may even occur more frequently. As a rule we do not make judgments about concepts and classes, about conceptual contents and the extensions of concepts, but rather about conceptual objects.

In any case, from this point of view we can only hold that:

All men are mortal.

[1] From *Husserliana XXII*, pp. 400–406. These are Husserl's working notes related to his 1891 article, "The Deductive Calculus and the Logic of Contents," translated above. [D.W.]

= Man is mortal.

= If something falls under the concept *man* (possesses the properties connoted by the name "man"), it also falls under the concept *mortality*.

The five relationships which pairs of conceptual objects can present are:

I. The concepts are compatible.

1) The object of the concept A (as such) is also object of the concept B. Being-A conditions being-B. If it is correctly asserted of something that it is A, then it is also correctly asserted that it is B. Consequently being-B is a "property" of being-A, the word taken in the sense in which it is understood, for example, in the *nota notae* principle. The converse relationship is *excluded*.

2) The converse relationship: Being-B conditions being-A, but not conversely.

3) Mutual conditioning; logical equivalence of the conceptual objects *in abstracto*.

4) No conditioning relationship.

II. The concepts are incompatible.

5) If something falls under the concept A, it does not fall under the concept B, and if it falls under the concept B, it [401]
does not fall under the concept A. A and B are mutually exclusive.

Symbolism

a = (= b

Whatever falls under the concept A also falls under the concept B. Gold is yellow.

Being-A conditions being-B.

With this is left open whether or not at the same time

b = (= a.

If a = (= b, but *not* b = (= a, then we write:

$$a (b.$$

If

$$\left.\begin{array}{l} a (b \\ b (a \end{array}\right\} \quad \text{then} \quad a = b$$

Definition of the identity – or, better, logical equivalence – of the conceptual objects.

Schröder's Principle I: $a = (= a$

Principle II: If $\left.\begin{array}{l} a = (= b \\ b = (= c \end{array}\right\} \quad \text{then} \quad a = (= c$

Definition of the Sum a + b

a + b = whatever "is either a or b," or is object of the concept *a or b*.

$$\text{If } \left.\begin{array}{l} a = (= c \\ b = (= c \end{array}\right\} \quad \text{then} \quad a + b = (= c$$

Gold is yellow
Brass is yellow

"Gold or Brass" (in general, inasmuch as something falls under this concept, it) is yellow.

Every virtuous man is praiseworthy.
Every wise man is praiseworthy.

Every man who is either virtuous or wise is praiseworthy.

Definition of the Product a · b

a · b = Whatever is a and b (summation of properties).

$$\text{If } \left.\begin{array}{l} c = (= a \\ c = (= b \end{array}\right\} \quad \text{then} \quad c = (= ab$$

Every perfect man is virtuous
Every perfect man is wise.

Every perfect man is virtuous and at the same time wise. [402]

Definition of the Null

The object of the concept of non-existence is called the *null*, something of which it is true that it is not, in the sense of truth. Accordingly, for any *a* it holds:

$$0 = (= a$$

Whatever has the property of non-existence also has any other arbitrary property.

= For something of which it can be asserted that it is not, it also may be asserted that it is *a*, whatever *a* may be.

Definition of the 1

The object of the concept of existence is called the 1.

Accordingly, for any *a* it holds:

$$a = (= 1$$

Whatever falls under the concept *a* also falls under the concept of the existing.

Of whatever it can be asserted that it is an *a* (whatever has the property *a*), it also may be asserted that it is something existing.

Negation

a_1 signifies non-a: that is, something to which the property belongs, or of which it can correctly be asserted, that it is not-a. (The negative judgment here functions as a relation, and grounds the relative determination, not-a.)

$$a \cdot a_1 = 0 \qquad a + a_1 = 1$$

$a \cdot a_1 = 0$. That is, whatever is simultaneously *a* and not-*a* does not exist; and whatever does not exist is simultaneously *a* and not-*a*.

In fact, A = 0 signifies: Whatever falls under the concept A, falls under the concept of the not-existing; and therefore A is a non-existent, which is equivalent with *A does not exist.*

$a + a_1 = 1$. That is, whatever is either *a* or non-*a* is something; and conversely: whatever is something (any arbitrary thing, anything we can make judgments about) is either *a* or not-*a*.

The distributive law is in its first part, as *Schröder* demonstrates, a mere consequence of what has been assumed to this point (negation excepted).

As to the second part:

a (b + c) = (= ab + ac,

Jevons has proven this, adapting it directly to conceptual objects.

By means of a line of argument which is immediately obvious:

Whatever is *a* and either *b* or *c* – that is, if it is *b*, it is ab, and if it is *c*, it is ac – is also either ab or ac. [403]

Question.

For the concept of the not-existing we have the symbol null. Why not also a symbol for the concept of existence?

1 is in fact not the sign for existence, since "0 = (= 1" holds. [Whatever falls under the concept of the non-existent – thus, what does not exist – also falls under the concept of the existing?!]

Particular Judgments:

Some M are P

We think the concept M as supplemented or delimited in such a way that it includes the M's which are P, and these alone. The unknown properties which must supervene, and indeed obviously multiplicatively, we call *v*.

v · M = (= P.

Inference: Some M are P*v* M = (=P
All P are Q P = (= Q
*v*M = (= Q.

The case enumerated as I. sub 4) above was "no conditioning relationship." That was obscurely expressed.

If we consider a pair of objects of the concepts *a* and *b*, then it can be that neither the relationship, *if* something is an *a*, it is a *b*, obtains, nor the converse. But instead we can have the categorical judgment, An *a* is a *b*; or else, A *b* is an *a*.

There exists here a conditionality, but it is indeterminately expressed. Through the determinative factor *v*, the particular judgment (just as above, where the indeterminacy was linguistically expressed by "some") is transformed into a hypothetical-general one.

It probably is not appropriate to speak of "relationships" which "any two conceptual objects" can exhibit, but rather of *relationships of any two concepts with regard to the objects which fall under them*, or of relationships of dependence of properties, conceptual determinations as such, or of relationships of concrete concepts.

In the class calculus, the cases where the class has one or no object constitute boundary cases. In the property calculus these are no boundary cases. Even if we have an individual concept designated by means of a proper name, the concept of the conceptual object still finds its appointed use. Something which is Socrates, i.e., something which falls under the concept *Socrates*. Something which is a round square?

In the case of the individual representation not mediated by general properties, the individual representation conforms to the fact (*der Sache*). The representation is the concept *of* that fact.

In the case of a general representation, mediated by general properties, the representation corresponds to something which concretely [404] possesses these properties.

I. Extensional Interpretation

All A are B = Class A (the totality of the A's)
= (= Class B (the totality of the B's).

II. Content Interpretation
a). *Hypothetical Interpretation.*
If something is an A (has the properties A), it is a B (has the properties B).

The properties A (as such, that is, when they occur in an object) condition the properties B.

A = (= B.

b). *Negative Interpretation.*
It is not the case that something is an A and not a B.

Since deductions are easier with hypothetical judgments than with these negative categorical ones, we can – in virtue of the indubitable equivalence of the two – base the calculus on the former.

III. Mixed Interpretation

It seems that not both the subject *and* the predicate are quantified, but rather only the subject or only the predicate.

All men are mortal. According to the wording we have here a certain relation between the totality of men and the attribute of mortality. There is first formed the class concept of the totality of men, and then there is brought to it the thought: Inasmuch as something belongs to this totality, it has the attribute of mortality (Or the negative: There is none from this totality which would not be mortal.)

["Man belongs to the mortal beings." Here the plural indicates the quantification of the predicate, whereas the subject is not quantified. Inasmuch as something is a man, it belongs to the class of mortal beings.]

Let us further consider the first example. We can obviously regard the following proposition as the essential logical thought in that wording:

Inasmuch as something belongs to the totality of men, it is a mortal being. With this we have a judgment that we can designate as a content judgment, not as a judgment of extension. It expresses not merely the relationship of two classes, but rather the relationship of two representations of properties.

Syllogisms constructed from such mixed judgments are dealt with according to the principle *nota notae*, which requires a prior transformation *via* substitution of identicals for identicals. [405]

E.g., If something is a member of the totality of A, it is a B.
If something is a member of the totality of B, it is a C.

Here we need for mediation: a B = whatever is member of the totality of the B's. With the aid of this identity we can transform the major, or if we wish the minor premiss, through which the sameness of the middle concept becomes Evident. ["Sameness of the middle concept" in the common sense of the word. The one time the concept stands as subject: something which has the properties X, inasmuch as it has them. The other time in the predicate position: something which has the properties X. The one time universal, the other time particular. The mediating properties X are the same in the two cases; but, since they are not what the assertion bears on, identity is not substituted for identity.]

Inasmuch as something is a member of the totality of A's, it has properties (it is a) B.

<u>If it a B, it is a C.</u>

Here there occurs an ordinary content (or property) inference.

If a judgment with mixed quantification were not in the first, but was in the second position, then the transformation discussed above would be necessary.

Of course any class judgment can also be interpreted as a property judgment: "Class a *sub* class b" says –

Inasmuch as something falls under the concept "class A," it also falls under the concept "subordinate to class B."

If a categorical judgment includes the existence of the subject, we always need two formulae in the calculus.

Socrates is ill:

S = (= I and S = (= e,

where e = existing.

A Difficulty:

One certainly cannot imagine all judgments actually brought into the forms stated.

E.g., Bismarck exists.

"If something has the properties of Bismarck, it has the properties of existence." But what about the represented judgment: "Something has the properties of Bismarck"? Are we to phrase it: "Inasmuch as something has the properties of being a certain something, it has the properties of Bismarck?

This in fact leads to an infinite regress.

This objection would have a foundation if we were to require or to claim that each judgment actually includes a conditionality.

What we claim is only that each judgment produced can be brought into the conditional form, which is necessary for the purposes of the calculative procedure. [406]

If we succeed in obtaining an interpretation for some uninterpretable symbol of a calculus, by means of re-interpretation of the principles of the calculus, then it is already demonstrated that the symbol is usable in the first domain of application. And conversely: If I can prove that one algorithm includes the totality of the general rules of the other, then the entire first algorithm can also be regarded as an "extension" of the second.

INTUITION AND *REPRÄSENTATIONEN* [1]: TOWARD A CLASSIFICATION OF *REPRÄSENTATIONEN*

Repräsentationen divide into:

1) Those in which the Representing contents do *not* belong to what is intended.

2) Those in which they do. Among these we also number those Representations where the Representing contents are more-or-less equivalent standins for the Represented ones. Thus, they are analogues of the latter, or are contents which stand in relationships to them, or the like. *The main point is that the Representing content itself is simultaneously intuited. It is a distinction of essence whether the Representing content is* MERELY *a means of Representation* – where the intuition directed upon it is a mere transition point for an annexed Representation – or whether the intuition exists in combination with the Representation and whether (if that is not possible) the currently intuited itself then becomes the objective of a Representation.

If I see a *thing*, then the intuited in this case is at the same time fundament of a Representation: fundament, but not mere means, not mere sign. Intuition is blended with Representation. If I wander over the intuitive image, then intuition and Representation are modified. But what was just now intuition then immediately turns into content of the Representation; and, conversely, what was merely Represented is now intuited.

Intuition is the activity of an interest (intention) founded through the Present content, directed upon it, and at the same time resting or satisfied in it.

[1] From 1883–1884. *Husserliana XXII*, pp. 406–411. Where stylistically possible, the word "*Repräsentationen*" is translated by the English "Representations," with capital "R." Similarly for "*präsent.*" [D.W.]

Repräsentation is the activity of an interest incited through the Present content – should we say here also, "founded through it"? – but *not* directed upon it. Nevertheless it is a determinately tinged[2] and simultaneously satisfied interest, with an Ideal direction upon a content not given. (Satisfaction is resolution of an intention which has an Ideal direction upon a content.)

Intention is a tensed interest that is incited by a given content but is not directed upon it (at least not primarily directed upon it). It has an Ideal direction upon a content not given, tinged in conformity with [407]
that content's special character; and it does not have the character of the satisfied interest – which, rather, is the corresponding positive feeling.

Apprehension is the activity of an interest which is partly directed upon a given content – restfully engaged with it or finding its satisfaction in it – but also partly reaching out beyond it, merely incited by it, but not directed upon it. This "reaching out beyond" part has its determinate tinction (in virtue of an Ideal direction upon a content not given) and its character of satisfaction.

In any case we have here a complex process essentially consisting in the unity of the interest as it activates itself in various forms.

If I apprehend a *thing*, or intuitively wander over all its parts and Moments, then the unity of the thing lies not merely in the contents – which, changing from moment to moment, yet in a steady transition develop out of one another – but rather in the acts also, and above all in the unified interest which encompasses the whole course. Even though the intentions and satisfactions (tensions and resolutions) change, they are still only different forms or specifications of a unified and continuous act of interest. This *act* grounds the consciousness of the unity of the entire course. Each new step is intimated and presignified in the previously given, and thus finds it fulfilment. This constitutes the consciousness of identity. It is not an identity of the act in the strong sense of the word, where one is an exact replica of the other, but identity in the sense it exists between intention and fulfilment. The acts are different; but they have a uniform intention, we can say, if we take the concept of intention somewhat more broadly. Thus the identity continuously perpetuates

[2] "*Tinktion*," "*tingieren*": – Terms used by Husserl to indicate the character of meaningfulness in an event or symbol. Probably from Stumpf or James. See the first sentence of the IInd "Logical Investigation." [D.W.]

itself through the total course; and our intention bears, not on what is given and intended at an instant, but rather upon the total course and what it may yield, so that we have a complex disposition of a wholly definite character, which determines the feeling at each given moment. The unity of the interest linking the intuited and intended from moment to moment conditions the characteristic distinction between apprehension (*Apprehension*) and mere Representation.

In the case of *mere* Representation, a sign serves as a support of the understanding; but the sign, to which we are also intuitionally turned in the moment when it is noticed – certainly only momentarily – does not require for itself identically the same interest as does that which manifests itself through the intention. If the sign, as sometimes occurs, is viewed for a while before an understanding of it sets in, or before it functions as a sign, and if *then* understanding ensues, there is in that case a break in the interest. The new condition does not take the intuition up into itself as an inherent part of the continuing [408] interest, but rather excludes it.

Yes, but what grounds the "unified interest"? It is surely the unity of the apprehension, unity of "the attention," which has its correlate in the unity of the object or of the event. If I observe a *thing* from various sides, bringing all of its parts and Moments to my intuition, then step-by-step I indeed have new apprehensions. But all of these apprehensions form an inner unity, *one* continuous "consciousness," one continuous apprehension, which perhaps itself sets up further apprehensions.

This unity of the intuition, this over-arching unity – which binds together the apprehensions steadily passing over into each other, and makes them into an intuition of the one object that, for all successive identifications, stands over against us in constant unity and is the intuited unity – is to be sharply distinguished from the mere temporal adjacency of the various acts of apprehension. A *different* object can be apprehended in temporally adjacent apprehensions.

Indeed, the same "content" admits of different "interpretations" ("*Auffassungen*") – *Schröder*'s stairs diagram, picture puzzles, etc – or it can be apprehended at one time as an object, and at another time as a sign for a series of objects. If I glance over a sign as a figure, then I have a unified apprehension, a unified intuition. If I then take it as a sign, I have a totally new mode of interpretation, a new kind of consciousness: possibly a consciousness of the same genus, but of quite another *differentia*. I do not have a continuous unity of apprehension in which one and the same object is grasped, just under different aspects.

Attempt at a Classification

Descriptive Differentiation

2) Contents which do not have the character of Representations[3] (intuitions, authentic representations).

1) Contents which do have it (inauthentic representations).

But that distinction does not coincide with intuitions and non-intuitions.

1) Contents which have a Representative function: a) with, b) without consciousness of the Representative function.

2) Contents which have no Representative function.

1) = *Repräsentationen*	2) = Intuitions
(inauthentic representations)	(authentic representations)

One decides with difficulty to speak of 1) as "concepts." For then [409] the indirectly seen would in fact – since the directly seen Represents it to me – have to be a conceptual representation of the same. Likewise my representation of father would have to be a conceptual representation of the father *as such*, if I have concepts with the consciousness of the mere representation. Certainly the explicated (*explizierte*) consciousness always leads as a matter of course to a conceptual representation which also refers to the object. It may also be that the primitively Representing representation also includes an intention upon this object even from the outset. But in the more usual sense we do not have a *conceptual* representation here. Perhaps one should say: it is a fragment, the rudiments of a concept. It logically Represents a concept, but that does not therefore enter into the psychology of the matter.

Classification of Repräsentationen

A.1) Direct Representations, the intention of which bears directly

[3] But there are no such contents. What is noticed is always *already* apperceived, in the developed human being. [Husserl's marginal note]

upon the object.

2) Indirect, the proximate intention of which bears upon a completely different Representation (that is, one more closely related to our interest), and, by means of it, directly or indirectly upon the object.

B.1) Adequate Representations (that is, those whose content possesses the same *habitus*, the same parts and Moments, and in the same combination, as the object intended) – that is, those which make their objects (inasmuch as, and as, they intend them) present by means of corresponding contents adequate to them. All of that which is intended in the object, and just as it is intended, also occurs analogously in the Representing content. Between the Representing and Represented there obtains content identity or content similarity in the direction of the intended. And this circumstance provides the basis for also speaking of intuitive Representations. These Representations are, in fact, those which are sometimes called "intuitive representations" in contrast to intuitions.

2) Inadequate Representations, which do not accomplish this.
a) Dispositions, in virtue of which the corresponding adequate Representations can come forward.
b) Not.

C.1) *Repräsentationen* which intend their object by means of some sort of intrinsic or extrinsic properties. The relation between property and subject serves, in virtue of fixed dispositions, to characterize the latter as the intended, e.g., this red thing, a blind horse.

2) *Repräsentationen* which intend their object without mediation through representations of properties (without mediation through parts, attributes, determinations of content). Thus, proper names and representations of individuals. Thus, also, "general representations" like *house*, *tree*, and so on, inasmuch as these names do not serve as deputies for more or less defi- [410]
nitely demarcated groups of properties.

CB. Conceptual and non-conceptual Representations can be adequate or not. *Adequate concepts* thus are those which, as is often said, are based upon a corresponding "intuition." More correctly stated, they are Representing representations that put forth adequate Representants for parts and properties which they intend.[4]

Adequate non-conceptual representations can be called *image representations*. Thus, the phantasm which I now have of that schoolhouse over there is an adequate Representant of the schoolhouse itself: that is, the corresponding intuition, as I can have it from out my window.

Adequate representation. Also, representation for which we have a *corresponding* intuition.

Thus adequate concept: One in the case of which we have intuition of an object falling under it, *as such*.

D.1) *Repräsentationen* through signs, natural / technical } conventional / non-conventional

2) Not.

Repräsentationen

1) Through analogy: similar for similar.

2) *Pars pro toto*: Beginning of a melody for the whole melody.

3) Through contiguity: the adjoining as representation of the adjoined, the symbol for the object. The beginning of the melody (clothed with a feeling of lack) as Representation for the lacking fragment.

4) Through properties. (Here parts are given in their relationship to the whole, but not the intended whole – rather, there is given only a part which is the same with respect to the possession of the properties. Thus, this belongs to 1), or it is a mixture of 1) and 2).)

* * * *

Repräsentationen[5]

1) Through analogy: similar for similar.

---

[4] These are 'intuitionalized' concepts. But "to have an adequate concept of an object" commonly has another signification. Of color I have an adequate concept, the blind do not. An adequate representation is, in general, an intuition. A more adequate concept is a direct representation whose corresponding intuition is available to us. [Husserl's note. See p. 475 of *Husserliana XXII*.]

[5] Quite likely a second attempt at a classification under the concept *Repräsentation*. (*Husserliana* editor's note.)

2) *Pars pro toto*: Beginning of the melody for the whole melody. One side of the thing for the whole thing. The appearance of a thing from one standpoint for the complete thing.

3) Through contiguity.

There really already belongs here 2), by means of signs. E.g., associated feelings serve as such . . .

4) Through conceptual determinations.

The Represented can be:

Something non-conceptual

Something conceptual. [411]

4) has a totally different character. It is a totally new group, wherein belongs: an A, an A which is B, ...

The A in general (the species of the individual, perhaps the genus of the species, and so with higher levels)

The grasp of *this* as an A, and the like.

Intuitions

INTEREST (TENSION) AND SATISFACTION (RESOLUTION): ON THE ROOT OF THE CORRELATIVE CONCEPTS: INTENTION/FULFILMENT[1]

In the transition from image to object – yet before the image is represented as such, as Representation of the object – there occurs a heightening of interest in what is known to be similar or identical, a heightened desire to observe. If this happens frequently, or if a habit has formed in virtue of the force of involvement, then upon the appearance of the image there is awakened, so to speak, a tendency toward the flowing of the interest over to an object identified, even where it is not present. More accurately stated: The interest will not remain at rest in the image, but rather pulls away from it. It becomes an awakening, a tensed interest, which strains toward the object. And if the object does not show up, then we feel stretched, unsatisfied; if it does, then satisfied. If the habit has become very powerful, then the tension can immediately pass over into resolution, even without the object showing up. Where the two are used to following quickly upon one another, there this rhythm is as a rule reproduced, and thus the following succession will occur: Image – tensed interest, but not turned toward the image – resolution of this tension into gratification, but one which does not here pass over into desire to observe, where that to be observed is lacking. The lack of this desire will not assert itself in the case of things very familiar to us. There the reproduced satisfaction precisely fills us out; and the stimulus to further activity, which would have to announce itself in a new tension, is lacking.

Similarly in the transition from sign to object. Here again the habitual succession from something of little interest to something of greater interest is the foundation. Thus the intuition of the sign will

[1] Probably prior to 1894. *Husserliana XXII*, pp. 411–414.

pass over into a feeling of tensed interest pulling away from it – if you will, into a prescient seeking. One is perhaps inclined to say: Something faintly reminds us of the *customarily* ensuing heightening of interest, of the elevated desire to observe, and puts us in tension; and that can only mean: excites our craving for realization. However, I hold this to be an incorrect exposition. It is already ruled out because it entangles us in a circle. The desire represented is not identical with the possibly actual one. It is only a Representation of the latter; and [412] how Representations are Represented, and wherein they consist, is here first to be explained. If merely a shadow, an echo, of the earlier desire is reproduced, and no more, how can this awaken a craving for more, for a correspondingly stronger, fuller desire (in case the entire distinction is to consist therein), if we have no inkling that the 'minus' Represents a plus? Thus, craving for "realization," for another stronger and fuller desire, cannot occur, if the non-being is not indicated in a certain fashion in the present content. But this indication is nothing absolutely primitive. It is the effect of habit, and indeed an original effect of it. We can set up the following principle: If to the transition A^B there is linked a (not insignificant or wholly unnoticeable) heightening or enrichment of our feeling tone, then there comes into play the habit that upon the appearance of A a deflection of the interest, and indeed in the form of a tension "toward B," ensues – even when B still has not turned up, or even *remains* absent. The deflection and tension of the interest is a peculiar phenomenon that cannot be further analyzed. "Deflection," "tension toward B" are non-literal expressions. They are only supposed to intimate that the feeling is determinately tinged, depending upon the special character of B.

Commonly one speaks in such cases of a *tensed expectation.* But distinct phenomena are then covered by the same name. If the handbell of the stage manager sounds, we look with tensed expectation toward the stage: We can therewith have a determinate representation (Representation) of the expected, and simultaneously the judgment that what is represented will appear. But often, and in the first instance, we have no such representation and expectation, but rather only the determinately tinged intention. The handbell at first evokes tension and nothing more.

The B can be one term, or a dispositional complex of terms interlinked in many ways. From A various series of terms can radiate outward, and various series from each of these terms again. Everything then depends upon the degree to which the dispositions are stimulated. If A stands in the same relation to B as to C and D, then, if B is primarily stimulated, the 'tinction' of the feeling of the tensed interest will be determined through B: We are then stretched toward B and not C. From the beginnings of this interweaving an entire series (e.g., a melody) can stand forth – that is, be unifiedly and primarily stimulated. Then the tension can preserve its tinction throughout that entire series. (It is a different matter whether upon a tone *one* other, or a tone configuration, follows; and again whether upon a configuation there follows *one* other, or a far-ranging melody.)

Again, it makes a significant difference whether the feeling dispositions that are co-excited through the terms stimulated are strong or weak. Or better: whether or not to the terms there also belong strong [413] feelings. Such feelings (more precisely, feeling dispositions) are especially excitable, and transfer the excitation over to their contents, near and remote, as far as their propogation customarily reaches. The stronger these feelings are, the stronger the tension actually stimulated in the interest. The tension is resolved through the appearance of the object, and possibly also by the appearance of an equivalent placeholder or image. It is "equivalent" precisely owing to this resolution, owing to the harmonizing satisfaction which it evokes. The resolution, which is not perhaps something negative – not perhaps mere suspension, annihilation or dissolution of the tension – but rather a positive feeling, then constitutes that which we call understanding or comprehension. First, the following observation. The resolution can also be an incomplete one. The tension relaxes momentarily, but tension is still left over. Perhaps after the resolution there follows a new tension, since the lack of completeness is tangible. E.g., the city of Cambridge in America is mentioned. The phantasm of the outline of the map of America occurs to me: resolution. But the resolution is incomplete: new tension. (Where is the city situated?) The influence of habit shows up in each case through the fact that, in virtue of the frequently or forcefully taken transition from tension to resolution, *along with the former the latter also is reproduced* – reproduced precisely as is usually the case in reproductions, as a stronger or weaker after-effect, as a more or less approximative copy of the original, of what was lived through (*Erlebten*) earlier. Depending on the power and type of the feeling reproduced, its relationship to the tension

is naturally a different one. If it sets in, then the tension feeling is resolved; that means, the feeling of satisfaction is not one which merely comes after, *post hoc*, but rather it is precisely a feeling of the satisfaction of *this* tension. But depending on the effective power of the habit, the result can be a more or less complete one. The tension T can pass over into its resolution R, and we, resting in R, can remain peacefully fixed. This is the case whenever the reproduced resolution is adequate to the tension, and accordingly completely removes it. Thus, for example, when a word evokes a complete understanding, even though it may be that no significational representation is reproduced. The tension stimulated through the word immediately finds the harmonization adequate to it. Sometimes the effect is not a sufficient one. We feel ourselves stretched, but the tension does not come immediately to the corresponding harmonization. We have perhaps only an "intimation" of the signification, only a weaker echo of the earlier relationship rings on, slackening the tension a bit, but not dispersing it. Later a full understanding may "light up," and we come to rest in the resolved tension. When the stage manager rings in the theatre, the first scenes from the part [of the play] seen earlier may occur to me: the tensed interest finds momentary, but only partial, discharge. Immediately, new tension emerges, and possibly one heightened through the intervention of the understanding: This was, precisely, not the adequate resolution, the phantasm evoked was no sufficient substitute, it served merely for a clear fixation of the [414]
direction in which the interest is stretched. If the curtain is raised, then we first feel full resolution of the tension. At least its most proximate goal is attained, and we have the fullest consciousness that the viewed is the intended. Thus we are in the moment of fullest understanding, but one which immediately dissolves into intuition, which was lacking in the previous examples.

The incompleteness of the understanding, or of the satisfaction of tensed interest constituting it, can also be of other types. A word "sounds familiar to me." Yet I have no inkling of its signification. Or it is totally unfamiliar to me; but I "feel" that it is an English word, without making this explicit judgment. The tensed (even though perhaps not highly tensed) interest which is stimulated by the word finds a certain resolution. But it is not the intended resolution. The tension springs up anew, and is only become stronger through the contrast. Or, to interpret the situation otherwise: The word was imbued with various intentions, and the one was resolved, the others not. And, depending upon the direction of the prevailing interest, we feel the lack, or else not, or we feel it more weakly.

APPENDIX VI

ON THE TRANSITION FROM THE MERE REPRESENTATION TO PERCEPTION. *REPRÄSENTATION* THROUGH ANALOGY [1]

The child hears it said, for example, "Papa is coming." It has the intuitive representation of Papa. Then Papa himself comes up. It senses a certain heightening of interest. A representation of greater fullness, vivacity, freshness, stability is there. It stimulates an incomparably stronger interest. Phantasma come and go. They do not retain their stability in the struggle against the co-temporous sensations. They do not offer such a fullness of perceptible details. They do not stimulate such strong satisfactions, appease the hunger, quench the thirst, and so forth. The emergence of the genuine presentation after the Representing representation (even though an intuitive one), is therefore sensed as a heightening of feeling. In the succession, then, the phantasm which has just been supplanted appears, in virtue of this heightening of feeling, as an imperfect image of the intuited thing. They seem related to one another, the imperfect as weaker reflection of the perfect, of the actually viewed. Then in addition the following comes into consideration: If a content stimulates a habitual interest and *satisfies* it, the appearance of a *similar* content which does *not* satisfy it will cause the lack to be sensed as immediately instinctive. It is once again the interruption of the habit which comes into consideration. The *mere representation* of Papa is very similar to the actual intuition of Papa (especially with the greater freshness and vivacity of the child's phantasy), etc. Thus the representation will show up as pervaded with lack. In the intuiting, then, the genuine interest [415 turns up again, heightened through the contrast. The sensation of satisfaction arises from the deficiency, etc.

[1] From 1893. *Husserliana XXII*, pp. 414–416.

So all *representations* of familiar things and their characteristics will therefore acquire the character of Representations. In further consequence phantasy representations in general become Representations – at least they easily do unless they, just as they are, satisfy the interest. This is the case not merely in dreams, but rather in the free play of phantasy, especially where it finds its basis in sensible intuitions (child playing with the headless doll).

It is on my view absolutely nothing primitive, not to mention essential, that representations appear to be something merely subjective over against the objects "which they represent," the pertinent objectivities. The relation [of the representation] to the ego which happens to be representing – that is, which "has" the content, or even has produced it in free phantasy – and which brings it to consciousness, as a rule only comes forward upon reflection, even in the mature mental life. In living thought, where representations nevertheless abound, nothing of that relationship is to be found.

Only deficient analysis keeps one from finding the characteristic property of the Representation in the content, and makes us think it necessary to fall back upon the representing *act*. But since this act also is not lacking in the intuition, one would in further consequence see the distinction between the two in a judgment that posits the object as existing, holding such a judgment to be absent in the representation. I must firmly deny that a judging is already present in simple intuiting. Simple intuiting is an act of exactly the same type as representing in phantasy. It is the taking in of the content. If we are wholly absorbed in the phantasies, then there certainly is no difference. If we subsequently deny existence, then we do not set aside a judgment previously there: at least not one of which we are conscious. To assume judgments of which we are unaware would require the most compelling of reasons. They are still hypothetical. An explanation which does not have recourse to things unconscious will always claim the decisive advantage.

Certainly representations and intuitions will often enough be accompanied by representations of existence and by assertions. The origin of the concept of existence has given the psychologist plenty to worry over. I believe the matter stands thus:

If, after we have had a mere phantasm, the corresponding object comes to intuition, then not merely satisfaction, but rather intuition of *the relation* of the intuited content to the represented also occurs.

This relation is recognized as the fulfilment of the intention; and here that means, above all, that the relation is intuited. The transition from the representation to the object, the intuition of the fulfilment of the intention, the overlaying and coinciding as to Moments of interest, the flowing of the representation over into the object with the feeling of relief, the satisfaction of the intention: these are nothing but characteristic Moments of the undivided phenomenon. (At first the representation is not altogether similar to the object. Still, the [416] similarity will be there in essentials. Or an intermediate term will be reproduced in virtue of which the original representation is so modified that the conformity comes about. E.g., Papa was represented as standing, but he is sitting. There the representation of being seated intervenes, and the like.) Thus existence is intuition of the conformity between representation and object; and, indeed, it is the object which, *as the intended*, has that property assigned to it through this process. Also the representation, as *the intending*, receives a property, if we will attend to the fact that through this process there is produced the transition into the satisfactory, into release. What is so represented is there. It corresponds to something in the representation. And this "something" must be expressed if one wishes to communicate to another his judgment: *There's Papa*!

But how can the judgment be represented? To the representation there is linked a shadow or residue of that judgment process with its affirmation, with the "belief."

CIRCUMSTANCES OF SEEING[1]

1) The content modifications in case the head is unmoved (and obviously also with the rest of the body unmoved), thus exclusively with movement of the eyes, possibly only with the movement of the inner gaze, wherewith the modifications by means of movements of the outer gaze can be reproduced, or also not. In any case one can abstract from the latter and attend to the appearing, the perceived alone.

2) The content modifications in case the head is moved while an object point is fixed or while the glance is changing.

In these cases the object can either be at rest or in motion.

The modifications of the first type, although considerable enough, are yet not so great by far as those of the second type. In the first case the matter stands thus: If we suppose in the first viewing the object to be analyzed into (A, B, C, D, ...), then the effect of the changing view is that the various partial contents become now clear, and now unclear. We can do this in such a way that in succession each becomes clear. Often, with the clarity, further analysis of one part occurs.

We thus have reciprocal and univocal coordination of the parts or groups of parts with the particular intuitions in the intuitional process. The wholes as well as the parts stand in the relation of similarity, and beyond that in the relation of relative clarity or unclarity. Such a reciprocal, univocal coordination, and such a relation of the clear to the unclear, are lacking in the second case. As a rule, totally modified intuitions occur, although ones proceeding through

[1] Probably from 1894. *Husserliana XXII*, pp. 416–417.

continuous modifications out of the original. And it is not merely the object considered that changes, but rather also the entire visual field in a similarly significant manner and in similar directions. (The [417] same happens also in the previous case, but merely as to the aspect under clarification and possibly expansion or constriction.) A plane is transformed into a line, a line into a point. Great into small and small into great. Shapes transform themselves, etc. Further, total novelties emerge with regard to the qualities. The concealed, the reverse side, etc.

Thing	Objective Unity	Group of Things
Table	Melody	Inkstand, Clock, etc.

Assume that the training of the eye is already developed to the point that there exists a certain command over the movements of the eye; and let us imagine a child which with fixed body moves its gaze over an object that stands out sharply from the background and is at rest – perhaps a colored form on a light ground. If the interest is satisfied in the first position of the gaze, then those portions of the image off to the sides which possess some outstanding Moment – a sharp boundary, a luminous hue standing out more strongly, and the like – will force themselves upon the awareness. If then movement occurs, it may not lead immediately to this part, but rather to another. But if it does strike this part, then in virtue of the similarity recognition sets in, and simultaneously therewith satisfaction of interest, precisely as clarity gives it. If the gaze turns back to the old position, whether directly and by choice or accidentally, then there occurs recognition again, with respect to the unclear parts to the sides – but with the feeling of a certain deficiency, of unsatisfied interest. This will certainly be the case with numerous types of change. And so, with a sufficiently diverse movement of the gaze, each of the bordering parts of the momentary image finally receives the intention appropriate to it and therewith Representation. At the same time, with sufficient command over the movements of consciousness, free availability of the intended contents arises. Any arbitrary movement leads the desired part into its clarified state. And so the fulfilment of the intention is lived through at will, and at the same time the identity consciousness of the total contents, flowing over into each other, occurs. If no command over the movements exist, then also, first of all, no unity of the type of *the thing* can arise. But command over the movements mainly originates under the influence of the satisfaction which the optical axes, set on identical points, bear within themselves through the fact that they make possible a seeing that is clearer and is free of struggle. And likewise command over the coordinated movements of the two eyes arises through the satisfaction which is produced by the transition of the marginally noticed into a fixed focus. A certain sensation of movement comes with this satisfaction. That case will stand out there where the marginal is brought near in the most direct way possible. For in the case of greater indirection, instead of stronger clarity there at first occurs obscurity.

CIRCUMSTANCES OF SEEING: VISUAL FIELD AND FIELD OF SIGHT [1]

1) The 'circumstances' consist – disregarding the change in noticing and attending – in the gradations of muscle feelings, which, proceeding in sequence, can reverse themselves and within their varieties change arbitrarily. We have at times lifted one dimensional varieties out of a two-fold variety of possible muscle feelings, and indeed in the first place with the head at rest and in the primary head posture: – To each possible position of a distinct point in the visual field there corresponds, given the fixed posture of the head, a *determinate* muscle sensation, and possibly a (no longer quite so determinate) sensation of transition. Given the same circumstances we always see the same in the object. Given the same change, the same change. The content of the object is something existing, and in each moment something intuitable. Consciousness is able, from each given term of the group, to derive that every other is there. That is, we have a certain feeling of satisfied assurance, at least after we have repeatedly run through the group. And likewise in the novel case, when the object is still completely unfamiliar to us: with the new object we have the feeling of being able to *freely* activate the interest. The circumstances do not belong to the content, for they are only means for emphasizing and intending it, perhaps arriving at it. Inasmuch as the content is actually there, the interest is turned upon it alone. The muscle sensations fall entirely outside of that interest.

2) We have to distinguish: a) the movements which bring a marginal image to the center of the retina, b) those which bring the double image to coincidence through convergence, c) those which,

[1] *Husserliana XXII*, pp. 418–419.

under otherwise identical circumstances, sharpen the contours: accomodation.

Objects which occur in the marginal parts of the visual field show a remarkable indeterminacy. With regard to the *very* marginal parts of the visual field, one can hardly still speak of contours or of boundaries. In any case they are most indistinct, and a grasp of the distinction between object and background is scarcely possible. The object forces itself forward in a manner as if it had no background at all. And it itself can at no time be observed as at rest, but rather only as moved or as movement. For all that, certain formal distinctions come to light. One can probably roughly recognize the elongation, thinness, the breadth and roundness. The more toward the center of the retina, the more the form gains in determinacy. The contours become graspable, and the bounded content itself gains, so to speak, a body. Further, the background itself becomes graspable. The seen appears on the background. Lack in accomodation also comes to light in a similar manner. Whatever is reflected in the retinal center without accomodation stimulates an image with unclear contours. Perhaps like an image pushed to the side, but not *too* far. Double [419] images trouble us mainly because of the struggle that comes with them. The unified image is also in a certain manner more complete than the doubled image.

Moreover, the taste sensations which the movement of the eyeball produces on .... {omission from MS} also come into question.

Field of sight and field of view.

The child originally has no field of sight in the authentic sense, but rather a total sensation, whose content indeed exhibits in itself relational orders, but relational orders which the child cannot recognize. It originally has nothing more than an indistinctly bounded intuition, tinged in a certain manner by the background. It recognizes nothing of boundaries, relationships, etc. To that extent the child does not see the same as we, even if we had the same position and accomodation of the eyes, since to it the seen signifies nothing, refers to nothing, intends nothing. *Our* field of sight is what is seen with eyes at rest, and indeed the total content: thus the directly as well as the indirectly noticed, thus also what is bound up with the primarily noticed. And this content is charged with intentions by us – which we also can realize in a certain manner without eye movement, through the wandering of awareness. But at the same time there exists intentions which reach onward. The field of the glance is the intended of second level, inasmuch as the intention is realized by means of the glancing. It is thus an Ideal content which only becomes intuitive in a sequence of intuitions. Each stage of this sequence includes a visual field for each stage of the observing.

Descriptive analysis of the field of sight and the field of the glance.

Each analysis already presupposes a certain objectification. Namely, we assume all of that as objectively existing in the content which we can single out in it through a separate act of noticing, correspondingly imposing on the content no other modifications than those introduced through the noticing itself, to be re-identified as often as is wanted.

DRAFT OF A LETTER BY HUSSERL TO MARTY[1]

Halle, July 7, 1901

Dear Professor:

It was a great joy to learn, from your letter of June 7, that you have taken an interest in my *Logische Untersuchungen*, and consider it worthy of closer study. A series of chance interruptions, and the (often painful) after-effects of last winter's over-exertions, made an immediate response impossible, for which I ask your kind forgiveness.

With reference to your observations concerning the IVth Investigation, I have nothing to add. We are, happily, totally in agreement [420] on the matter. But it would have been of great interest if you had also expressed your views concerning my conception of "pure grammar". Your agreement would be of quite special value to me. That the exposition in the IVth Investigation is somewhat too narrow, in the light of the results of the VIth Investigation, or that it would require a somewhat revised concept of *signification*, has been noted by me in the Appendices of pp. 718 and following.

The unfavorableness of the circumstances under which the work was brought to completion often makes itself painfully felt in such unevennesses.

So it is now a matter of settling our differences about the distinction between immanent and real objects.

[1] *Husserliana XXII*, pp. 419–426.

1. The need for this distinction lies above all in the fact that every representation refers to an object, whereas there does not to each representation correspond an object in actuality. A representation without an object represented is not conceivable: there are, therefore, *no* objectless representations. On the other hand, however, actually existing objects do not correspond to all representations: there *are*, therefore, objectless representations. The contradiction is to be avoided, it seems, only if one distinguishes between represented and actual objects. Thus: There are no representations without immanent objects, but there are representations without actually existing objects.

2. A second argument invokes the fact that the representation is a relationship of the ego to the represented content. If the relationship *is there*, then the relatum must also be – and thus there must be the represented content – so often as I represent. Sometimes one existing in actuality, but not always.

3. The distinction can also be directly verified. The phantasy representation projects (*vorstellt*) an image which nevertheless is not the actually existing object. The symbolic representation has the sign content immanently within it, which is obviously different from the designated object.

Thesis: With each representation a distinction is to be made between the intentional (or immanent) object and the actually existing object. And this is to be done in such a way that – disregarding the exceptional case – the intentional object is *other* than the actually existing one. The first is a (necessarily reel (*reelles*)) constituent of the act, and is therefore existing at all times in or upon the act, "immanent in it." The latter is, in general, transcendent to the act, and so now existing and now not existing.

Antithesis: Every representation *R* represents an object; and *this same* object, which it represents, is actually *there* (*wahr*) if the affirmative existential judgment *E*(*R*), whose subject representation is *R*, is true; and it is not "actual" if it is false. "Actual" does not mean exterior to consciousness, but rather *there*; and the object there is immanent if the representation intends something immanent. It is not immanent if it intends something external – whether it is directed upon something physical or upon something psychical.

There is *no* distinction between intentional and actual or non-actual objects. The customary talk of merely intentional objects, or even of merely intentional existence, does not signify the existence of objects residing, in any literal sense, in the intention, but rather *the mere existence of the respective representations* together with the non-existence of their objects. That the representation has not merely an intentional, but also an actual object, means that it does not merely present the object to us, but rather that the object also *is*. To represent = there exists a minding of the object in a representing fashion; but this does not require the object itself to exist, either within the representing act or outside of it.

a. It is clear that the "actually existing" objects of the representation – and, in the opposite case, the non-actual ones – are the objects which are represented in it. If I speak of this house, then this house is represented. If I speak of this centaur (which is painted here in the work by Böcklin), then this centaur is represented. That object is represented which the representation means, which is judged of in the 'corresponding' judgment, loved in the corresponding love, desired in the corresponding desire, etc. If I see the Charles Bridge, or if it is present to me by means of a picture or in phantasy, and it evokes delight in me, then the delight is not delight in my sensation or representation. Rather, as I may also quite correctly state it, I am delighted with the Charles Bridge. If I desire a certain dish, then it is not my phantasm or the inner image or something else psychical that is desired, etc. What is desired, loved, hated and judged of is that which is represented; and obviously the represented is, in general, not something immanent (in the literal sense) in consciousness, but rather is the object itself, the object which we speak of as true or real whenever the existential judgment is valid, and which after all is the identical point of reference for all of these possible acts or modes of reference to an object.

If there should be present in each representation an additional something-or-other, which I will call "immanent object," in no wise does that eliminate the difficulty posed by the apparent contradiction under 1. above. That difficulty is: How can a representation refer to an object when *that* object does not exist at all? And how can consciousness direct itself upon something that is not even present to it, or that is not immanent in it?

If, now, the intentional reference of the act to the object is in its nature (*Sinn*) a reference to an object that may exist in actuality, if love is love of the object actually existing (or held to be actually existing), and so forth, but not love for something immanent – disregarding the exceptional case where the act *does* bear, precisely, upon an inner state – then the second argument proves too much. It proves the necessary existence of the actually existing object.

As to the third argument, it is already dispatched by this: The phantasy representation does not represent the phantasm or the phantasy image, but rather the object which is phantasized. And the symbolic representation is not a representation of the sensations of [422] the symbol and of the object functioning as a symbol, but rather of what is symbolized.

b. There is represented in a representation just that which a corresponding adequate perception would set before our eyes – that thing itself, and precisely in the manner it was represented. A representation can be composite, and when it is it then has parts which are themselves representations, and which therefore themselves, in turn, have objects. But if, in a certain fashion, these objects are already encompassed by the intention of the total representation, yet it is not they that are the "represented," in the genuine sense, but rather the *total* object of the representation is. The representation, "the pen-knife on the table," represents a pen-knife, and it does so precisely in the mode of "the ." That representation certainly is also inclusive of the table; but it is not a representation *of* the table. Rather it is precisely a representation of the pen-knife as being on the table. Not the table, but rather the pen-knife – and in this mode of being represented – is the subject for the respective judgments, emotion, etc. Should one therefore advance the theory that every representation, in this inclusive fashion, also has immanent objects – i.e., that no representation can represent what is transcendent without also including representings of what is immanent – then we certainly would have here a fact worth noticing. But these immanent objects would not be, in the literal sense, the intentional objects. They would not be the objects represented by the total representation.

However illuminating these arguments may seem, I still regard them all as mistaken. Their basic error consists in the confusion of two very different things. Representation is not the mere being-there (*Dasein*) of a content in consciousness – that is, its presence in the real context of psychical processes (*Erlebnisse*). Rather, representation is an intentional process, a certain minding (*Meinen*), whereby an *object appears* (whether it appears "itself," or appears in image) or whereby an *object* is even just *designated*, or meant *via* significations.

But then an object is meant in virtue of the fact that a psychical content lived through (*erlebte*) is "interpreted." In that such a content is subjected to an interpretation (*Auffassung*), in that it functions as Representant in an act of "Representation" – *therein* comes to pass what we call the appearing of an object, and, in general, its being represented.

The confusion consists in this: that (i) the phenomenological relationship (of purely descriptive psychology) between the act-character of the interpretation and the psychical content (belonging to the actual ego) which functions as the subject of the interpretation, is confused with (ii) the relationship between the act – i.e., the psychical experience that we call "the representation" – and the represented *object*.

And this confusion is grounded in the fact that in a narrow range of *exceptional* cases psychical contents really lived through (*reell erlebte*) do themselves become objects of representations; and, indeed, they do so in such a direct fashion that the represented objects are simultaneously the Representing contents of the representation. [423] Since, then, in all cases of psychological reflection upon Representing contents those contents are transformed into such objects, one assigns to *every* act of representing a "direct" representing of the Representing contents, and regards those contents as objects immanent in the representation.

In my opinion, the theory that all representations of transcendent objects (understood simply as objects that are not immanent) must involve representation of immanent objects stems from *apriori* considerations, and not from descriptive analyses themselves. *Repräsentation* by means of images passes itself off as something obvious: the object is brought before us by means of the image, while the image is the immanent object and makes the object itself present in virtue of its similarity to the object. Where no image-like Representative is to be found, representation by means of relationships appears as an intelligible means of referring to what cannot be represented through an image. Thus one says: Either our representations are authentic, in which case everything is clear; or they are inauthentic, and then must always be founded upon an authentic representation, in such a way that what is authentically represented is content for a thought that perhaps could be expressed in words – though it need not be – as: the such-and-such which stands in relations ϕ to the represented object A.

However, it still must be emphasized that the image is not the thing itself (*die Sache selbst*), and that to represent something through an image is not to represent the image, but rather to represent the thing in the image. Hence the view must be modified as follows: The authentically represented is only that which is *itself* given, and indeed in its genuine selfness. The inauthentically represented is that which is represented through an image, *as well as* that which is symbolically represented – the former mediated through relations of resemblance (picturing relations), the latter through other sorts of relationships. And in each case a conceptual thought is presupposed. Now it certainly would be questionable to claim that a plain and simple act of imaging must already involve conceptualization of a relationship. Still worse, however, is the reflection that these conceptualizations of relationships surely are themselves representations, and that they could not possibly be "authentic" representations, since authenticity would presuppose the existence of the indirectly represented. But if they are inauthentic representations, then there must be still further authentic ones (these again would have to have the character of relation representations) whereby the inauthenticity would be possible. And so on *in infinitum*.

I believe that a more fundamental distinction is to be made:

1) Representations which – whether through an image or through a symbol – do not in any case represent what is immanent in the literal sense. Between image and thing there obtain relationships. But these relationships are nothing within the act of representation – neither as actualized nor as conceptually represented relations. They obtain objectively. That is, if we placed the representation into relationship with the thing itself – given in an adequate perception – and thus were to produce a synthesis of judgment which brought the representation and adequate perception to union, then we would find with Evidence, in reflection upon the fulfilment experience, the analogy between the [424]
image (or shadowy picture [*Abschattung*]) and that which is imaged, etc.

2) Representations through relationships – that is, attributive representations – e.g., the thing to the right of the corner, and so forth. These contain relational concepts.

Assuming that the *usual inadequate representing*, the representing of *things*, is a complex act in which the genuinely immanent content is authentically represented, and in which this authentic representing is the foundation, there is yet built upon the underlying authentic representing a new act of representation, precisely that one which intentionally refers itself not to the immanent, but to the transcendent object. This new act of representation has no immanent object, for the immanent object is object for the underlying and distinct act of representing. This immanent object is not the referent of consciousness in the representing here at issue, but rather the new or additional act of representing relates to the transcendental object. The underlying representing could be considered for itself, and *then* we would have a representing which has an immanent object, and which "means" that object and it alone.

If it is said that the authentically represented, the immanent *object*, is *a surrogate*, my objection to this deputization theory has been that being a surrogate cannot merely amount to the objective fact that, instead of the transcendent object, the immanent object is lived through and, within the context of my representing activities, can stand in place of the transcendent object – perhaps because for the psychical effects, which alone are of interest, the one object serves as well as the other, so that the final outcome would be objectively the same. Rather, on my view, being a surrogate has the character of a psychical, a phenomenological, a spiritual (*inneren*) fact that modifies the character of the phenomenon. It is a matter of *meaning* something different. When I speak of Paris, I do not in place of it mean the map or the word "Paris" brought forth. Rather, I mean the city, Paris. Moreover, in my opinion talk of a surrogate is not appropriate here. Neither the explicit conceptualization nor the objective fact of deputization comes into play, but the relationship between the novel manner of meaning and possible future fulfilment and disappointment does. If in speaking of representing through relations one is saying that the one term of the relation and the relation itself aid us in bringing the other term before our minds, that again is something which I find unclear. Conceptualization of the relation comes forth only upon reflection, only for judgmental activity after the fact. The pure and simple "inauthentic" representation is not a relational representation, and is no relating act, but rather is a characteristic type of consciousness of the object – possibly built upon yet another representation, as when the representation of the sign, as a perceptual representation, undergoes that modification, bears that novel mode of being-conscious, which turns it into the sign of something. Only the understanding of the inauthentic representation as representation-through-relation has conferred apparent obviousness on the view, which is nonetheless unthinkable, that something [425] is inauthentically represented in cases where *nothing* is authentically represented. But of course I can realize a relational thought which gains *determinateness* of objective reference only by means of relative attributes grounded in objects that are known.

Even if I am wrong in thinking that sensation does not require sensing (representational) acts to be directed upon it, it is important to have dissipated the alleged obviousness of the view that it does.

That every representation is, in its essence, interpretation, and, as such, requires contents to be interpreted, is a view which I myself have sought to work out. But the relationship of the interpretation to the interpreted (the Representing content) is one essentially different, from the ground up, from the intentional relation as relationship of the total interpretation (or relationship of the sense-animated sign, of the content insouled by the interpretation) to the Represented or intended object. The former relationship is a real one, while the latter is Ideal or logical. The interpretation does not represent the content to be interpreted; but rather in that it interpretively grasps that content it represents (generally speaking) a content different from it. 3). The general object is the object of a general representation. If we understand by *general object* a species, e.g., the number π or the color red, then this object – and I have had to emphasize this point – is not the fleeting red-Moment, which is an individual aspect in the intuition, and *real* in the fashion of a non-independent Moment of a concrete whole. (We need not dispute about whether to call such things "real" or not.)

Now the representation of a universal requires a Representing content, just like any representation; and the intuition of the universal is constituted upon the basis of the interpretive grasp of a corresponding individual. But this individual, the red-Moment, whose immanence in the intuition of the universal (*the* red) I never deny, but instead have always stressed, is not the red *in specie*, which rather is a merely Ideal intentional unity that is "itself" given or intuited in carrying out "authentic" abstraction – or *Ideation*, as I also say in one word, in order to distinguish it from the usual concept of abstraction. Ideation is perception of the universal, "perception" understood in the broadened sense. Naturally the universal does not exist after the fashion of a particular, but it nonetheless exists – namely, in such a manner that the relevant act of "general intuition" can actually be carried out.

When we include in the essence of the "authentic" representation of the universal the actual and adequate execution of the consciousness of the universal on the basis of a corresponding particular (the red-Moment) which itself is lived through, this does *not* amount to a psychological hypostatization of the universal. For this "Moment" is not the universal, but rather the universal is the identical factor there for the repeated execution of the abstraction. This identical factor is neither the momentary red, nor the abstracted general red; for those [426] are contingent particulars.

When I represent a general object, then *it* is the intentional object – by which I always understand that which is meant. That this representation necessarily involves an immanent content in its sense, *viz.*, the red-Moment, is something I have never denied. But this is not to make the universal into a real psychical element; for the realized is not the universal, but rather it is the immanent content, which is precisely nothing general, but a fleeting aspect of my experience (*Erlebnisses*). The red in general, the subject of the general judgment, is identically the same whoever it is that may be judging. But each person who judges has his own red-Moment.

IDENTITY OF REPRESENTED OBJECTS[1]

a) Intuitive representations:

I represent a silkworm in phantasy. It turns and twists. Yet it is always the same creature. The different representations in this continuous synthesis of acts represent the same object ("under different aspects").

Or else I represent a silkworm, and, after a period of time in which I am engaged in totally different thoughts, "it" occurs to me again; and I recognize that what is now represented is the same as what I earlier represented.

In neither case do I believe in its actual being.

What does the identity of the represented object signify here?

The represented or imagined object is the same. The two representations indeed do not have totally the same content, but they have contents which contain a certain identical factor – namely, that which we speak of as the represented object. "To mean the one object *A*" in representing is something in the act, and *that* is the same in the various acts. The different acts in an intuitive synthesis differ from each other not only temporally, but also in their content. But they are one in "that they mean the same object." And each one means it *via* distinctive Presenting (*präsentierender*) "contents." In each one "other sides of the object come to primary appearance." Already with this we are considering something Ideal (a genuine identity).

But in all of this the following is not to be overlooked: In intuitive identification there of course resides no judgment, as it does in the explicit sense of the assertion, "The two representations represent the same thing," in which the representations, therefore, are the subjects. In that an intuitive synthesis is actualized – as continuously varying

[1] Possibly from around 1898. *Husserliana XXII*, pp. 426–429.

representations follow upon one another, or else, fixed or partially fixed, they pass over into one another with a shift in their content – we live through the unity of consciousness as to object. This is not a judgment *about* a constant identity, but an experience (*Erlebnis*) of identity. [The Gestalt quality that constitutes the unity is a case of [427]
the Gestalt qualities (unity forms) which are based upon sameness (contrasted with difference). Things which are the same naturally flow together into a unity; and because they do so here, we have a steady "consciousness" of the same thing.]

Living through the identity serves as basis for the Evident judgment of identity: that the two representations represent the same object, or that I mean the same thing in both of them.

When I have seen a man earlier, and I now once again see a man and recognize him as the one seen earlier, what are the facts of the case here?

Here I can be deceived about whether the actual object is the same. I have perceived the object earlier. Presumably I grasped an actual existent. I now once again grasp something presumed to exist. I believe that I am grasping the same object. But it is in general not necessary that, now or earlier, an object of the kind intended must exist. Nor is it necessary, if they did exist, that the object should be the same. Only this holds true: I remember that I perceived an *A* at some earlier time. I have the perception that I now perceive something – and, to be sure, I do so in such a manner as to believe, besides, that the past and present objects are the same thing. The two representations which underlie the acts represent the same object. But the real entity which corresponds to them is not the same. That is, the intuitive synthesis which belongs to the second representation – insofar as it actually has an object for its basis – is not the synthesis presumed (which would belong to it in the sense of the intention of the representation); thus it is not that synthesis coinciding with the synthesis belonging to the first representation. We have here rather complicated relationships.

b) Conceptual representations or significations.

i) The two significations obviously represent the same object, the Ideal content of the respective acts. The signifying content is not the same, but it obviously possesses a common character, which we understand to be the identity of the object. This identity of the "representations" (significations), to which corresponds a *sameness* in the acts, is either an immediate identity, and then we have an *immediate* equivalence with respect to the conceptual objects; or else it is a *mediated* identity. In the latter case the identity rests *a priori* upon an encompassing Ideal interconnection, yielding a mediated equivalence – and, to be sure, an *a priori* one.

It must be emphasized that Evidence of an identical objectivity presupposes two significations, and that we here have to do with general or with mixed significations. Purely individual significations (proper names) are necessarily different as to their individual content.

This brings us to judgments of the following sort: "The equilateral triangle is identical with the equiangular," or "The shortest connection between two points is identical with the straight line" (specified by [428] another definition), and so on. The two representations do not mean the same in the sense that they have the same signification, but in the sense that they refer to the same objects.

These judgments are equivalent judgments about the respective individual objects, and therefore they have the character of hypothetico/*a priori* interrelationships.

It certainly must be asked whether in the formulation of such identities we make a judgment concerning the representation. Indeed we do not. My judgment concerns the relevant "general objects." But in this case the objects also exist. Where we have equivalent concepts that are objectless, there our judgment can only be about the concepts, if it is to be an *authentic* judgment.

There simply are two distinct modes of judgment: the one assumes that the respective representations have objects, and hence for it the respective existential judgments are valid. The other type does not assume this, and if we make use of modes of expression of the same type, then their sense is changed (to: The intended S as such is identical with the intended P). Supplementation by the relevant existential judgments transforms the merely hypothetical connections into "causal" or existential ones. Hence there is a broad sphere of thought which is indifferent to the question of existence.

ii) Now we proceed to the case of non-existence. Here the identity rests upon empirical (only probable) connections in judgment. In each case we in fact judge that an A as such is identical with a B as such, or that the individual I is the same as the individual I' given in other circumstances and specified through them, and so on.

Here the identity does not rest solely upon the content of the significations and upon the Ideal interrelationships of those significations with others, but rather upon what we call the unity of experience (*Erfahrung*). On the basis of the assumption of empirical facts and laws we identify objects which are merely represented. That is, we assert that, given the assumption, the objects would be the same; or that the respective representations are, with the aid of the relevant assumption, equivalent to each other. The factors we have discussed under i) here come into play. For it is not in themselves that the representations have the character of sameness in objectivity, but only as they are conjoined with the assumptions. That character does not attach to the representations taken by themselves, but it accrues to them with necessity under presupposition of the assumption.

The unity of the object points to the unity of an Ideal, closed totality of judgments bearing upon *that* object. To say that the object exists is to say that some of the representations *A*, univocally representing it, authorize the judgment, "A exists." If this judgment is not valid, then the representations have no object in the authentic sense (no object corresponds to it). But we still have an indeterminate number of judgments which "are about this object" – that is, which would [429] turn into judgments about this object if it did exist. And if the existence is not decided, we still have *this* valid judgment: that *if* it is decided affirmatively, then we have judgments authentically about the object. In those judgments is expressed what is true of the object: its characteristics – and the intrinsic ones, or properties, as well as the extrinsic ones (thus its relations to other objects also). The properties and other sorts of characteristics truly belong to it in case it does exist. (In other words: The categorical judgment presupposes the existence of the object.) In the other case we have hypothetical properties: they *would* belong to the object *if* it existed, if the representation of it grounded a true existential proposition.

The property of the representation which determines that it shall represent this and no other object does not belong to it in such a way as to involve an intentional object. Rather, it is in the "minding toward" (*Vermeinen*) that the essence of the representing act (whose Ideal form is the essence of the "one" signification) is constituted. This "intending" is an aspect which, precisely in its Ideal form, is the foundation for the validity of the objective interrelationship within the judgment (or sentence) that constitutes the correlate of the unity of the object.

APPENDIX XI

INTUITION AND THE GEOMETRIC AXIOMS[1]

The consistency of axioms is also not a matter of "axiomatic" insight, but is certified only by freedom from contradiction in the consequences derived. This of course does not preclude insight into the consistency of particular propositions or sub-sets of axioms. But such insight will, then, not rest upon the intuition of geometric relationships, since we lack that with reference to Ideal concepts, but rather upon insight into certain extra-geometrical relationships: namely, universal and categorial ones.

When it is claimed that the space of intuition is a geometrical space, little can be said against that. It is just as when one says that actual motions proceed according to the Ideal laws of mechanics. Naturally they do this within the limits of precision imposed upon our observations, within which we can comprehend appearances under concepts. Likewise here. Geometrical phenomena cannot be comprehended in Ideal concepts, but we can, nonetheless, effectively deal with them by means of such concepts. It is taken for granted that the Ideal relationships are preserved when the concepts are carried over to approximations, and therefore that rationality and continuity reign in the domain of the spatial. (Similarly, the physicist assumes that the functions which show up in mechanics are rational.) However, it is overlooked that these details also can be subjected to mathematical treatment, inasmuch as the empirical straight line can be defined as a segment of a plane, and the angle as a magnitude varying within the appropriate limits, and so forth. [430]

But in each case the assertion, that perceptual space is Euclidean and geometrical space, rests upon the same type of logical justi-

[1] *Husserliana XXII*, pp. 429–430.

fication as the assertion that motions occurring in nature proceed according to the laws of gravitation and the parallelogram of forces. The verification in application is always stronger the more precisely the [Ideal] relationships in question are specified, in which the degree of exactitude and approximation is, in turn, governed by these conceptual structures.

THE ASSUMPTION OF AXIOMS IN GEOMETRY, THEORY OF MANIFOLDS AND PURE MECHANICS[1]

Matters stand in geometry, therefore, exactly the same as they do in the formal algebras – say in theories of manifolds. The difference consists merely in the fact that the delimitation of the latter is a free act of the mathematical mind, which is not so in the case of Geometry. There is an incalculable number of possible ways in which to delimit algebras – by conventions which formulate mutually consistent, formal, primitive concepts and axioms – or to define manifolds and then investigate the laws implicit in them. In geometry, by contrast, we deal with what is given, a domain of deductions arising naturally. Its starting point – indeed, we can almost say, its fundamental principles – consist in certain obviously true judgments, arising out of everyday experience and phantasy shaped by that experience. These are judgments about empirically given space. They operate with certain basic forms to which we come through analogizing idealizations on the basis of the empirical peculiarities of the spatial domain, as in the domain of colors we come to pure red, pure blue, etc. Corresponding to the character of the concepts involved, the judgment represents what we might call the ideal average of judgments of external experience and phantasy, whose epistemic value – as is to be expected in light of their origination – remains problematical in each case, so long as a scientific analysis has not been carried through. How necessary such an analysis is is sufficiently shown by the debates through the centuries concerning the character and epistemic status of the geometric axioms. But the geometer does not enter into this investigation. He simply assumes the axioms and draws his consequences.

[1] *Husserliana XXII*, pp. 430–431.

At the very most he seeks to restrict the number of primitive concepts and axioms, and to establish what the minimum number of logically independent concepts and axioms is from which the entire system of deductions can be derived. That geometrical space and its theories is a means of mastering the various spatial relationships and providing the basis for the knowledge of nature: this places geometry on a par with pure mechanics. But the one has as little as the other to do with considerations relevant to the clarification of nature. That is, on the one side, a matter for physical mechanics, and, on the other side, a matter for physical geometry – the lack of which we must lament [431] on behalf of the interests of the theory of knowledge, although they would be completely useless in broadening the range of knowledge in the natural sciences. The task for pure theory is pure deduction. It does not inquire into the source of the axioms. It just assumes them. There may be debate over the epistemic status of the basic concepts and the axioms of geometry, but the geometric theorems stand above all debate. And quite naturally, since validity in their case means only the cogency of their derivation from the fundamental assumptions of geometry.[2]

[2] This concludes *Husserliana XXII* and its Appendices. [D.W.]

"PERSONAL NOTES"[1]

September 25, 1906

Since the beginning of this month I have been deeply immersed in the work. But have I gone about it in the right way? First I studied *Meinong*'s book[2] on assumptions, which repeatedly required me to look back into my own earlier articles and think my way back into them.

I did a lot of reading in the *Philosophy of Arithmetic*. How immature, how naïve and almost childlike that work appeared to me! Well, it was not without reason that I was conscience-stricken upon its publication. Actually, I had already gone beyond it as I published it. Indeed, it was drawn in essentials from the years 1886 and 1887. I was a novice, without a correct understanding of philosophical problems, without proper exercise of philosophical abilities. And while
laboring over projects concerning the logic of mathematical thought, [443]
and of the mathematical calculus in particular, I was tormented by those incredibly strange realms: the world of the purely logical and the world of actual consciousness – or, as I would say now, that

[1] Some of Husserl's reflections on his own career and works. Excerpted from a black notebook ("X, x 5" in the Archives at Louvain) in which he kept notes and quotations from works and people who impressed him, along with diary-like entries. The text here translated was prepared by Walter Biemel and first published in *Philosophy and Phenomenological Research*, March 1956, pp. 293–302. Second edition in *Husserliana XXIV*, pp. 442–449. All of the footnotes which follow, other than mere references to publications, appeared in the 1956 edition, and are Biemel's, unless they are followed by "H.S." for "Herbert Spiegelberg." Dr. Ullrich Melle has given me invaluable advice on the translation. [D.W.]

[2] A. Meinong, *Über Annahmen*, Leipzig, 1902.

of the phenomenological and also the psychological. I had no idea of how to unite them; and yet they had to interrelate and form an intrinsic unity. So I wracked my brains concerning, on the one hand, the essence of representation and judgment, the theory of relations and so on, and, on the other hand, concerning the elucidation of the interrelationships between the formalisms of mathematics and logic. Extension of my efforts to the whole domain of the purely logical was no doubt occasioned, more than anything else, by engagement with the logical calculus during the Winter of 1890. Then there came the lectures on psychology in 1891/1892, which brought me to look into writings on descriptive psychology with eager anticipation. Some reading in *James' Psychology* – of which I was able to read only a very little – yielded a few flashes of insight.[3] I saw how a bold and creative man was able to free himself of bondage to all tradition, and how he sought genuinely to hold onto and describe what he saw. The influence of this was, I suppose, not without significance for me, although I was able to read and understand only a very few pages. Indeed, to describe and to be faithful – *that* was absolutely necessary. But only *after* the appearance of my treatise of 1894[4] did I read and make excerpts from larger sections of his work. That treatise (a working out of ideas which came to me in the lectures mentioned) I have now read once again. It is a first sketch of the *Logical Investigations*, especially of the IIIth and Vth.

Unfortunately I can no longer judge to what extent *Meinong*'s

---

[3] Husserl's copy of the *Principles of Psychology* shows signs of intensive reading. Marginal notes are especially present in Volume I, Chapters IV (Habit), V (The Automaton Theory), VI (The Mind-Stuff Theory), VII (The Methods and Snares of Psychology), VIII (The Relation of Mind to Others Things), IX (The Stream of Thought), XI (Attention), XII (Conception); and in Volume II, Chapters XX (The Perception of Space), XXI (The Perception of Reality), XXII (Reasoning), and XXVI (Will). – See also the reference to James in the second note to the Appendix to Chapter 5 of the IInd "Investigation" in *Logical Investigations*, Volume II (1901): "How little James' brilliant observations in the domain of the descriptive psychology of the representational experience force one into psychologism is to be seen from the present writing. For the assists which I owe to this outstanding researcher in descriptive analysis have only favored my disassocation from the psychologistic standpoint." According to a remark of Husserl's to Dorian Cairns, in the year 1931, it was Carl Stumpf who called Husserl's attention to James. [H.S.]

[4] The reference here obviously is to the "Psychologischen Studien zur elementaren Logik," which appeared in *Philosophische Monatshefte*, Vol. XXX, (1894), pp. 159–191, and which in two footnotes invokes James as a corroborating witness. [H.S.] {Translated in this volume. [D.W.]}

theory of relations influenced me. I had already been reading in it around 1890, but it was correspondence with him in 1891 that first led to a thorough study.[5] I find it hard to believe that his theory provided me anything in the way of method, beyond a few ideas of an insignificant nature.

I have also read anew my reviews of recent years,[6] along with sections from my *Logical Investigations.*

Meinong's book could no longer offer me much with respect to the study of the representation and the judgment, beyond the great stimulus which always comes when a not insignificant person works through the same problems that have engaged one for many years. I find in this book only one important idea that was left unexpressed in my *Logical Investigations.* Although it occurred to me and was considered during the elaboration of that work, yet I did not venture to adopt it at the time: namely, the idea of carrying the modification of the judgment into a "mere representation" on over to apply to wishes and all other types of acts. I still have the dated page dealing with this question (1894), in which I took precisely *Meinong*'s position. [444]
But certainly I saw great difficulties, which *Meinong* does not see, and which prevented me from proceeding along this line. *Meinong*'s concept of the representation is totally incomprehensible to me. A discussion with *Meinong* will, for obvious reasons, be necessary and unavoidable – not to mention the fact that it must at some point be shown that in actuality the domains of investigation and the most essential points established are common to both sides.

We are like two people traveling in one and the same dark continent. Naturally we frequently see and describe the same things, but often differently, corresponding to our differing apperceptive masses. This can be shown paragraph for paragraph, except for his chapter on emotional acts, as well as the one on hypothetical judgments and inferences from assumptions. I certainly hold the latter to be wholly mistaken.

After *Meinong* I went on with the arrangement and review of my own manuscripts. I saw with amazement how much is contained in them, and how much is begun but unfortunately not completed. They clearly testify to how powerfully I was gripped by deep, and by

[5] In the Husserl-Archives at Louvain there is to be found no such correspondence from this period. [H.S.]

[6] *Husserliana XXII*, pp. 162ff. Translated in this volume. [D.W.]

the deepest, problems. And how I was gripped anew by them, once again, in looking through the manuscripts! This much is certain: I must *never* forsake these areas of investigation, leaving uncompleted the bore-holes and foundations begun. That would be to give up myself. These were my life for many years, and my life must and shall never fall into fragments. How much time, how much life, spiritual labor and first insights (beginnings of valuable knowledge) I have frittered away! How many times have I begun to build and then allowed all to fall to ruin again! Since the publication of the *Logical Investigations* my life has taken on an inner stability. And it should and must exhibit inner unity from now on. Unfortunately, my personality can no longer become full and whole. It can no longer attain unity of outlook upon its world, the unity of a freely developed, beautiful and natural, organic formation. But, thank heavens, results have not been lacking, and further fruits are ripening. The peculiar varieties of fruit that can thrive upon this trunk – which, alas, is so much broken and scarred – shall now, God help me, be brought to maturity. That is henceforth my life, the range of life-tasks proper to me. I will not despair, but will live in hope. I will be satisfied if I do work which genuinely benefits those who come after me. However much it may pain me to relinquish the joy of a harmonious unity and freedom present in a naturally pleasing development of the soul, I must do it. I can only admire the beauty and harmony of others. But I must abide in my own tasks, and seek in their solution my worth and my inner security. "By their fruits shall ye know them!" And by my fruits I will come to know myself. I shall be able to respect myself if, with great labor, I bring them to maturity, in ordered sequence.

Above all there is required the greatest of inner concentration and the best use of time. All drafts composed up to now must be arranged and systematically brought to completion. The three weeks which I have given to such arranging were not wasted, but they have far from sufficed. I also have erred in not first studying my lectures, and *then* [445]
the appended pages, which often sought to clarify and solve the same problems in related – and then again in new – formulations. What writing projects must I carry out? Which problems take up?

1. In the first place I mention the general problem which I must solve if I am to be able to call myself a philosopher. I mean: *A critique of reason*, a critique of logical and practical reason, of normative reason in general.

Without getting clear on the general outlines of the sense, essence, methods and main points of a critique of reason, without having thought out, outlined, formulated and justified a general sketch of such a critique, I cannot live truly and sincerely. I have had enough of the torments of unclarity, of tottering back and forth in doubt. I have to come to an inner stability. I know that this concerns high, even the highest, matters. I know that great geniuses have run aground and failed here. And were I to compare myself with them I would have to despair at the outset. I will not compare myself to them, but I simply cannot go on without clarity. I will – I *must* – approach these sublime goals, through self-sacrificing labor and purely disinterested absorption in the work. I am fighting for my life, and because of this have confidence that I shall be able to make progress. The hardest conditions of life, the defense of oneself against threat of death, gives unsuspected, boundless strength. I do not strive for honor and fame. My aim is not to be admired. I take no thought of others, or of my own external needs. Only one thing will fulfil me: I must come to clarity! Otherwise I cannot live. I cannot endure life without believing that I shall attain it – that I myself can, with clear eyes, actually look into the promised land.

My many investigations into particular topics have provided handholds for me, and have acquainted me with methods. I first must get clear concerning the most general vantage points.

2. On the other hand: We need not only knowledge of goals, guiding principles, standards, methods, and of our posture toward other areas of knowledge and other sciences. We also need the actual execution. We must walk the paths themselves. We must step-by-step solve the particular problems. Thus there is necessary here, above all, a step-by-step treatment of the phenomenology of reason, and, based thereon, a genuine illumination (*Aufklärung*) of logical and ethical reason in the form of the basic principles and concepts of each.

And here there stand in first position the problems of a phenomenology of perception, of phantasy, of time, and of the thing.

I gave preliminary and most incomplete sketches for a systematic treatment of all of this in the lectures on "Main Points" during the winter of 1904/1905.[7] Yet earlier there were the treatises from the

[7] See *Husserliana X* and *XXIII*.

year 1898 – presumably already perfected for the press, and in any case completely worked out – which served as the basis for my lectures, and which must be looked into once again. What is of [446] value in these must be excerpted, and the remaining repudiated and discarded. In addition there is a group of supplemental pages, often dealing with difficulties.

In connection with this I have also made attempts at a *Phenomenology of attention.* But I lack a phenomenology of space – although I wanted to make a beginning on this latter in the year 1894 and have in my possession all sorts of attempts at it (but nothing of use).

Further, there is required a systematic realization of a *Phenomenology of significations.* The foundation for this is offered by my *Logical Investigations*, which are by and large of value here, and yet do not suffice, not being systematic enough.

In connection with this, there is required a phenomenology of empty intentions and symbolic representations.

Further, a theory of the judgment, that great desideratum on which I have already done so much work. On this I have the lectures, and ever so many more unprocessed manuscripts to be utilized.

In connection with the phenomenological theory of the judgment there stands the analysis of the essences of the various forms of propositions, which belong, from another viewpoint, to the domain of pure grammar. Here once again, then, is an area for a new and large work.

Investigations concerning pure logic (and pure grammar), the logical calculus – also the definite manifold – the logic of probability, and investigations concerning the essence of the categorical proposition as well as the existential.

A comprehensive work on hypothetical judgments and hypothetical inferences, as well as on the concepts *necessity*, *impossibility*, and so forth.

It seems to me that at this point the following projects are most advanced:

1. A work on *an introduction to the critique of reason*, in particular theoretical reason.

2. A very comprehensive work concerning *perception*, *phantasy*, and *time.*

I am not so clear on whether the beginnings of a phenomenology of the *thing representation* lie here – though it certainly seems to me that they do – and whether those beginnings are to a certain degree already pure enough. The phenomenology of attention (at least within the intuitive and sensible sphere) is to be placed along side that of the thing representation. This will be, it seems, a vast work, and one which must be brought to completion as soon as possible.

3. Treatises on pure logic, on the analysis of the essences of propositions. Of quite special value here (besides analyses of the existential and the categorical proposition) might be *theories contributing toward an account of hypothetical propositions and inferences*. This latter would by itself provide the basis for a significant work that would delight me.

4. Toward a phenomenology of the representation and the judgment, or, to begin with: *achievement of a theory of the judgment*. That again would yield a significant work. Is the preparation for this now sufficient?

Toward a theory of belief (already involved in #2).

5. A treatise against *Meinong*, and a discussion with him concerning the concepts of the representation, of the assumption (the hypothetical proposition), and of the judgment. [447]

6. Discussion with *Cornelius*.

7. Treatises on Apriori-Aposteriori,
Analytic-Synthetic,
The categories, and
Motivation and *Causation*.

What is to be done with the draft of a treatise on *intentional* objects? It cannot be published as it is. Perhaps the substance of it can be used in the context of the discussion with *Meinong*: – if, perchance, that discussion should take the form of several lectures.

This, of course, does not exhaust all of the tasks. (I have forgotten to mention abstraction, general formal ontology, and so forth, in the above.) But here I am speaking only of the plans which could come first to execution because the preliminary work is already far advanced. In choosing my lectures, especially those for advanced students, I must seek to benefit myself and obtain drafts suitable for publication.

Above all, I need assistance from Heaven. Favorable conditions necessary to the work, and an inner concentration, an inner oneness with the problems. Again and again to read, to improve, to transcribe the old manuscripts. Always to hold oneself in readiness for the great purpose. To be prepared. (Cf. *Amiel* and also the many fine words of *Carlyle*.) How weak I am, needing the assistance of the great spirits! Out of the fullness of their strength and their pure will they must strengthen me. I would fully nourish myself from them, and learn to avert my glance from the distracting impulses of the commonplace. O God! This last year! How *could* I have allowed myself to be so crippled by the disdain of my colleagues, by rejection of the faculty, by the disappointment of my hopes for higher position. Is *that* what I have always been working for? And if *ever* that – though I believe, really, never – then certainly not for the last decade. Pure inner conviction, purely inward life, inward gestation of the problems, being devoted purely to them and only to them: *that* is the hope of my future. If I do not succeed in this, then I can only live a life which is, rather, a death. I still may hope. But the hour has struck in which I must commit myself. The mere "act of will," as a momentary resolve, does not suffice. There is required inner renewal, or inner purification and firmness. Against all that is superficial, against all the temptations of the old Adam, I must arm myself with nine-fold layers of bronze.

I must follow my path with the certainty, the fixed resolve, and the seriousness of *Dürer*'s knight, in defiance of death and the devil. Ah, life has been grim enough for me. The cheer of the sensible pleasures of life has become foreign to me, and must remain so. I must not be passive (and pleasure *is* passivity!). I must live in work, in battle, in a passionately serious grappling for the garland of truth. Cheerfulness shall not be lacking. A cheerful heaven above me, as above *Dürer*'s knight, as I valiantly and steadily march forth! And God be with me, as with him, in spite of the fact that we are altogether sinners. [448]

November 4, 1907

So I have once again tasted the grimness of the life of aspiration. Tasted? As if it were a misery only once visited upon us, a bitter draught to be downed on one occasion! I have now been in Göttingen six years, and it will soon be seven years since the appearance of my *Logical Investigations*. After many years of confused struggling, lifting myself toward clarity, I published those attempts at conceptualizing problems, elaborating methods, opening and making useable certain possible routes toward solutions. How my heart swelled with pride as – something I never dared to hope – they began to exercise a rapid and strong influence, and especially upon the younger generation. How my hopes soared, how close at hand appeared the great goal of gaining genuine insight in logic, in the critique of knowledge, and in the critique of reason in general: of establishing the natural order among the problems, of discovering the natural sequence in investigations, of raising the problems themselves to the highest level of precision, of refining the methods to the point of total consistency and complete assuredness – and then, within this clarity of goals and methods, of doing what is in each case dictated, step by step.

Now almost as many years have passed as preceded the *Logical Investigations* from the time I began the elaboration of the sections which make them up, or began to devote myself exclusively to the range of problems they comprehend. And with what result? Serious work year by year, surely – even if I do not count one year, *the* year of external hindrances and anxieties. I am in fact well along, although, in periods of dejection, I am also inclined to underestimate my progress. In great part my teaching activity has been oriented toward my life-goals: How many new starts, how many attempts to penetrate deeper and farther, and then the ever new attempts to bring to clarity the sense of logic, of the critique of knowledge, and of the method. And after all of that, how far in arrears I am! Life flies, and the years of my strength pass away. Alas for me if I remain stuck in these studies and modes of work! That would be little less than to have wasted a life of impassioned struggle, of the most extreme labors. Live in vain! I will never do that. My intention is never to give up, and I never will. Henceforth my earnest endeavor is, above all, none other than to come into an absolutely firm pathway. At the moment all my efforts concern the question of the natural order of the investigations, as well as the manner in which the most fundamental of those investigations are themselves to be entered into afresh and set in order.

March 6, 1908[8]

In the following Winter semester of 1906/07 I worked very diligently, and not without results. In the first half of the lectures on logic (up to Christmas) I gave a general introduction to the theory of science.[9] I sought to lay out the possibilities and demarcations of essence required by the Idea of a theory of science. After Christmas I sought to develop briefly the various forms of objectivities. But there was then nothing of a more complete treatment, although those presentations [449] might not be wholly worthless. I greatly exerted myself, but was exhausted by February. At Easter (of 1907) I went to Italy, my first trip into the holy land of the Germans. During the summer, in a four-hour course of lectures on the main points of phenomenology <and critique of reason>, I made a great effort in the endeavor to produce a phenomenology of thinghood, and especially of spatiality.[10] This was a grand new beginning – unfortunately not so well understood and accepted by my students as I had hoped. Also, the difficulties which arose were far too great, and could not be overcome at the first approach.

During vacation, up to the beginning of the Winter lectures on *Kant*, I busied myself with the problem of the signification, as well as with that of the analytic judgment. I attempted an elaboration of the lectures of the previous Winter; and, dealing with the Idea of apophantic logic, I found it necessary to become clearer on the various questions about significations.

During the Winter of 1907/1908, my strength for work receded. I saw it best, under those conditions, to apply myself to the lectures, with which my students seemed not dissatisfied. In spite of all, I learned much through an in-depth study of *Kant*'s *Critique of Pure Reason.*

During the Christmas vacation I 'recovered' myself. Dear

[8] The text below follows in the notebook after the entry of 29, IX, 1906. The text from 1907 was glued into the note-book by Husserl at some later time.

[9] The MS of these lectures is in the Husserl-Archives under the catalogue number F I 25, "Introduction to Logic and the Theory of Knowledge." (Now in *Husserliana XXIV*. [D.W.])

[10] The MS of this series of lectures bears the catalogue number F I 13. These are the lectures which have since been published under the title, *Die Idee der Phänomenologie*, in *Husserliana II*.

Heinrich[11] came to visit me. Unfortunately, Malvine's condition worsened, and so I more and more lacked inner joy. Afterwards came Gerhart's accident and his concussion. And black clouds covered the heavens. I could almost say that this was, for reasons of which I will say nothing, the most miserable period of my life. Now is not the time to reflect on it. Easter vacation has begun. I am here all alone. I hope to pull myself together. I hope to overcome my inner fragmentation. I want to rebuild my life, to give my spiritual existence a unified reference to its great goals. I was and am in great "peril of my life." Well, whatever is pleasing to God! But I intend to triumph or die. To perish in spirit, to succumb in the battle for inner clarity, for philosophical integrity – and yet to live on physically: that, I hope, will not be my lot, not be possible for me. However, I must not dwell on this until I have come to a point of greater inner rest, quieter and more assured.

First a few days given to objective matters. I will keep a diary.[12]

---

[11] Presumably E. Heinrich, an earlier student at Göttingen, whose dissertation, *Investigations toward a Theory of the Concept*, appeared in Göttingen in 1910.

[12] Here the entries cease. There follows only one entry, of three lines, concerning family matters.

INDEX

Printed in the United States
136024LV00006B/6/A

9 780792 322627